ENCYCLOPEDIA OF ORGANIC CHEMISTRY

ENCYCLOPEDIA OF ORGANIC CHEMISTRY

Vol. 2

By
Sananda Chatterjee
DAE, DCA, CMCNet
Scientific Consultant
Institute for Natural Sciences
Kolkata
(West Bengal)

DISCOVERY PUBLISHING HOUSE PVT. LTD.
NEW DELHI-110 002

Published by:
Tilak Wasan
DISCOVERY PUBLISHING HOUSE PVT. LTD.
4383/4B, Ansari Road, Darya Ganj
New Delhi-110 002 (India)
Phone : +91-11-23279245, 43596064-65
Fax : +91-11-23253475
E-mail : parul.wasan@gmail.com
discoverypublishinghouse@gmail.com
web : www.discoverypublishinggroup.com

***First Edition:* 2012**

ISBN: 978-81-8356-875-3 (Set)

Encyclopedia of Organic Chemistry

Printed at:
Shree Balaji Art Press
Delhi

Preface

Organic Chemistry is that branch of chemistry, which deals with the reactions related to the life form, or one of the source of the life form, carbon (we can say the end of it).

This carbon mainly conjugates with different elements such as nitrogen, hydrogen and oxygen sometimes metals and other non-metals also). Organic chemistry studies mainly the extraction, structure, properties, of these compounds.

In this book chapters deals with the different methods of purification, extraction of compounds, some processes were used and devised by Alchemists.

The extraction preservation of natural dyes are examples of alchemical work, these been discussed in the chapter "dyes indicators and pigments".

Dyes is the mal-pronunciation of the damathi word dyum meaning colour, it was, the ancient Egyptians who use different natural pigments to extract different colours.

In this book the organic compounds has been mainly classified into two broad groups, the aromatic and the aliphatic compounds. The extraction of alcohol is definitely an alchemical process'.

The destructive distillation of wood is one of the finest processes devised by the alchemists for the preparation of methanol is one of the important topics of this book.

Looking ahead in to the modem world, the world of future, the subject comes in vision is the biochemistry, the base of which is organic chemistry, the chapter containing carbohydrates, and the chapter, chemical compounds of biological and biochemical interest is the best element of focus for the students stepping foot in the advanced staircase of biochemistry.

The last chapter, chapter 41, discusses the different organic reactions and their suitably mechanisms, this will help those students who are very much afraid of organic formulas and reactions. This chapter also deals with the chirality one of the interest of Dr. Palash Gangopadhyay (author's friend).

Organic Chemistry is not a mere subject, it is the rhythm of life tuned in the essence of carbon, which looks black but carries the dazzleness of diamond in its heart.

SANANDA CHATTERJEE

Contents

CHAPTER

8

Halogen Derivatives

Introduction

The compounds which are obtained by the replacement of one or more hydrogen atoms in hydrocarbons by a corresponding number of halogen atoms are called 'halogen derivatives' of the hydrocarbons. They may be divided into Mono, Di, tri, tetra etc. Halogen derivatives according to the number of the hydrogen atoms presents in the molecule.

MONO-HALOGEN DERIVATIVES

Alkyl Halides

General Formula: $C_nH_{2n+1}X$ or $R-X$

Some important members are:

Name	Formula	Melting Point (°C)	Boiling Point (°C)	Density as liquid
Methyl Bromide	CH_3Br	–93	–24	0.95
Methyl Chloride	CH_3Cl	–93	5	1.73
Methyl Iodide	CH_3I	–66	42	2.28
Ethyl Bromide	C_2H_5Br	–119	38	1.45
Ethyl Iodide	C_2H_5I	–105	72	1.92
Ethyl Chloride	C_2H_5Cl	–142	12.5	0.92

General Methods of Preparation

1. From Paraffins

Alkyl halides are prepared by the halogenation in sunlight

(a) $CH_4 + Cl_2 \xrightarrow{\text{Sunlight}} CH_3Cl + HCl$

(b) $C_2H_6 + Br_2 \xrightarrow{\text{Sunlight}} C_2H_5Br + HBr$

This method is not suitable for the preparation of the alkyl halides because the reaction proceeds further to produce the di, tri, tetra and other higher derivatives, which are very difficult to separate.

Iodo derivatives are, however, prepared by indirect methods.

2. From Olefines

By direct addition of halogen acids:

(a) $C_nH_{2n} + HX \rightarrow C_nH_{2n+1}X$

(b) $C_2H_4 + HBr \rightarrow C_nH_5Br$
Ethylene → Ethylene bromide

3. From Alcohols

By the action of the halogen acids or phosphorous halides.

(a) Halogen acids react with alcohols in the presence of anhydrous zinc chloride or concentrated sulphuric aid which absorbs the molecule of water formed in the process:

(i) $ROH + HCl \xrightarrow{ZnCl_2} R-Cl + H_2O$

(ii) $CH_3-OH + HCl \xrightarrow{ZnCl_2} CH_3Cl + H_2O$
methyl alcohol → methyl chloride

(iii) $C_2H_5OH + HCl \xrightarrow{ZnCl_2} C_2H_5Cl + H_2O$

The above reactions can also be effected catalytically by passing a mixture of alcohol vapours and HCl gas over Al_2O_3 at 400 - 450°C

(b) Phosphorus halides also react with alcohols to give alkyl halides

(i) $R-OH + PCl_5 \rightarrow R-Cl + POCl_3$

(ii) $C_2H_5-OH + PCl_5 \rightarrow C_2H_5Cl + POCl_3 + HCl$
Alcohol + Phosphorous Pentachloride → Ethyl Chloride + Phosphorous Oxychloride + Hydrogen Chloride

(For preparing alkyl bromides or ethyl chloride by this method, it is not necessary to empty previously prepared phosphorous bromide or iodide. The halogen is gradually added to a mixture of alcohol and red phosphorous and the whole is then warmed to complete reaction.

2P	+	$3Br_2$	→	$2PBr_3$
Red Phosphorous		Bromine	→	Phosphorous Bromide

Physcial Properties

1. The lower members such as CH_3Cl, CH_3Br and C_2H_5Cl are gases at the ordinary temperature. Higher members are mostly are pleasant smelling liquids. Members with higher molecular weights are solids
2. They are particularly in suitable in water, but readily dissolve in alcohol, ether, carbon di sulphide and other organic solvents.
3. Lower members produces cooling on evaporation
4. Most of them (CH_3Cl, C_2H_5Cl) are suitable
5. They burn with a green edged flame
6. The iodides are heaviest and chlorides are lightest
7. The iodides posses higher boiling points than bromides and chlorides

Chemical Properties

1. Non-polar character: In alkyl halides the halogen atom is more firmly bound to the carbon atom by non-ionizable or covalent bond. Hence alkyl halides are non-polar compounds and do not ionize in solution and therefore are not good conductors of electricity.

This is why carbon tetrachloride is non-polar.

2. Synthetic Reactions: Halogen atom of an alkyl halide is very reactive and can be carefully replaced by many atoms or groups. This makes alkyl halides very valuable synthetic agents in organic chemistry. Alkyl iodides and bromides are less volatile and react more readily than the alkyl chloride.

Alkyl halides are used for the synthesis of:

1. *Paraffin*: By reduction or by Wurtz's Reaction

(a) $\underset{\text{ethyl iodide}}{C_2H_5I} + 2H \xrightarrow{\text{Zn/cu couple and alcohol}} C_2H_6 + HI$

(b) $\underset{\text{ethyl bromide}}{2C_2H_5Br} + 2Na \rightarrow 2NaBr + \underset{\text{butane}}{C_4H_{10}}$

The reaction may be used to convert lower paraffins into higher paraffins.

2. *Olefines*: By heating with alcoholic caustic alkalies

C_2H_5I	+	KOH	→	C_2H_4	+	KI	+	H_2O
Ethyl Iodide		Potassium Hydroxide	→	Ethane	+	Potassium Iodide	+	Water

3. *Alcohols*: By treatment with aqueous caustic alkalies or moist silver oxide

(a) $C_2H_5I + KOH \rightarrow C_2H_5-OH + KI$

(b) $C_2H_5I + AgOH \rightarrow C_2H_5-OH + AgI$

Ethyl iodide Ethyl alcohol

4. *Ethers*: By the treatment with sodium alkoxide or silver oxide

$C_2H_5 I + Na-O-C_2H_5 \rightarrow C_2H_5-O-C_2H_5 + NaI$

Sodium Ethoxide Ether

This reaction is known as Williamson's Synthesis and may be used for preparing simple mixed ethers

$2C_2H_5I + AgO \rightarrow (C_2H_5)_2O + 2AgI$

ethers

1. *Amines*: By treatment with alcohols ammonia

$C_2H_5I + NH_3 \rightarrow C_2H_5NH_2 + HI$

Ethyl amine
Primary amine

This reaction proceeds further to give (i) secondary amine, (ii) tertiary amine and (iii) quaternary ammonium salt

(i) $C_2H_5NH_2 + C_2H_5I \rightarrow (C_2H_5)_2NH + HI$

Ethyl amine (pri) Ethyl Iodide → Diethylamine

(ii) $(C_2H_5)_2NH + C_2H_5I \rightarrow (C_2H_5)_3N + HI$

Diethyl amine(sec) → Triethyl amine + HI

(iii) $(C_2H_5)N + C_2H_5I \rightarrow (C_2H_5)_4NI$

Triethyl amine Tetraethyl ammonium Iodide

2. Alkyl Nitrates By the action of potassium nitrite, KNO_2

$C_2H_5I + KNO_2 \rightarrow C_2H_5NO_2 + KI$

Ethyl Iodide Ethyl Nitrite

3. Nitro-Paraffins By the action of silver nitrate

$C_2H_5I + AgNO_3 \rightarrow C_2H_5NO_2 + AgI$

Nitro-paraffins and alkyl nitrates are isomers of each other

4. Alkyl cyanides By treatment with potassium cyanide

$C_2H_5I + KCN \rightarrow C_2H_5-C\equiv N + KI$

5. Alkyl isocyanide and alkyl cyanides are isomers of each other*

$C_2H_5I + AgCN \rightarrow C_2H_5-NC + AgI$

6. *Esters*: By the action of silver salt of fatty acids i.e.

$C_2H_5I + CH_3COOAg \rightarrow CH_3COO\,C_2H_5 + AgI$

Silver acetate — Ethyl acetate(ester)

7. *Grignard's Reagent*: By heating with magnesium powder, in the presence of anhydrous ether

$$C_2H_5I + Mg \xrightarrow{\text{Dry Ether}} MgC_2H_5I$$

Magnesium Ethyl Iodide (Grignard's Reagent)

8. Mercaptans or Tio-alcohols: - By the treatment with potassium hydrosulphide KHS

$C_2H_5I + KHS \rightarrow C_2H_5-S-H+KI$

Ethyl mercaptan

9. *Homologous of Benzene*: By reacting with benzene in presence of anhydrous aluminium chloride

$$C_2H_6 + CH_3Cl \xrightarrow{AlCl_3} C_2H_5CH_3 + HCl$$

Ethyl Benzene
toluene

This reaction is called Friedel Craft's Reaction in characteristic to aromatic.

Individual Member

Methyl Chloride

Formula: CH_3Cl

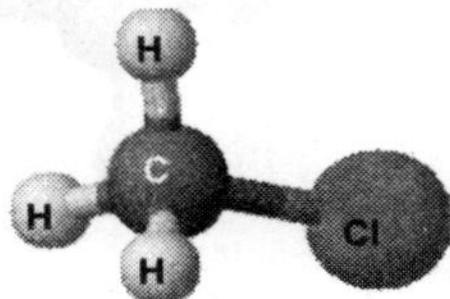

Preparation

1. By saturating methyl alcohol CH_3OH containing anhydrous zinc chloride with dry HCl gas

$$CH_3OH + HCl \xrightarrow{ZnCl_2} CH_3Cl + H_2O$$

2. By controlled chlorination of methane in the presence of suitable catalyst, such as $SbCl_3$ and $CuCl_2$

$CH_4 + Cl_2 \rightarrow CH_3Cl + HCl$

3. By heating trimethylamine hydrochloride with hydrochloric acid at 360°C

$(CH_3)_3N.HCl + 3HCl \rightarrow 3CH_3Cl + NH_4Cl$

Trimethylamine Hydrochloride — Methyl Chloride

*Alkyl cyanides yields acids *i.e.* $RCN + 2H_2O \rightarrow RCOOH + NH_3$.

Physical Properties

1. It is a colourless gas with a sweet ethereal smell
2. It is inflammable and burns with a green edged flame
3. On evaporation it causes cooling
4. It is sold as a liquid and its boiling point is - 23°C

Chemical Properties

It gives all the typical reactions

Uses

1. As a methylating agent for dyes in the colour industry
2. For the productions of the low temperature
3. As a local anæstatic
4. For extracting perfumes from flowers.

Ethyl Bromide or Monobromo Ethane

Formula: C_2H_5Br

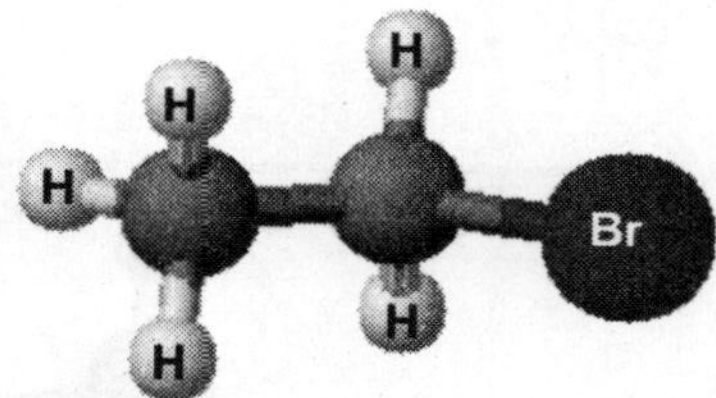

Preparation

Ethyl bromide can be prepared by all the general methods of preparation and gives all the typical reaction of alkyl halides.

It is commonly prepared by the following methods: -

1. By the action of red phosphorous and bromine on ethyl alcohol

2P	+	3Br	→	$2PBr_3$		
$3C_2H_5OH$	+	PBr_3	→	$3C_2H_5Br$	+	H_3PO_3
Ethyl Alcohol		Phosphorous Tribromide	→	Ethyl Bromide		Meta phosphoric Acid

Laboratory Preparation

Theory: By distillation of mixture of ethyl alcohol, potassium bromide and sulphuric acid.

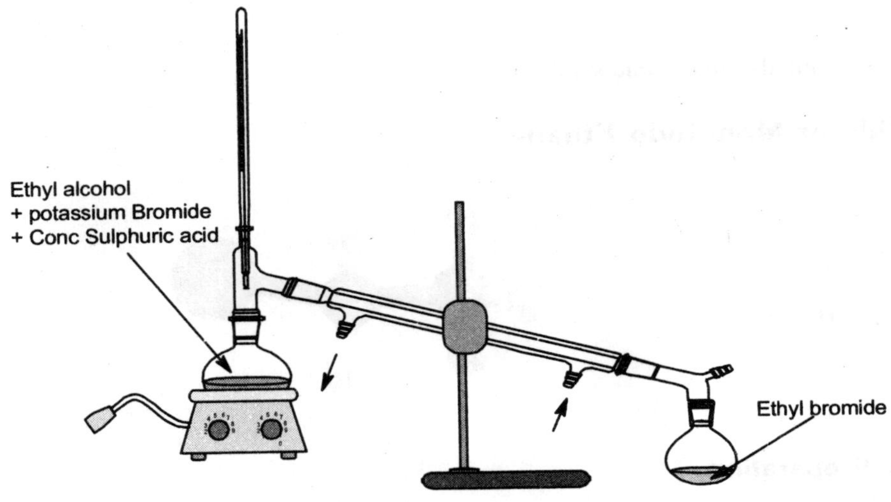

Fig. 8.1. The Lab preparation of Ethylene Bromide

Apparatus	Reagents	Catalyst	Temperature
Distillation apparatus, stands clamps and heating mantle	Potassium Bromide Ethyl alcohol Sulphuric acid (conc)	No	110°C

Reaction Equation

$$KBr + H_2SO_4 \rightarrow KHSO_4 + HBr$$

$$\underset{\text{Ethyl Alcohol}}{C_2H_5OH} + HBr \rightarrow \underset{\text{Ethylene Bromide}}{C_2H_5Br} + H_2O$$

Purification: It is first washed with sodium carbonate solution and then with water. It is dries with calcium chloride, $CaCl_2$ and finally purified by redistillation. Pure ethyl bromide is collected between 37 - 40°C.

***Precaution**: Do not use burners while experimenting with alcohols (the distillation process) always use temperature controlled heating mantles.*

Properties

It is colourless pleasant smelling and heavy liquid.

Melting Point	–119°C
Boiling Point	38°C
Refractive Index	1.42386
Density g/mol	1.430
Solubility	
Water	1.8
Alcohol	∞
Other solvents (ether)	∞

Uses

As an anæsthetic and also in organic synthesis

Ethyl Iodide or Monoiodo Ethane

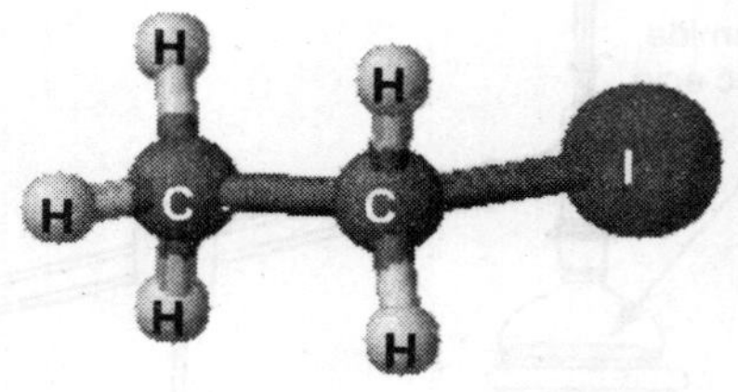

Formula: C_2H_5I

Laboratory Preparation

Theory: The reflux distillation of the mixture of red phosphorous iodine and ethyl alcohol.

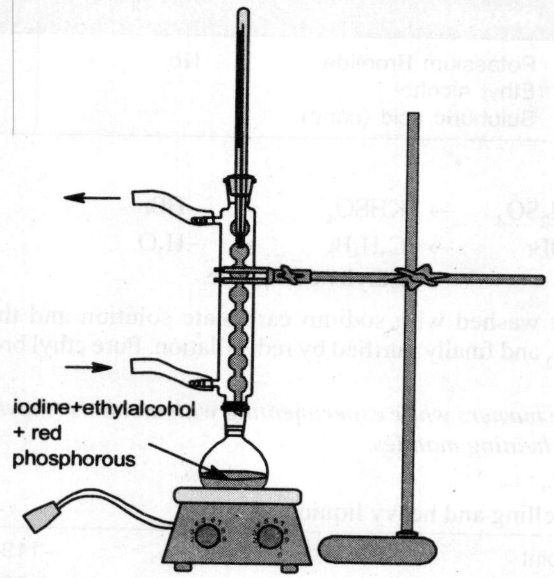

Fig. 8.2. The Lab preparation of Ethynl Iodide

Apparatus	Reagents	Catalyst	Temperature
Flask, reflux condenser, stand clamps, heating mantle	Ethyl alcohol, iodine and red phosphorous	No	85°C , reflux time 2 hours

Reaction Equation

2P	+	$3I_2$	→	$2PI_3$		
$3C_2H_5OH$	+	PI_3	→	$3C_2H_5I$	+	H_3PO_3
Ethyl Alcohol		Phosphorous Triiodide		Ethyl Iodide	+	Metaphosphoric Acid

Precaution: During reflux distillation heating mantle is must, do not use burner.

Properties

1. It is a colourless, pleasant smelling liquid
2. It boils at 72°C

Uses

Ethyl iodide is largely used in the synthesis of organic compounds

Dihalogen Derivatives

Two different types of dihalogenic derivatives are known: (i) Alkaline derivatives and (ii) Alkaledine derivatives.

1. ***Alkaline Derivatives***: These are the compounds in which the two halogen atoms attached to the different carbon atoms ethylene chloride as 1: 2 dichloroethane.

2. ***Alkaledine Derivatives***: These are the compounds in which both the halogen atoms are attached to the same carbon atom i.e. ethylidene chloride or 1:1 dichloroethane.

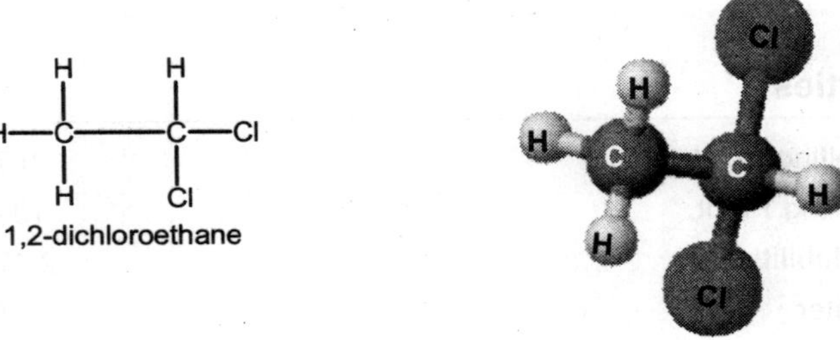

Ethylene chloride and ethylidene chloride are isomeric compounds.

Ethylene Chloride

Formula: $C_2H_5Cl_2$

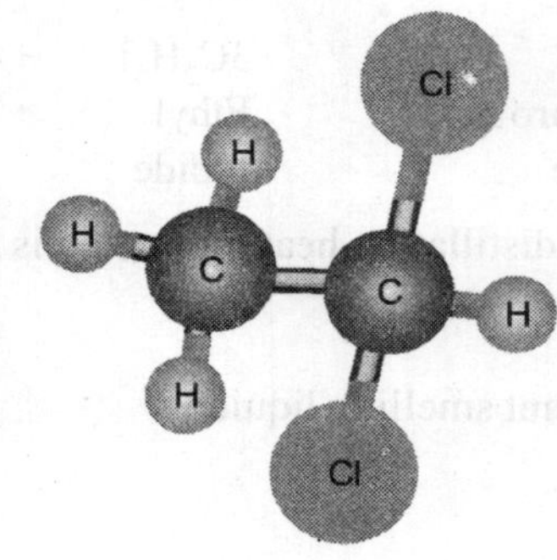

Preparation

1. By addition of chlorine to ethylene

$$\underset{\text{Ethylene}}{H_2C{=}CH_2} + Cl_2 \longrightarrow \underset{\text{1,2-dichloroethane}}{ClH_2C{-}CH_2Cl}$$

2. By the action of PCl_5 on ethylene glycol

$$\underset{\text{ethylene glycol}}{HOH_2C{-}CH_2OH} + 2PCl_5 \longrightarrow \underset{\text{1,2-dichloroethane}}{ClH_2C{-}CH_2Cl}$$

Properties

Physical Properties

Melting Point	9.7°C
Boiling Point	131°C
Solubility	2.1701
Water	0.431
Alcohol	∞
Others	∞ Soluble

Chemical Properties

In chemical properties it resembles alkyl halides:

1. ***Reduction***

   On oxidation it gives ethane

$$\underset{\text{1,2-dichloroethane}}{ClH_2C{-}CH_2Cl} \xrightarrow{+2H} \underset{\text{Chloroethane}}{CH_3{-}CH_2Cl} + HCl \xrightarrow{+2H} \underset{\text{Ethane}}{CH_3{-}CH_3} + HCl$$

2. ***Chlorination***

   On complete chlorination, it gives hexachloroethane as the final product

$$\underset{\text{1,2-dichloroethane}}{ClH_2C{-}CH_2Cl} \xrightarrow{4Cl_2} \underset{\text{Hexa chloroethane}}{C_2Cl_6} + 4HCl$$

3. ***Hydrolysis***

   On hydrolysis with aqueous sodium hydroxide it gives ethylene glycol

$$\underset{\text{1,2-dichloroethane}}{ClH_2C{-}CH_2Cl} + 2KOH \longrightarrow \underset{\text{Ethylene glycol}}{HOH_2C{-}CH_2OH} + 2KCl$$

4. ***Cyanization***

   On the treatment with KCN and subsequent hydrolysis it gives succinic acid (solid)

$$\underset{\text{1,2-dichloroethane}}{ClH_2C{-}CH_2Cl} \xrightarrow{2KCN} \underset{\text{Succinonitrile / Ethyl Cyanide}}{N{\equiv}C{-}CH_2{-}CH_2{-}C{\equiv}N} \xrightarrow{4H_2O} \underset{\text{Succinic acid}}{HOOC{-}CH_2{-}CH_2{-}COOH} + 2NH_3$$

Ethylidene Chloride

Formula: C_2H_3Cl

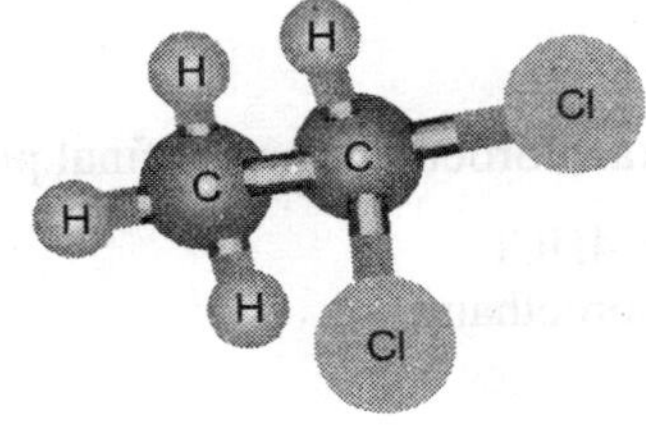

Preparation

1. By the action of hydrogen chloride on acetylene

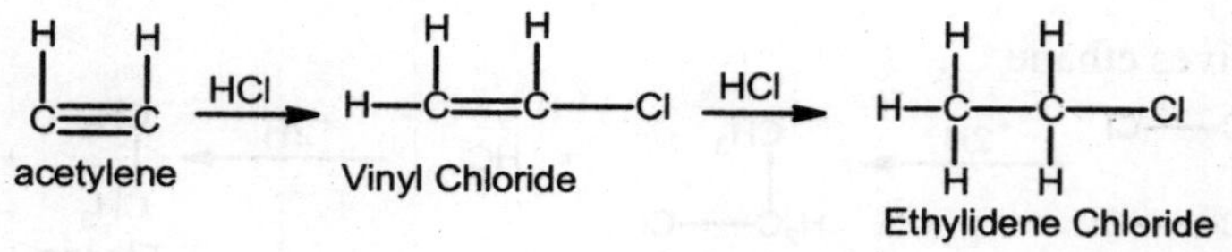

$$HC\equiv CH \xrightarrow{HCl} H_2C{=}CHCl \xrightarrow{HCl} CH_3CHCl_2$$

acetylene → Vinyl Chloride → Ethylidene Chloride

2. By the action of PCl_5 on acetaldehyde

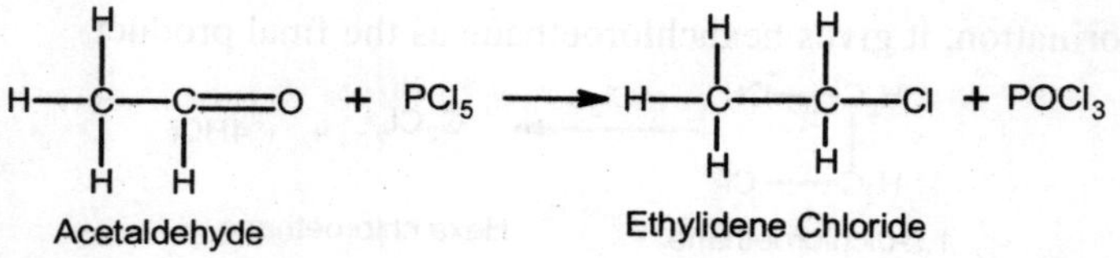

$$CH_3CHO + PCl_5 \longrightarrow CH_3CHCl_2 + POCl_3$$

Acetaldehyde → Ethylidene Chloride

Properties

Physical Properties

1. It is a colourless liquid.
2. Boils at 57°C.

Chemical Properties

It is similar in properties to alkyl halides halogen atoms easily replaceable by other groups.

1. Reduction

On reduction, it gives ethan

$$CH_3CHCl_2 \xrightarrow{+2H} CH_3CH_2Cl + HCl$$

Ethylidene Chloride → Ethyl Chloride

$$CH_3CH_2Cl \xrightarrow{+2H} CH_3CH_3 + HCl$$

Ethyl Chloride → Ethane

2. Chlorination

On chlorination, it gives hexachloroethane as the final product

$$CH_3CHCl_2 \xrightarrow{4Cl_2} C_2Cl_6 + 4HCl$$

Ethylidene Chloride → Hexachloroethane

3. *Hydrolysis*

On hydrolysis with aqueous potassium hydroxide, it gives acetaldehyde

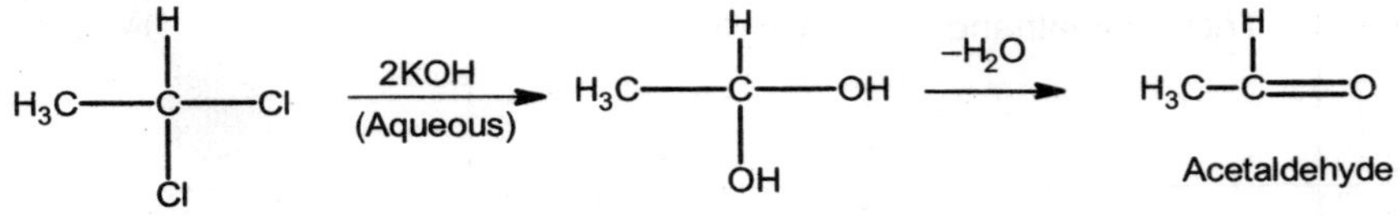

$$H_3C-CH(Cl)-Cl \xrightarrow[\text{(Aqueous)}]{2KOH} H_3C-CH(OH)-OH \xrightarrow{-H_2O} H_3C-CH{=}O$$

1,1-dichloroethane
Ethylidene Chloride

Ethane-1,1-diol

Acetaldehyde

4. *Cyanization*

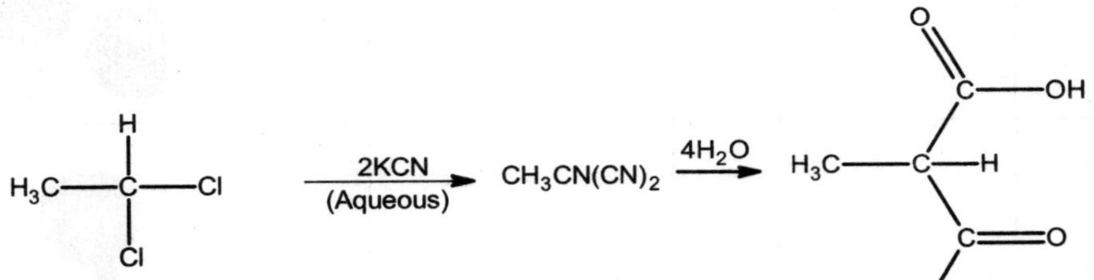

$$H_3C-CH(Cl)-Cl \xrightarrow[\text{(Aqueous)}]{2KCN} CH_3CN(CN)_2 \xrightarrow{4H_2O} H_3C-CH(COOH)-COOH$$

1,1-dichloroethane
Ethylidene Chloride
(1)

methylmalonic acid
(2)

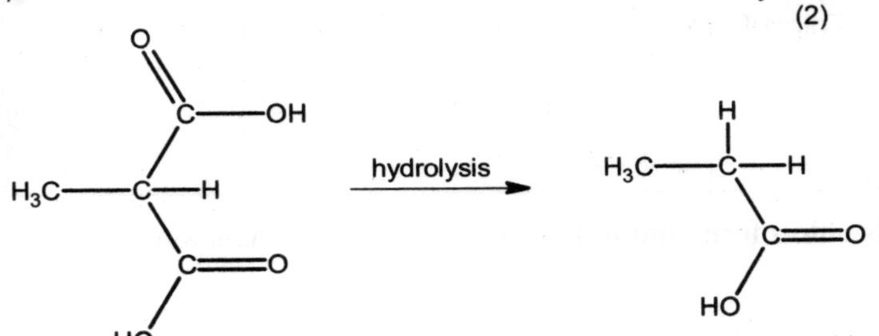

$$H_3C-CH(COOH)-COOH \xrightarrow{\text{hydrolysis}} H_3C-CH_2-COOH$$

Methylmalonic acid
(3)

Propionic acid
(4)

The above two reactions (No 3 and 4) may be employed to distinguish ethylidene chloride from its isomer, ethyl chloride. Constitution of both ethylene and ethylidene halide follows from their methods of preparation and their behaviour on hydrolysis with aqueous alkalies.

TRIHALOGEN DERIVATIVES

The trihalogen products of methane are the simplest halogen compounds in which the three halogen atoms are linked to the same carbon atom. They are:

Chloroform	Trichloromethane	$CHCl_3$	
Bromoform	Tribromomethane	$CHBr_3$	
Iodoform	Triiodoform	CHI_3	

The compounds chloroform and iodoform are more important since they are used in surgery and medicine.

Chloroform

Formula: $CHCl_3$

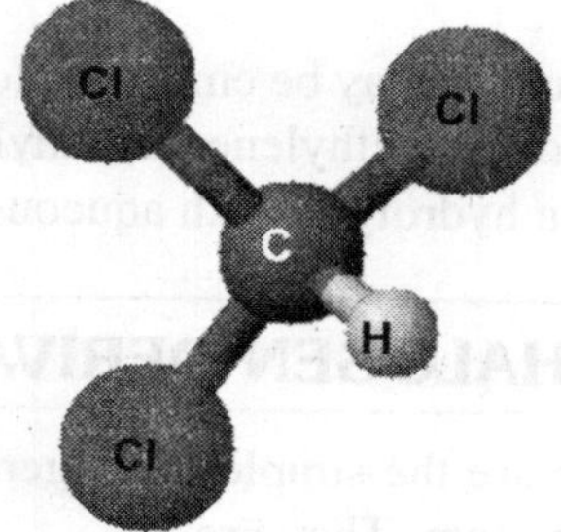

Preparation

1. By the action of chlorine on methane in sunlight

$$\underset{\text{Methane}}{CH_4} \xrightarrow{Cl_2} \underset{\text{Methyl Chloride}}{CH_3Cl} \xrightarrow{Cl_2} \underset{\text{Methylene Chloride}}{CH_2Cl_2} \xrightarrow{Cl_2} \underset{\text{Chloroform Trichloromethane}}{CHCl_3}$$

   The method is not of practical value.

2. By the partial reduction of carbon tetrachloride with zinc and sulphuric acid

$$CCl_4 + 2H \rightarrow CHCl_3 + HCl$$

The reaction was once used for the large scale preparation of chloroform, but at present has of a little technical importance.

Laboratory Preparation

Theory: Chloroform in the lab is prepared by the action of beaching powder paste and ethyl alcohol

(Prepare a paste of bleaching powder and water)

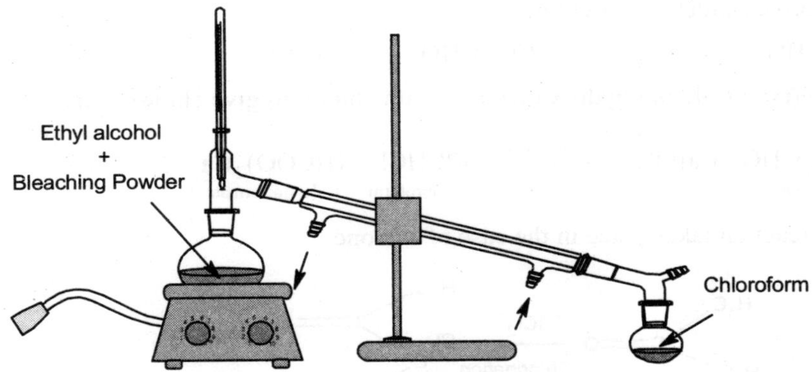

Fig. 8.3. The Lab preparation of Chloroform

Apparatus	Reagents	Catalysts	Temperature
Distillation apparatus, heating mantle, stands clamps flask etc.	Ethyl Alcohol, 25 ml Bleaching powder 200gm, water 100gms	No	70°C

Reaction Equation

$$\underset{\text{Ethyl Alcohol}}{C_2H_5OH} + O \rightarrow \underset{\text{Acetaldehyde}}{CH_3CHO} + H_2O$$

$$\underset{\text{Acetaldehyde}}{CH_3CHO} + \underset{\text{Hypochlorous Acid}}{3HOCl} \rightarrow \underset{\text{Chloral}}{CCl_3CHO} + 3H_2O$$

$$\underset{\text{Chloral}}{2CCl_3CHO} + Ca(OH)_2 \rightarrow \underset{\text{Chloroform}}{2CHCl_3} + \underset{\text{Calcium formate}}{(HCOO)_2Ca}$$

Expalnation of the Reaction Equation

(a) ***Bleaching Powder*** gives chlorine and chlorine and calcium hydroxide

$$\underset{\text{Bleaching Power}}{Ca(Cl)(Cl{:}O)} + H_2O \longrightarrow Cl_2 + Ca(OH)_2$$

(b) ***Oxidation:*** Chlorine acts as an oxidizing agent and converts ethyl alcohol to acetaldehyde

$$CH_3CH_2OH + Cl_2 \rightarrow CH_3CHO + 2HCl$$

(c) ***Chlorination:*** Chlorine also acts as a chlorinating agent and converts acetaldehyde into trichloroacetaldehyde or chloral

$$CH_3CHO + Cl_2 \rightarrow CCl_3CHO + 3HCl$$

(d) ***Hydrolysis:*** Calcium hydroxide reacts with chloral to give chloroform

$$\underset{\text{Chloral}}{2CCl_3CHO} + Ca(OH)_2 \xrightarrow{\text{Hydrolysis}} \underset{\text{Chloroform}}{2CHCl_3} + \underset{\text{Calcium formate}}{(HCOO)_2Ca}$$

A similar reaction takes place in the case of acetone

$$\underset{\text{acetone}}{(H_3C)_2C{=}O} \xrightarrow[\text{Chlorination}]{+3Cl_2} \underset{\text{1,1,1-trichloroacetone}}{H_3C{-}C(=O){-}CCl_3} + 3HCl$$

$$\underset{\text{1,1,1-trichloroacetone}}{H_3C{-}C(=O){-}CCl_3} \xrightarrow[\text{hydrolysis}]{Ca(OH)_2} \underset{\text{chloroform}}{H{-}CCl_3} + \underset{\text{Calcium Acetate}}{(HO)(O{=})C{-}Ca{-}C(=O)(OH)}$$

Pure Chloroform

It is prepared by the distillation of chloral with sodium hydroxide

$$\underset{\text{Chloral}}{CCl_3CHO} + NaOH \rightarrow \underset{\text{Chloroform}}{CHCl_3} + \underset{\text{Sodium Acetate}}{HCOONa}$$

Industrial Preparation

Chloroform in industry is prepared by the electrolysis of sodium chloride in the presence of ethyl alcohol in down cell

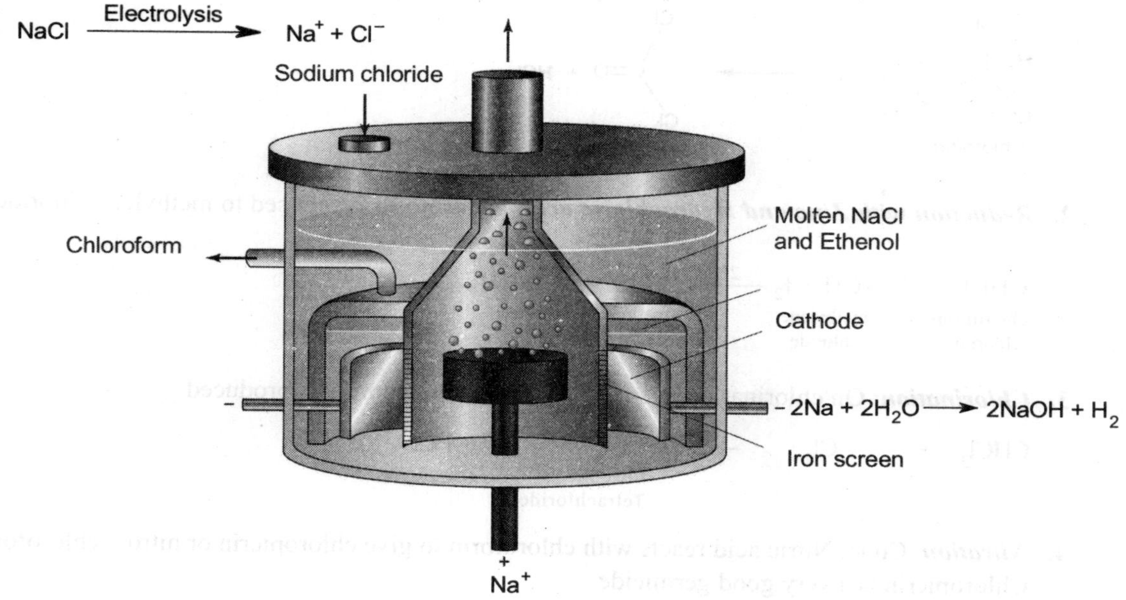

Fig. 8.4. The Industrial Preparation of Chloroform in Down's cell

Properties

Physical Properties

Melting point	-63°C
Boiling Point	61.26°C
Density (gm/mol)	1.44845
Refractive Index	1.44643
Solubility(g/100ml)	
Water	1.0
Alcohol	¥
Others	¥

Chemical Properties

1. ***Oxidation***: On exposure to light and damp air it is oxidised to carbonyl chloride or phosgene gas it is poisonous and attacks respiratory organs.

$$CHCl_3 + O \longrightarrow COCl_2 + HCl$$

Chloroform → Carbonyl dichloride

2. ***Reduction with Zinc and Hydrochloric acid:*** Chloroform is reduced to methylene chloride

$$CHCl_3 \xrightarrow{2H} CH_2Cl_2 \xrightarrow{2H} CH_3Cl \xrightarrow{2H} CH_4$$

chloroform chloride, methylene chloride, methyl, methane

3. ***Chlorination***: On chlorination in sunlight carbon tetrachloride is produced

$$CHCl_3 + Cl_2 \rightarrow CCl_4 + HCl$$

Carbon Tetrachloride

4. ***Nitration***: Conc. Nitric acid reacts with chloroform to give chloropicrin or nitro – chloroform Chloropicrin is a very good germicide

$$CHCl_3 + HO-NO_2 \longrightarrow Cl_3C-NO_2 + H_2O$$

Chloroform, Nitric acid → Trichloro(nitro)methane chloropicrin

5. ***Condensation***: With acetone in the presence of NaOH Chloropicrin is also produced

$$(CH_3)_2C{=}O + HCCl_3 \xrightarrow{NaOH} (CH_3)_2C(OH)CCl_3$$

Acetone, Chloroform → 1,1,1-trichloro-2-methylpropan-2-ol chloroetone

Chlroetone is used as hypnotic drug and also for the sea-sickness treatment.

6. ***Hydrolysis***: When heated with alcoholic caustic potash chloroform is hydrolyzed to potassium formate

$$\underset{\text{chloroform}}{H{-}C(Cl)(Cl)(Cl)} + KOH \longrightarrow 3KCl + \underset{\text{methanetriol}}{H{-}C(O{-}H)(O{-}H)(O{-}H)}$$

$$\underset{\text{Methanetriol}}{H{-}C(O{-}H)(O{-}H)(O{-}H)} \longrightarrow \underset{\text{Formic acid}}{H{-}O{-}C(=O){-}H} + H_2O$$

$$\underset{\text{Formic acid}}{H{-}O{-}C(=O){-}H} + KOH \longrightarrow \underset{\text{Potassium formate}}{H{-}C(=O){-}O{-}K}$$

7. ***Hofmann's Carbylamines Reaction***: When a drop of chloroform is added to aniline or any primary amine and alcoholic KOH and the mixture is gently warmed unpleasant and distinguished smell of phenyl carbylamine (phenyl isocyanide) is given out.

$$\underset{\text{Aniline}}{C_6H_5NH_2} + CHCl_3 + \underset{\text{Chloroform}}{3KOH} \rightarrow \underset{\text{Alcoholic Carbanyl Aniline}}{C_2H_5N{=}C} + \underset{\text{Phenyl}}{3KCl}$$

The reaction is used as a delicate test for the identification of chloroform and also for primary amines.

8. ***Heating with Silver Metal***: On warming with silver powder it gives acetylene

$$2CHCl_3 + 6Ag \rightarrow CH.CH + 6AgCl$$

Chloroform as Anæsthetic

Chloroform used as an anæsthetic must be pure, when exposed to light and damp air it is rendered highly poisonous due to formation of phosgene gas.

(i) chloroform is prepared from chloral

(ii) It is stored in closed bottle to avoid the action of light

(iii) The bottles containing chloroform are kept filled full to the brim to expel all the external air

(iv) One present of ethyl alcohol is added to chloroform to destroy phosgene by combining with it to form the neutral carbonate.

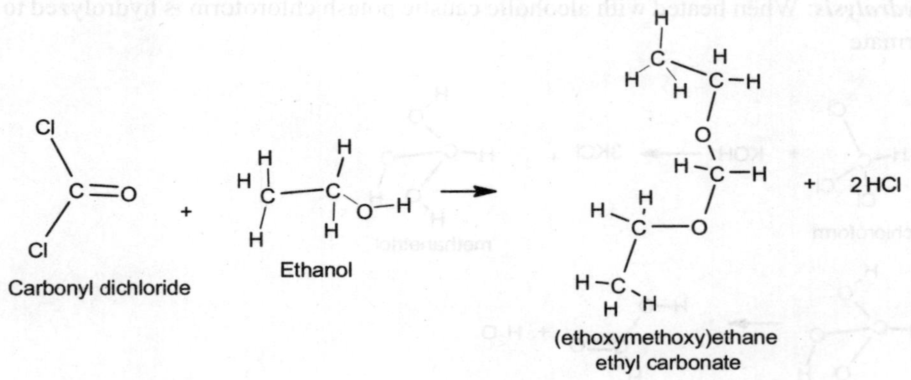

Carbonyl dichloride + Ethanol → (ethoxymethoxy)ethane ethyl carbonate + 2 HCl

Modern Anæsthetics

In the 1970s chemists produced the compound halothane (2-bromo-2-chloro-1,1,1-trifluoroethane), which gives deep yet safe anæthesia. Soon afterwards two similar compounds were developed – enflurane and isoflurane. These new anaesthetics have allowed surgery to progress so that not only are operations much safer for patient, but an enormous variety of new surgical procedures have become possible. These range from operations carried out through minute openings made in the body wall to massive transplant operations involving both patients and doctors in many hours in the operating theatre.

The fact that modern anaesthetics are not poisonous makes surgery lasting more than 24 hours a possibility. They also make it possible for patients to have surgery and go home the same day, because the anaesthetic itself wears off fairly quickly and has relatively few side effects.

1. The structural formulae of three of the most important modern anaesthetics are shown below. Enflurane and isoflurane are isomers. The structural formulæ of some of them are given below:

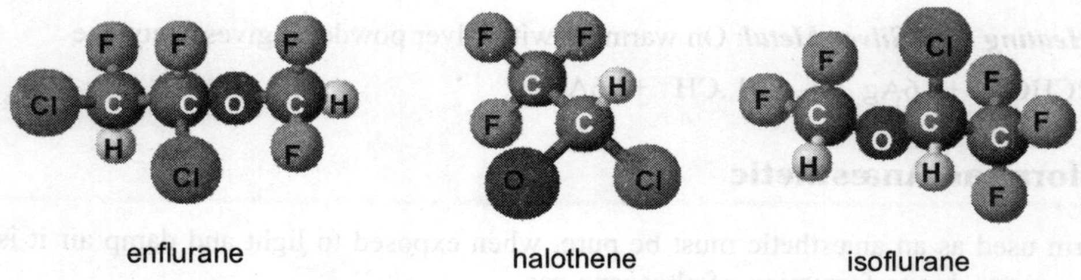

enflurane halothene isoflurane

The other uses of the chloroform are as follows:-

(a) As a solvent,
(b) As a lab reagent
(c) As a preservative
(d) As a flavouring agent to distinguish test of nauseous medicines

BROMOFORM

Formula: $CHBr_3$

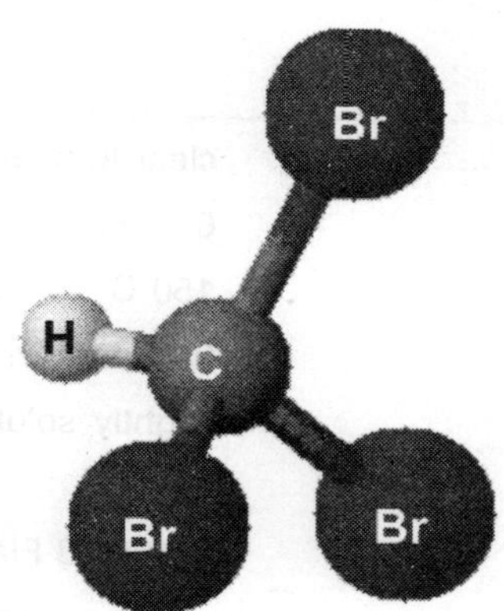

Preparation

Bromoform is prepared by the action of liquid bromine (prepared by the action of potassium bromide and conc. Sulphuric acid see inorganic chemistry chapter 12 halogens.) on alkaline alcoholic mixture ($C_2H_5OH + NaOH$ or Na_2CO_3). Reaction equation are similar to iodoform

(i) Oxidation

$$\underset{\text{Alcohol}}{C_2H_5OH} + Br_2 \rightarrow \underset{\text{Acetone}}{CH_3CHO} + 2HBr$$

(ii) Halogenation

$$\underset{\text{Acetone}}{CH_3CHO} + 3Br_2 \rightarrow \underset{\substack{\text{Tribromoacetaldehyde}\\ \text{Bromal}}}{CBr_3CHO} + 3HBr$$

(iii) Hydrolysis

$$CBr_3CHO + NaOH \rightarrow CHBr_3 + (COOH)Na$$

Other Methods of Preparation

From 2 - butanone

$$\underset{\text{butan-2-one}}{H_3CCH_2-\overset{\overset{\large O}{\|}}{C}-CH_3} \xrightarrow[-OH]{\text{excess } Br_2} \underset{\text{1,1,1-tribromobutan-2-one}}{H_3CCH_2-\overset{\overset{\large O}{\|}}{C}-CBr_3} \xrightarrow{-OH} \underset{\text{propionate}}{H_3CCH_2-\overset{\overset{\large O}{\|}}{C}-O^-} + \underset{\text{bromoform}}{HCBr_3}$$

Properties

Physical Properties

Physical State	clear to pale yellow liquid with a sweet odor
Melting Point	6 - 7 C
Boiling Point	150 C
Specific Gravity	2.89
Solubility in Water	slightly soluble
Vapor Density	8.8
Nfpa Ratings	Health: 3 Flammability: 0 Reactivity: 0
Flash Point	nonflammable
Stability	Stable under ordinary conditions

Chemical Properties

It is similar to chloroform and iodoform

Uses and Application of Bromoform

Bromoform is mainly used as a reagent for the extraction of minerals, the process is explained below:

Diamond Extraction by Caustic Dissolution

Processing for Micro diamond Analysis

Introduction

Caustic dissolution of exploration samples efficiently produces a concentrate from which diamonds can readily be extracted during microscopic examination. The process uses diamond's property of high resistance to caustic soda (NaOH). Use of this method eliminates diamond size reduction and losses that often occur during extraction procedures that rely on crushing and attrition milling.

Procedure

Very few minerals survive the harsh attack by caustic soda; therefore weight reductions commonly exceed 99% of the initial sample weight. Samples are divided into equally sized charges of less than 8 kg. Smaller charge sizes are necessary if the sample contains a high proportion of carbonate minerals that are vigorously reactive with NaOH (evaluated by an acid test completed prior to charge preparation). If a high proportion of the sample is composed of greater than 8 cm fragments, simple breakage, crushing or attrition milling may be required, or the length of the dissolution process increased. Client consultation and approval is necessary before any size reduction of the sample is initiated.

After digestion in molten caustic soda, the sample is poured onto a large diameter 150 mesh screen. The + 150 mesh residue is liberated from the NaOH by washing the sample in a series of water and acid leach (HCl) baths. Once all of the NaOH is dissolved and removed, the concentrate is dried and screened on a 6 mesh screen to remove undigested material. The undigested material is examined microscopically by a mineralogist. If the + 6 mesh material is significant or consists of possible diamondiferous rock fragments, further digestion may be required. If the undigested material is of insignificant size or not considered as a possible source of diamonds, the - 6 mesh residue is further processed by a two (possibly three if the residue is large) stage magnetic separation procedure utilizing a permanent magnet and a Frantz Barrier Magnetic Separator.

The magnetically characterised residue is then submitted for microscopic examination and diamond selection. (In addition to diamonds, the residue may contain partially undigested indicator minerals, colourless to opaque spinel, garnet, ilmenite, graphite, moissanite, zircon and kyanite.) Each of the magnetic fractions is examined at a magnification of 400x using a binocular microscope. Grains of questionable mineralogy are examined using a scanning electron microscope equipped with an energy dispersive spectral (SEM-EDS) analyser. Although each magnetically characterised fraction is examined, particular emphasis is given to the diamagnetic portion.

The X, Y and Z dimensions of selected microdiamonds are measured in millimetres. Macrodiamonds are weighed individually while microdiamonds are weighed in groups of 20 or 30 and the milligram weight, in each case, converted to carats. The colour, clarity and morphology of each diamond are determined and all observations reported in a Certificate of Analysis.

Quality Control

At Lakefield routine quality control tests are utilised to evaluate the efficiency of the caustic dissolution processing technique by running blank samples spiked with "Congo Rounds". The chance of diamond or indicator mineral contamination is evaluated by running caustic soda blanks between client's samples and examining the residue for microdiamonds and indicator minerals. Recovery of the diamond spikes typically ranges from 97 to 100%. 1998 statistics showed that, on average, only a single indicator mineral grain was carried over into the caustic soda blanks run between different client's samples.

Each residue is picked twice by separate diamond pickers. Questionable grains are examined by SEM-EDS for verification.

Every effort is made at each stage of sample handling during caustic dissolution, residue preparation and diamond picking to eliminate any possibilities of contamination. These steps include:

- A rigorous sample tracking procedure.
- Dedicated screens and equipment for each sample during sample processing.
- Replacement of screens between each sample after pouring caustic soda.
- Thorough washing and scrubbing of all sample containers.
- Thorough cleaning of equipment used to prepare caustic residues between each processed sample.
- Sandblasting of each kiln pot once a month to remove any scale build-up that might entrap micro diamonds or indicator minerals.

Kimberlite Indicator Minerals

For each micro-diamond sample interval, a kimberlite indicator mineral (KIM) sample of 5 to 10 kg will be collected. This sample will be a representative composite section collected from the remaining split portion of the core approximately every 2m.

The KIM sample material is put into well labelled poly sample bags (double bagged) and sealed using locking ties. A tag displaying drill hole number and depth interval is placed into each sample bag.

Prior to shipping, the KIM samples are packed into five gallon pails and sealed. The samples are shipped to the Lab for processing and analysis. Upon each shipment, a fax is transmitted to the lab and to Diamonds North's head office listing the samples shipped; date shipped, means of transport, and expected arrival date. Upon receipt, the lab conducts a full inventory of the shipment and reports the information to Diamonds North for confirmation.

KIM Sample Processing and Analysis

Kimberlite indicator mineral data for a given kimberlite provides supporting information with the micro-diamond data for the assessment of the economic potential of a kimberlite. The abundance of indicators and chemical signature of the mineral grains indicate the nature and composition of the mantle (how diamondiferous) and how well it was sampled by the kimberlite. Indicator minerals characterize each kimberlite and help to identify undiscovered kimberlites represented in nearby till.

Processing of the KIM Samples

Kimberlite samples requiring indicator mineral analysis will be sent to an experienced heavy mineral processing lab. The kimberlite material is concentrated for the extraction and analysis of indicator minerals. The method obtains a quantitative indication of the abundance of mantle-derived garnet, chromite, ilmenite and Cr-diopside as well as the mineral compositions of representative populations of these minerals.

The kimberlite is subjected to careful multistage crushing and screening to liberate indicator minerals using jaw and rotary crushing methods maximizing the yield of material in the 0.3 to 2.36 mm size range. The 0.3 to 2.36 mm size kimberlite fraction is concentrated using dense liquid processing (bromoform - SG = 2.9). The concentrate is meticulously sorted to extract kimberlite indicator minerals by methods that provide a quantitative indication of mineral abundance as well as a representative suite of grains for chemical analysis.

The extracted mineral grains are analyzed by electron microprobe to provide major and minor element compositions. Trace element data for a selection of peridotitic garnets is obtained using well-calibrated laser ablation ICP-MS methods. The numbers of grains of each indicator mineral for analysis varies and if sufficient grains are present in the sample the following number of grains are analyzed:

High-Cr garnet (2 wt% Cr_2O_3)	100
Low-Cr garnet (<2 wt% Cr_2O_3)	50
Chromite	40
Ilmenite	40
Cr-diopside	40

Trace element analysis (by LA-ICP-MS) for selected grains ideally, all G10-type high-Cr garnets (up to about 30), as well as a compositionally representative population of G9-type high-Cr garnets may be conducted.

Petrography

In addition to the micro-diamond and KIM samples, representative petrology samples (likely 1 or 2 samples per interval) corresponding to each micro-diamond sample interval is collected and sent to the lab for thin section description and photo microscopy. The thin section descriptions provide insight into the diamond carrying capacity of the kimberlite.

Representative Sample Suite

During the core logging procedure, the geologist retains a small representative suite of reference samples from each hole (collected from the split portion of the core). The core samples are labeled and retained in a core box that is shipped to Diamonds North's head office at the end of the programme.

Core Storage

The remaining split core is sealed with a lid and screws and shipped to Yellowknife or Vancouver at the end of the programme. Secure storage is arranged by Diamonds North. This core is retained for follow-up work and re-analysis if needed.

IODOFORM

Formula: CHI_3

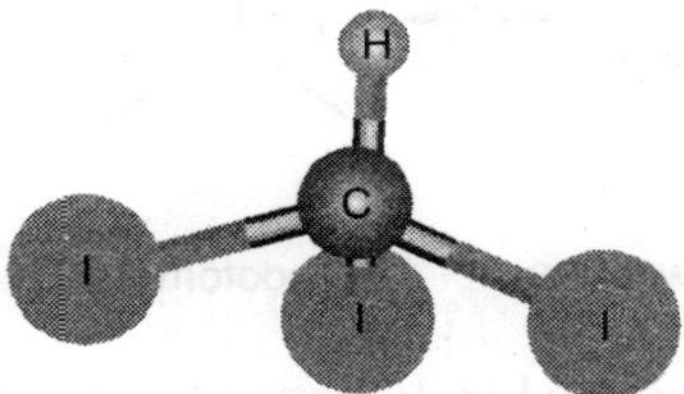

Iodoform closely resembles chloroform in the mode of preparation and chemical process reaction:-

Preparation

By warming a mixture of iodine and ($NaOH$ or Na_2CO_3) and ethyl alcohol or acetone

The process is probably similar to the formation of chloroform. The reaction are:

(i) ***Oxidation***: -

$$\underset{\text{Alcohol}}{CH_3CH_2OH} + I_2 \rightarrow \underset{\text{Acetaldehyde}}{CH_3CHO} + 2HI$$

(ii) ***Halogenation: -***

$$\underset{\text{Acetaldehyde}}{CH_3CHO} + I_2 \rightarrow \underset{\text{Iodal}}{CI_3CHO} + 3HI$$

(iii) ***Hydrolysis: -***

CI_3CHO (Iodal) + NaOH → CHI_3 (Iodoform) + HCOONa (Sodium Formate)

Or [$2CI_3CHO$ (Chloral) + Na_2CO_3 (Sodium Carbonate) → $2CHI_3$ (Iodoform) + HCOONa (Sodium Formate) + CO_2]

(iv) 5HI + 5NaOH → 5NaI + $5H_2O$

The reaction with acetone are: -

(1) Halogenation

$H_3C-CO-CH_3$ (acetone) + $3I_2$ → $CI_3-CO-CH_3$ (1,1,1-triiodoacetone) + 3HI

(2) Hydrolysis

$CI_3-CO-CH_3$ (1,1,1-triiodoacetone) $\xrightarrow{NaOH}$ HCI_3 (iodoform) + $H_3C-C(=O)-O-Na$ (Sodium Acetate)

Two parts of Na_2CO_3 are dissolved in 10 parts of water, one part of ethyl alcohol or acetone is pored in the solution. This is warmed to about 60 - 70°C on an heating mantle and one part of iodine is gradually added to it till a reddish tinge is obtained. A little NaOH or Na_2CO_3 solution is also added to discharge the red colour of the liquid and the temperature is kept at 60-70°C when yellow crystals of iodoform separates out. The crystals of iodoform are followed washed and purified by recrystallization from alcohol dilute alcohol.

Manufacturing of Lodoform

By Electrolysis of a dilute alcohol or acetone solution of KI containing Na_2CO_3 at 65°C.

$$KI \xrightarrow{\text{electrolysis}} K^+ + I^-$$

at cathode: $2K + 2H_2O \rightarrow 2KOH + H_2O\uparrow$

at anode: Iodine is set free

or $C_2H_5OH + 6KOH + 4I_2 \rightarrow CHI_3 + HCOOK + 5KI$

Ethenol — Iodoform — Potassium Formate

Properties

Physical Properties

Melting Point	119°C
Boiling Point	210°C
Density g/mol	4.008
Solubility(g/ 100ml)	
Water	0.01
Alcohol	1.3
Others	13.06

1. It is a yellow solid.
2. It has a characteristic unpleasant smell.
3. It is soluble in water but readily dissolves in alcohol.
4. it sublimes readily and is volatile in steam.

Chemical Properties

It its chemical characteristic it behaves like chloroform

1. ***Oxidation***: Methylene iodide is formed.

$$CHI_3 + 2H \rightarrow CH_2I_2 + HI$$

Iodoform — Methylene Iodide

2. ***With Silver Nitrate***: A light yellow precipitate is produced, AgI. This indicates that CHI_3 is less stable than chloroform.

$$2CHI_3 + 6AgNO_3 + 5H_2O \rightarrow 6AgI + 6HNO_3 + C_2H_5OH + 2O_2$$

Iodoform — Silver nitrate — Silver Iodide — Ethanol

3. ***Hydrolysis***: With alcoholic KOH potassium iodide and potassium formate is produced

$$\underset{\text{Iodoform}}{CHI_3} + 4KOH \rightarrow 3KI + \underset{\text{Pot Formate}}{HCOOK} + 2H_2O$$

4. ***Carbylamine Reaction***: With aniline and alcoholic KOH phenyl isocyanide is formed

$$\underset{\text{Aniline}}{C_6H_5NH_2} + \underset{\text{Iodoform}}{CHI_3} + 3KOH \xrightarrow{-3H_2O} \underset{\text{Phenyl Isocyanide}}{C_6H_5Na.C} + 3KI$$

5. ***Heating with Ag Metal***: Acetylene is formed

$$2CHI_3 + Ag \rightarrow HC.CH + 6AgI$$

Uses

It is used in medicine and surgery as a strong antiseptic and disinfectant, but on account of its unpleasant smell. Other organic iodine compounds (such as Iodole C_4I_4NH) has been proposed substitutes, under the commercial name betadine.

The important member of this class is carbon tetrachloride

Carbon Tetrachloride or Tetrachloromethane

Formula: CCl_4

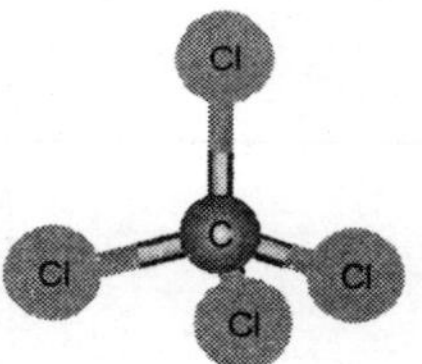

Preparation

1. By passing chlorine gas into boiling chloroform in sunlight

$$\underset{\text{Chloroform}}{CHCl_3} + Cl_2 \xrightarrow{\text{Sunlight}} \underset{\text{Carbon Tetrachloride}}{CCl_4} + HCl$$

2. By the action of sulphur monochloride in carbon disulphide in the presence of $FeCl_3$ etc. as catalyst. Carbon tetrachloride obtained by any of the above process is purified by re distillation.

Properties

Physical Properties

1. It is colourless pleasant smelling heavy liquid.
2. It is insoluble in water but dissolves in organic solvents.
3. It is non-inflammable and non-explosive.
4. Much inhalation of the vapour can cause lot of sneezing.

Chemical Properties

It does not decompose in sunlight or air.

1. ***Reduction***: On treatment with nascent hydrogen it is converted to chloroform, methylene chloride, and methane.

$$CCl_4 \xrightarrow[(-HCl)]{2H} CHCl_3 \xrightarrow[(-HCl)]{2H} CH_2Cl_2 \xrightarrow[(-HCl)]{2H} CH_3Cl \xrightarrow[(-HCl)]{2H} CH_4$$

tetrachloromethane → chloroform → dichloromethane → chloromethane → methane

2. ***Hydrolysis***: On hydrolysis with hot alcoholic it gives potassium carbonate and potassium chloride

$$CCl_4 + KOH \rightarrow K_2CO_3 + 4KCl + 3H_2O$$

4. ***Stability***: Carbon Tetrachloride is fairly stable at high temperature. It is however, reacts with steam above 500°C giving phosgene gas ($COCl_2$)

$$\underset{\text{Carbon Tetrachloride}}{CCl_4} + \underset{\text{Steam}}{H_2O} \rightarrow \underset{\text{Phosgene}}{COCl_2} + 2HCl$$

5. ***With fuming sulphuric acid***: It reacts to give phosgene and pyrosulphuryl chloride

$$\underset{\text{Carbon Tetrachloride}}{CCl_4} + \underset{\text{Fuming Sulphuric acid}}{SO_3} \rightarrow \underset{\text{Phosgene}}{COCl_2} + \underset{\text{Pyrosulphuryl chloride}}{S_2O_5Cl_2}$$

Uses

In the past, carbon tetrachloride was widely used as a cleaning fluid (in industry and dry cleaning establishments as a degreasing agent, and in households as a spot remover for clothing, furniture, and carpeting). Carbon tetrachloride was also used in fire extinguishers and as a fumigant to kill insects in grain. Most of these uses were discontinued in the mid-1960s. Until recently, carbon tetrachloride was used as a pesticide, but this was stopped in 1986. Until 1986, carbon tetrachloride was also used as a pesticide.

For Revision (Conversion)

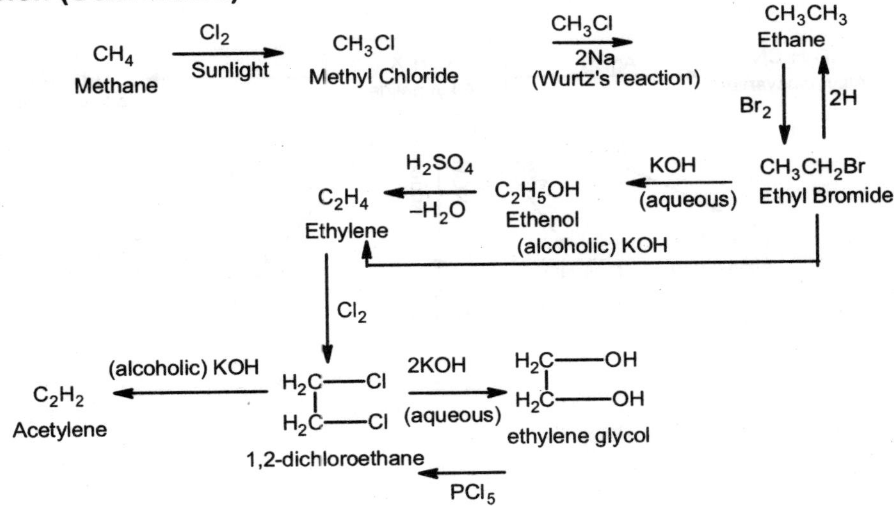

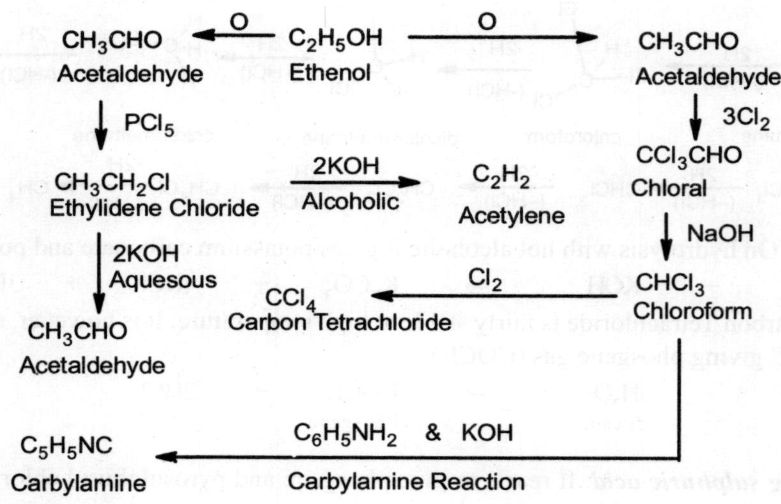

CH_3CHO
Acetaldehyde
O
C_2H_5OH
Ethenol
O
CH_3CHO
Acetaldehyde
PCl_5
$3Cl_2$
CCl_3CHO
Chloral
CH_3CH_2Cl
Ethylidene Chloride
2KOH
Alcoholic
C_2H_2
Acetylene
NaOH
2KOH
Aquesous
Cl_2
$CHCl_3$
Chloroform
CCl_4
Carbon Tetrachloride
CH_3CHO
Acetaldehyde
C_5H_5NC
Carbylamine
$C_6H_5NH_2$ & KOH
Carbylamine Reaction

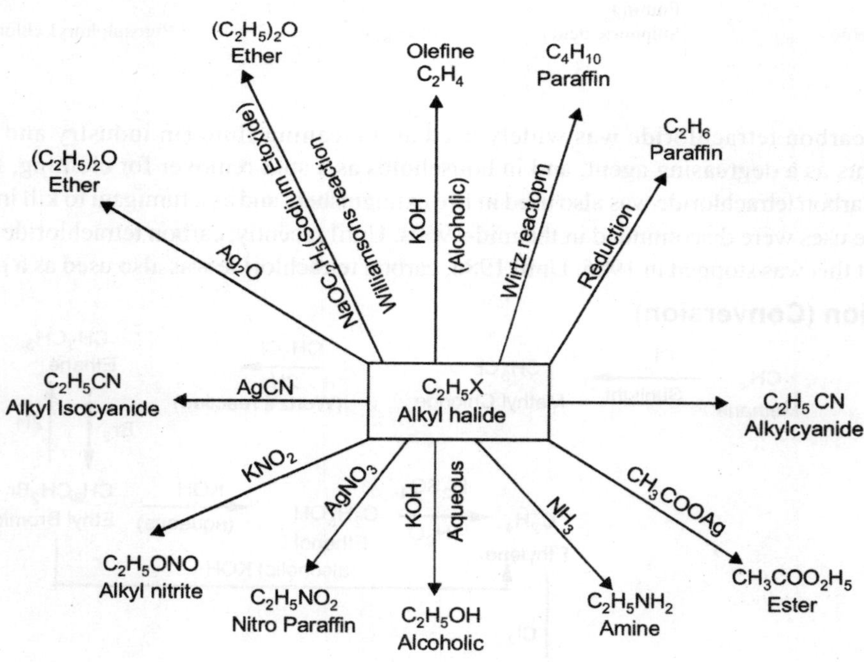

$(C_2H_5)_2O$
Ether
Olefine
C_2H_4
C_4H_{10}
Paraffin
C_2H_6
Paraffin
$(C_2H_5)_2O$
Ether
Ag_2O
$NaOC_2H_5$ (Sodium Etoxide)
Williamsons reaction
KOH
(Alcoholic)
Wirtz reactopm
Reduction
C_2H_5CN
Alkyl Isocyanide
AgCN
C_2H_5X
Alkyl halide
C_2H_5 CN
Alkylcyanide
KNO_2
$AgNO_3$
KOH
Aqueous
NH_3
CH_3COOAg
C_2H_5ONO
Alkyl nitrite
$C_2H_5NO_2$
Nitro Paraffin
C_2H_5OH
Alcoholic
$C_2H_5NH_2$
Amine
$CH_3COO_2H_5$
Ester

CHAPTER 9

Alcohols

Introduction and History

Alcohols have a long history back since as early as 6000 B.C. The best element of leisure and amusement was wine. We find the word "*som rass*", the chronology of the history of wine has being shortly discussed for the interest of the readers.

The word 'alcohol' comes from the Arabic word "*al-khwl* or *al-ghawl*", previously wines were generally prepared by the fermentation of carbohydrates.

History of Wines

> "The peoples of the Mediterranean began to emerge from barbarism when they learnt to cultivate the olive and the vine." —Thucydides, Greek Historian, 5th Century BC.

Throughout time, wine has had a special influence on western culture, a sentiment which is evidenced in the above quote. From its earliest development, wine has had a special place in our customs, diet and social gatherings. From France to Australia, wine is produced and enjoyed in large quantities. Few areas of the world remain untouched by its many virtues. The growth of wine's popularity closely resembles the development of the western world. As the wine trade began to spread throughout the world, western civilization also advanced and grew. This is evident throughout the study of wine's history and diversification. In today's modern world wine is widely discussed and consumed globally. The wine industry has undergone a massive boom in the last few decades resulting in many varieties of wine from which to choose. This boom should continue into the near future as we discover even more technologies and benefits that wine brings us.

> "I think wine has taken over from the toys of the old days like watches and cars. Wine shows you have money, but it also shows you have taste." —Thomas Matthews.

List of Alcohols

Name	Formula	M.P. °C	B.P. °C	Relative density (at 20°C)	Solubility, g/100 g H_2O
Methyl	CH_3OH	–97	64.5	0.793	∞
Ethyl	CH_3CH_2OH	–115	78.3	.789	∞
n-Propyl	$CH_3CH_2CH_2OH$	–126	97	.804	∞
n-Butyl	$CH_3(CH_2)_2CH_2OH$	–90	118	.810	7.9
n-Pentyl	$CH_3(CH_2)_3CH_2OH$	–78.5	138	.817	2.3
n-Hexyl	$CH_3(CH_2)_4CH_2OH$	–52	156.5	.819	0.6
n-Heptyl	$CH_3(CH_2)_4CH_2OH$	–34	176	.822	0.2
n-Octyl	$CH_3(CH_2)_6CH_2OH$	–15	195	.825	0.05
n-Decyl	$CH_3(CH_2)_8CH_2OH$	6	228	/829	
n-Dodecyl	$CH_3(CH_2)_{10}CH_2OH$	24			
n-Tetradecyl	$CH_3(CH_2)_{12}CH_2OH$	38			
n-Hexadecyl	$CH_3(CH_2)_{14}CH_2OH$	49			
n-Octadecyl	$CH_3(CH_2)_{16}CH_2OH$	58.5			
Isopropyl	$CH_3CHOHCH_3$	–86	82.5	.789	∞
Isobutyl	$(CH_3)_2CHCH_2OH$	–108	108	.802	10.0
sec-Butyl	$CH_3CH_2CHOHCH_3$	–114	99.5	.806	12.5
tert-Butyl	$(CH_3)_3COH$	25.5	83	.789	∞
Isopentyl	$(CH_3)_2CHCH_2CH_2OH$	–117	132	.813	2
active-Amyl	(–)-$CH_3CH_2CH(CH_3)CH_2OH$		128	.816	3.6
tert-Pentyl	$CH_3CH_2C(OH)(CH_3)_2$	–12	102	.809	12.5
Cyclopentanol	*cyclo*-C_5H_9OH		140	.949	
Cyclohexanol	*cyclo*-$C_6H_{11}OH$	24	161.5	.962	
Allyl	$CH_2{=}CHCH_2OH$	–129	97	.855	∞
Crotyl	$CH_3CH{=}CHCH_2OH$		118	.853	16.6
Methylvinyl-methanol	$CH_2{=}CHCHOHCH_3$		97		
Benzyl	$C_6H_5CH_2OH$	–15	205	1.046	4
α-Phenylethyl	$C_6H_5CHOHCH_3$		205	1.013	
β-Phenylethyl	$C_6H_5CH_2CH_2OH$	–27	221	1.02	1.6
Diphenylmethanol	$(C_6H_5)_2CHOH$	69	298		0.05
Triphenylmethanol	$(C_6H_5)COH$	162.5			
Cinnamyl	$C_6H_5CH{=}CHCH_2OH$	33	257.5		
1, 2-Ethanediol	CH_2OHCH_2OH	–16	197	1.113	
1, 2-Propanediol	$CH_3CHOHCH_2OH$		187	1.040	
1, 3-Propenediol	$HOCH_2CH_2CH_2OH$		215	1.060	
Glycerol	$HOCH_2CHOHCH_2OH$	18	290	1.261	
Pentaerythritoal	$C(CH_2OH)_4$	260			6

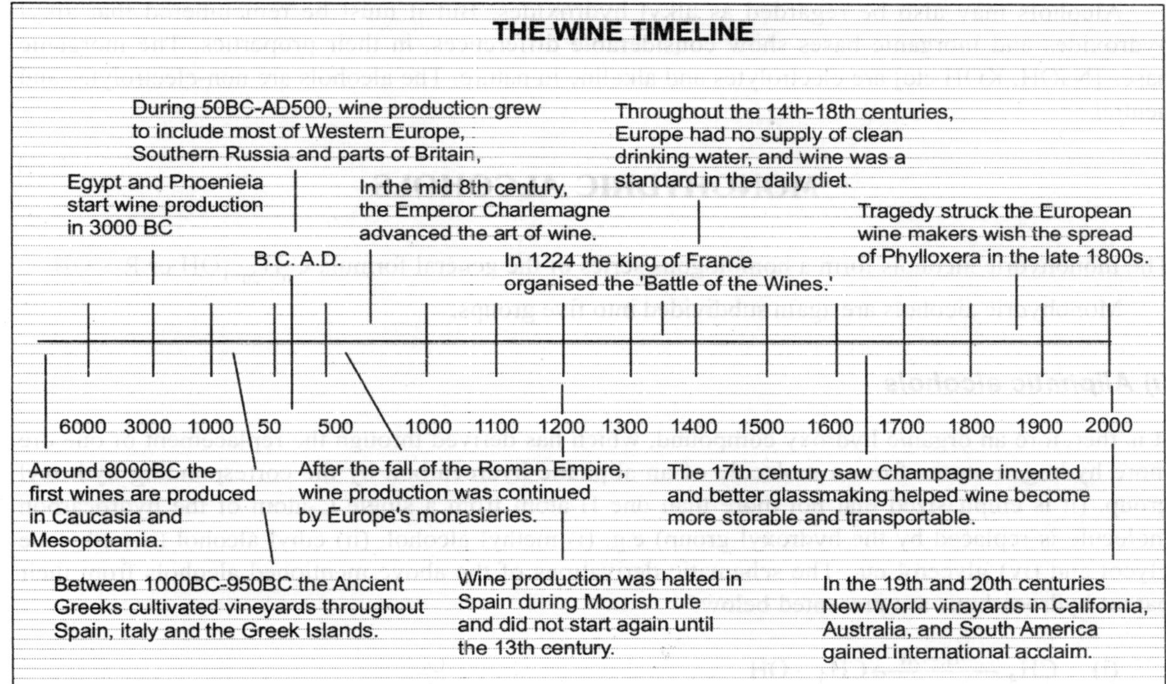

Chemistry of Alcohols

We shall now discus the chemistry and physical profile and different ways of synthesis and action of alcohols including their classification and derivations and conversions.

The alcohols have derived form the aliphatic hydrocarbons by the substitution of the monovalent hydroxyl (-OH) groups for one or more atoms of hydrogen. The important classes of alcohols are being stated below:

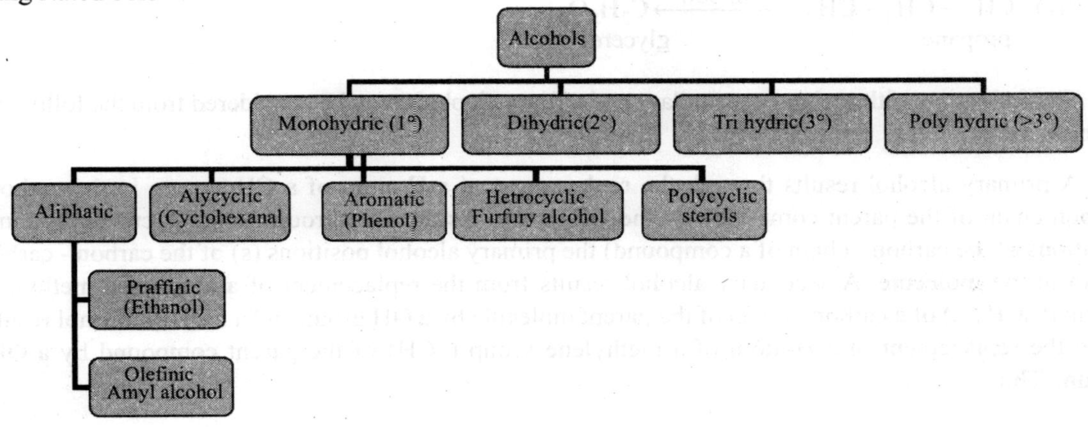

Alcohols may also be regarded as alkyl hydroxides. But it must be remembered that alkyl hydroxides and inorganic bases show considerable differences, in their properties. The inorganic bases [NaOH, KOH etc] are electrolytes and alkaline in nature. The alcohols are non-electrolytes and neutral.

MONOHYDRIC ALCOHOLS

The monohydric alcohols form a homologous series of the general formula $C_nH_{2n+1}OH$ or R – OH.

Monohydric alcohols are again subdivided into five groups:

(i) Aliphatic alcohols

It is therefore an organic hydroxy compound, which has derived through the replacement of one ore more hydrogen atoms from a molecule of an aliphatic hydrocarbon by the corresponding hydroxyl group. (it is emphasized that not more than one H-atom from a single C-atom of the hydrocarbon molecule is replaced by the hydroxyl group) e.g. (i) methyl alcohol, (ii) ethyl alcohol (iii) ethylene glycol and (iv) glycerol etc. The schematic derivations of the above mentioned alcohols from their parent hydrocarbons are presented below:

(i) $\underset{\text{methane}}{CH_4} \xrightarrow{-H,+OH} \underset{\text{methyl alcohol}}{CH_3-OH}$

(ii) $\underset{\text{ethane}}{CH_3-CH_3} \xrightarrow{-H,+OH} \underset{\text{ethyl alcohol}}{CH_3CH_2OH}$

(iii) $\underset{\text{ethane}}{CH_3-CH_3} \xrightarrow{-2H,+2OH} \underset{\text{ethylene glycol}}{C_2H_6O_2}$

(iv) $\underset{\text{propane}}{CH_2-CH_3-CH_2} \xrightarrow{-3H,+3OH} \underset{\text{glycerol}}{C_3H_8O_3}$

The formation of the primary secondary and tertiary alcohols may be considered from the following point of view:

A primary alcohol results through the replacement of a H-atom of a CH_3 group in the carbon-carbon chain of the parent compound by the OH group. As the CH_3 group cannot occur in the inner positions of the carbon - chain of a compound) the primary alcohol positions (s) of the carbon - carbon chain in the molecule. A secondary alcohol results from the replacement of a atom of a methylene group (%CH_2%) of a carbon - chain of the parent molecule by a OH group and a tertiary alcohol results from the replacement of a H-atom of a methylene group (–CH) of the parent compound by a OH-group. Thus

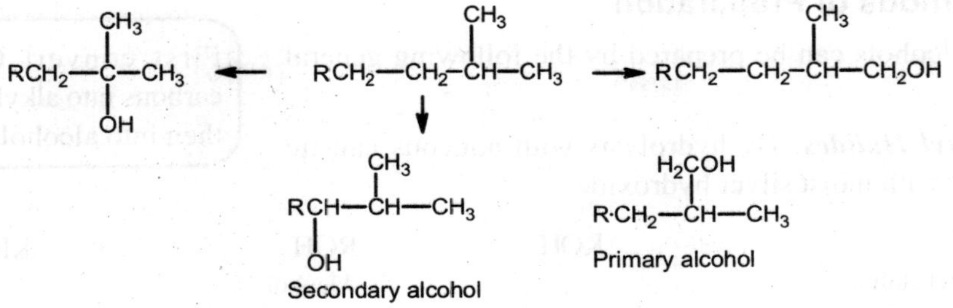

$$RCH_2-\underset{OH}{\overset{CH_3}{C}}-CH_3 \leftarrow RCH_2-CH_2-\overset{CH_3}{CH}-CH_3 \rightarrow RCH_2-CH_2-\overset{CH_3}{CH}-CH_2OH$$

$$RCH(OH)-\overset{CH_3}{CH}-CH_3 \qquad R{\cdot}CH_2-\overset{H_2COH}{CH}-CH_3$$

Secondary alcohol   Primary alcohol

The examples of primary alcohols:

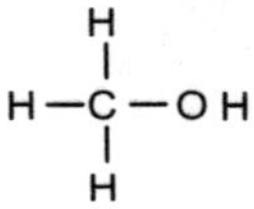

CH_3OH Methanol, Methyl Alcohol

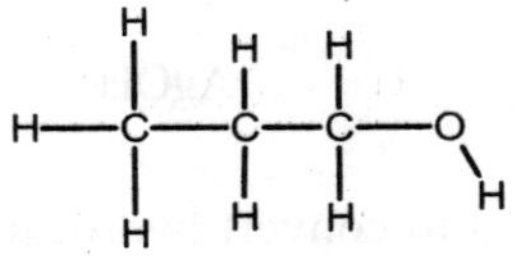

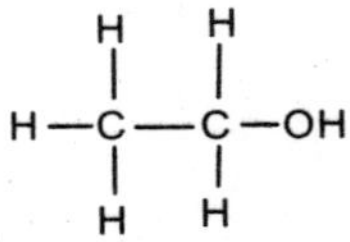

C_2H_5OH Ethanol, Ethyl Alcohol

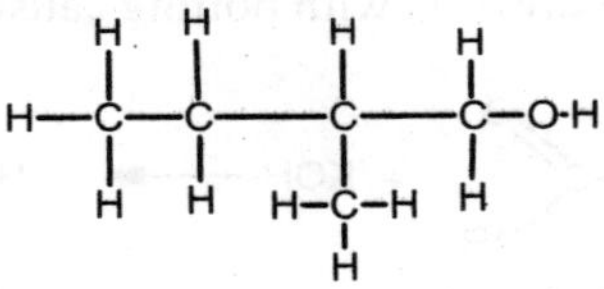

Propyl Alcohol, propan-1-ol, C_3H_7OH

Amyl Alcohol, $C_5H_{11}OH$

The examples of Secondary alcohols

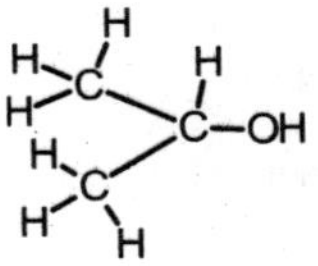

C_3H_7OH propan-2-ol, Isopropyl Alcohol

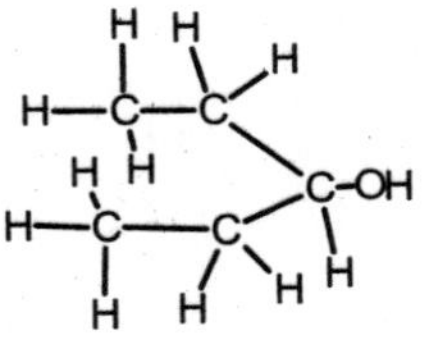

$C_5H_{11}OH$ pentan-3-ol Isopentyl Alcohol, Diethyl Carbinol sec-amyl alcohol

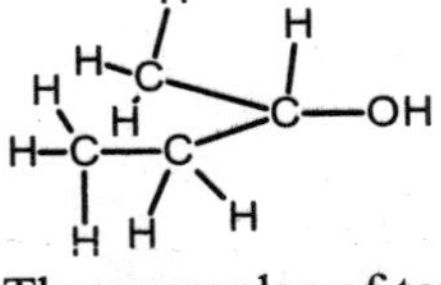

C_4H_9OH butan-2-ol Isobutyl Alcohol, Methyl Ethyl Carbinol

The examples of tertiary alcohol

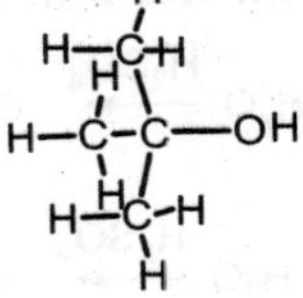

C_4H_9OH

2-methylpropan-2-ol
Trimethyl carbinol

General Methods of Preparation

> First convert the hydrocarbons into alkyl halides and then into alcohols

Monohydric alcohols can be prepared by the following general methods

(1) ***From Alkyl Halides***: By hydrolysis with aqueous caustic alkalies or with moist silver hydroxide

(i)	RX Alkyl Halide	+	KOH	→	ROH Alcohol	+	KBr
(ii)	CH_3Br Methyl bromide	+	KOH	→	CH_3OH Methyl Alcohol	+	KBr
(iii)	C_2H_5I Methyl Iodide	+	AgOH	→	C_2H_5OH Ethyl Alcohol	+	AgI

The methods can be used to convert hydrocarbons into alcohols.

(2) ***From Esters***: By hydrolysis with boiling caustic alkalies

$$R{-}C(=O){-}OR + KOH \longrightarrow H_3C{-}C(=O){-}O{-}K + R_1{-}OH$$

Methyl acetate; Potassium Methylate; Methanol

$$H_3C{-}C(=O){-}OCH_3 + KOH \longrightarrow H_3C{-}C(=O){-}O{-}K + R_1{-}OH$$

Methyl Acetate; Potassium Methylate; Methanol

(3) ***From Ester***: By hydrolysis with boiling caustic with boiling water

(i) $RHSO_4 + HOH \xrightarrow{Boil} R-OH + H_2SO_4$

Alyl hydrogen Sulphate

(ii) $C_2H_5HSO_4 + HOH \rightarrow C_2H_5OH + H_2SO_4$

ethyl hydrogen sulphate

(4) ***From Ethers***: By heating under pressure, with steam in the pressure of sulphuric acid

$$H_3C{-}O{-}CH_3 + H_2O \xrightarrow{H_2SO_4} C_2H_5OH + H_2SO_4$$

$$H_3CC\,H_2{-}O{-}CH_2CH_3 + H_2O \xrightarrow{H_2SO_4} C_2H_5OH + H_2SO_4$$

diethyl ether; Ethyl Alcohol

(5) ***From Aldehydes***: By relation with sodium amalgam and water or zinc and hydrochloric acid

(i) $R-CHO+2H \rightarrow R-CH_2OH$

(ii) $H-CHO+2H \rightarrow HCH_2OH$

(iii) $CH_3CHO+2H \rightarrow CH_3CH_2OH$

(6) ***From Ketones***: By reduction with sodium and dilute sulphuric acid secondary ketones are obtained:

$$(CH_3)_2C{=}O + 2H \longrightarrow (CH_3)_2CHOH$$

Acetone → Propan-2-ol (isopropyl alcohol)

(7) ***From Acid Chloride***: By catalytic reduction with hydrogen in the presence of platinum as catalyst

$$H_3C{-}COCl + 4H \longrightarrow CH_3CH_2OH$$

Acetyl chloride → Ethanol

(8) ***From Primary Amines***: By the action of nitrous acid (mixture of $NaNO_2$ and H_2SO_4 or HCl) on ethyl amine and methyl amine*

$$H_3CNH_2 + HO{-}N{=}O \longrightarrow H_3C{-}OH + N{\equiv}N + H_2O$$

Methanamine + Nitrous acid → Methanol + Nitrogen + Water

$$H_2C{=}CH{-}NH_2 + HO{-}N{=}O \longrightarrow H_3C{-}CH_2{-}OH + H_2O$$

Ethylenamine + Nitrous acid → Ethanol + Water

(9) ***From Grignard's Reagent***: Primary Alcohols: (a) By the action of oxygen and subsequent hydrolysis

$$CH_3{-}Mg{-}I + O \longrightarrow CH_3{-}O{-}Mg{-}I \xrightarrow{H_2O} H_3C{-}OH + Mg(OH)I$$

Magnesium Methyl Iodide → Magnesium Methoxy iodide → Methanol + Magnesium Hydroxy iodide

$$CH_3{-}CH_2{-}Mg{-}I + O \longrightarrow CH_3{-}CH_2{-}O{-}Mg{-}I \xrightarrow{H_2O} CH_3{-}CH_2{-}OH + Mg(OH)I$$

Magnesium Ethyl Iodide → Magnesium Ethoxy iodide → Ethanol + Magnesium Hydroxy iodide

*On reactioin with nitrous acid NH2 group is replaced by OH group always.

(b) By the action of formaldehyde and subsequent hydrolysis

$$H-CH=O + H_3C-Mg-Br \rightarrow H-CH(CH_3)-O-Mg^+$$

formaldehyde; Magnesium Methyl Bromide; Magnesium Ethoxy Bromide

$\downarrow H_2O$ Hydrolysis

$$Mg(OH)Br + H_3C-CH_2-OH$$

Magneisum Oxybromide; ethanol

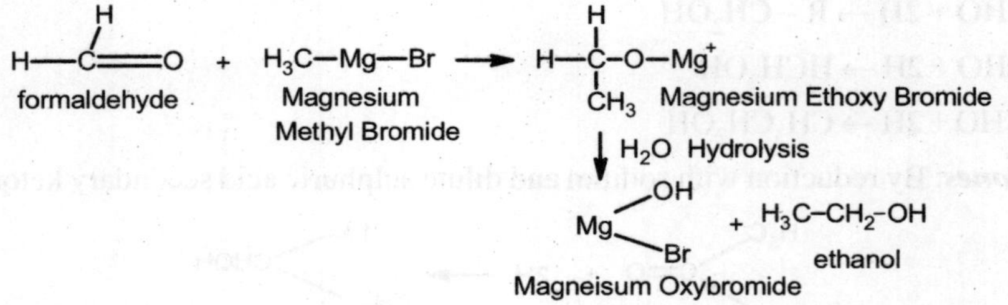

The reaction can be used for assuming the homologous series

Secondary Alcohol

By the action of aldehydes (except HCOH) and subsequent hydrolysis

$$H_3C-CH=O + H_3C-Mg-I \rightarrow H_3C-CH(CH_3)-O-Mg-I \xrightarrow{H_2O} Mg(OH)I + (H_3C)_2CH-OH$$

acetaldehyde; Magnesium Methyl Iodide; Magnesium Propyl Oxide Iodide; propan-2-ol Isopropyl Alcohol

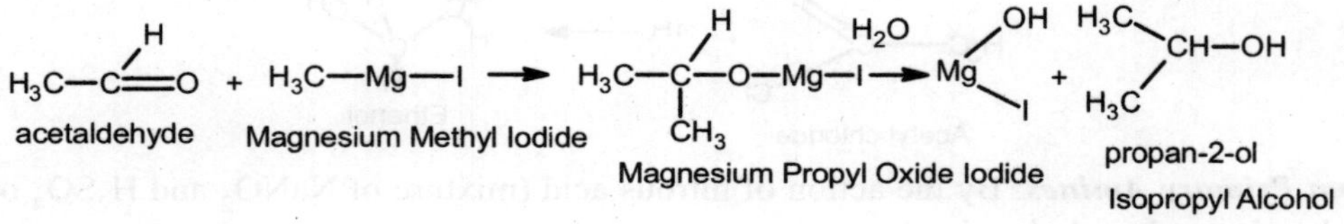

Tertiary Alcohols

By the action of ketones and subsequent hydrolysis

$$H_3C-Mg-I + (H_3C)_2C=O \rightarrow (H_3C)_2CH-O-Mg-I \xrightarrow{H_2O} (H_3C)_3C-OH$$

Magnesium Methyl Iodide; Acetone; Magnesium Propyl Oxide Iodide; 2-methylpropan-2-ol Trimethyl carbinol

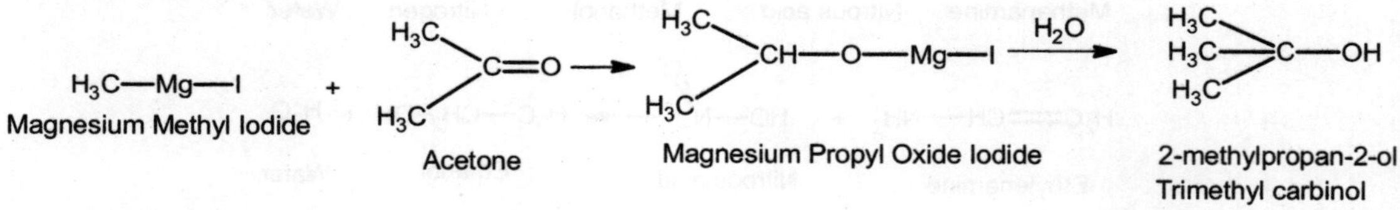

10. From Carbohydrate by fermentation

$$C_6H_{12}O_6 \rightarrow 2C_2H_5OH + 2CO_2$$

This is the special method for preparation of ethyl alcohol

Properties

Physical Properties

1. The alcohols are colourless neutral compounds.
2. Most of the members (up to $C_{12}H_{25}OH$) are liquid. Higher alcohols are waxy solids (gaseous alcohols are not known).
3. The lower members have a burning taste.

4. Methyl ethyl and propyl alcohols are miscible with water, but with the increase of carbon atoms the solubility of the alcohol rapidly decreases.
5. Specific gravity and boiling points increase as the molecular weight increases. Primary alcohols boil at a higher temperature than the corresponding secondary alcohols and the latter at a higher temperature than the tertiary isomerides.

Chemical Properties

The chemical behaviour of the H-atom in the –OH group (a group common to all alcohols) is very different from all that of the H-atoms in the alkyl group. The important chemical reactions are:

(1) Action of Sodium and Potassium: Alcoholetes are formed and hydrogen gas is given out

(I) $2CH_3OH + 2Na \rightarrow 2CH_3ONa + H_2$

Methenol — Sodium Methoxide

(II) $2C_2H_5OH + 2Na \rightarrow 2C_2H_5ONa + H_2$

Ethanol — Sodium Ethoxide

(2) Action of Phosphorous Halides: Alkyl halides are formed

(I) $C_2H_5OH + PCl_5 \rightarrow C_2H_5Cl + POCl_3 + HCl$

Ethyl chloride — Phosphorous Oxy chloride

The above reaction may be applied for a detecting the presence of an hydroxyl (OH) group in an organic compound.

(3) Action of Dehydrating Agent: Olefines or the ethers are formed

(I) CH_3OH cannot be dehydrated to give olefines, it however gives di methyl ether

$$CH_3OH \xrightarrow{\text{Conc } H_2SO_4} CH_3OCH_3$$

Methenol — Dimethyl Ether

(II) Ethyl alcohol on hydration gives ether ethylene or diethyl ether, according to the experimental conditions

$$C_2H_5OH \xrightarrow{H_2SO_4} C_2H_5HSO_4 + H_2O$$

Ethyl Hydrogen Sulphate

Heated with excess H_2SO_4 → C_2H_4 (Ethane)

Heated with excess C_2H_5OH → $(C_2H_5)_2O$ (Diethyl Ether)

(4) *Action of Acids*: Esters are formed, when an alcohol is treated with an acid(inorganic or organic) in the presence of a suitable dehydrating agent (like H_2SO_4 $ZnCl_2$ or P_2O_5) in yields an ester. The reaction is known as esterification

(I) $CH_3OH + HCl \rightarrow CH_3Cl + H_2O$
Methyl Chloride

(II) $C_2H_5OH + HNO_3 \rightarrow C_2H_5NO_3 + H_2O$
Ethyl Nitrate

(III) $C_2H_5OH + CH_3COOH \rightarrow CH_3COOC_2H_3 + H_2O$

(5) *Oxidation of Alcohols.* The structure of an alcohol determines the products formed on oxidation. For instance, in the primary alcohols $R-CH_3-OH$, the functional carbon atom is in the oxidation state – 1 it is oxidized to an aldehyde $R-CH==$, in which the carbon is in the oxidation state +1. The aldehyde is farther oxidized to a carboxylic acid

$$R-\underset{\underset{O}{\parallel}}{C}-OH$$

having the functional carbon in oxidation state +3 the highest with the carbon skeleton remaining intact. In a secondary alcohol, R_2CH-OH, the functional carbon in oxidation state O; on oxidation to a ketone $R_2C==O$, it reaches the oxidation +2. As there is no hydrogen attached to the functional carbon in a ketone, it resists further oxidation, which may break the carbon skeleton in a tertiary alcohol, R_3C-OH, the situation is similar to a ketone, it has no hydrogen on the functional carbon. So it is not normally oxidized. Under severe conditions the carbon skeleton is snapped. A primary alcohol is in the first stage of oxidation in the series

$$\underset{\text{Ethane}}{R-CH_3} \xrightarrow{O} \underset{\text{Ethanol}}{R-CH_2-OH} \xrightarrow{-2H} \underset{\text{Acetaldehyde}}{R-CH{=}O} \xrightarrow{O} \underset{\text{Acetic acid}}{R-\underset{\underset{O}{\parallel}}{C}-OH}$$

Oxidation State of C

$$\underset{\substack{\text{Methane}\\-4}}{CH_4} \longrightarrow \underset{\substack{\text{Ethanol}\\-2}}{H_3CCH_2-OH} \longrightarrow \underset{\substack{\text{Formaldehyde}\\0}}{H_2C{=}O} \longrightarrow \underset{\substack{\text{Formic acid}\\+2}}{H-\overset{\overset{O}{\parallel}}{C}-OH} \longrightarrow \underset{\substack{\text{Dioxomethane}\\+4}}{O{=}C{=}O}$$

Since the aldehydic hydrogen is linked with a more highly oxidised carbon than hydrogen atoms in the alcohol, it is removed by oxidation more easily than the hydrogens in the primary alcohol. Hence, a carboxylic acid results on an oxidation from a primary alcohol until and unless the aldehyde is removal as soon as formed. This is usually done by carrying out the oxidation at a temperature above the b.p. of the aldehyde but below the b.p. of the alcohol (the aldehyde boils lower than the alcohol because it cannot associate by forming inter-molecular hydrogen bond). For example

$$\underset{\text{propan-1-ol}}{H_3C-CH_2-CH_2-OH} \xrightarrow[\text{distils at 70°C}]{Na_2Cr_2O_7 + H_2SO_4} \underset{\text{propionaldehde}}{H_3C-CH_2-CH=O}$$

The primary alcohol on oxidation yield aldehydes as the initial products and these on further oxidation yield carboxylic acids. As a preparative method for aldehydes the method has the distinct disadvantage that the aldehyde is more susceptible to oxidative attack than is the alcohol so that the useless aldehyde is removed from contact with excess oxidizing agent essentially as rapidly as it is formed further oxidation to the carboxylic acid results. Fortunately, aldehydes (because of the absence of hydrogen bonding) are more volatile that the corresponding alcohols and may be preferentially removed by steam distillation or by a stream of carbon dioxide from the reaction mixture.

Dichromate of potassium in dilute sulphuric acid solution has been most frequently used as a laboratory-oxidizing agent.

Oxidation of higher primary alcohols leads to the formation of ester, acids and acetals as by products. If the concentration of sulphuric acid is sufficiently high and the reaction is run at the low temperature to prevent escape of the intermediate aldehyde, the ester becomes the principle product. Thus *n*-butyl and *n*-butyrate may be prepared in 41-47% yield by oxidation of *n*-butyl alcohol with potassium dichromate in 50% sulphuric acid at a temperature of 20-35°C

The oxidation of secondary alcohols to Ketones after little experimental difficulty since the product is not susceptible to oxidation except under quite vigorous conditions. For example 1-menthane (1-methyl 4-isopropyl –cyclohexanone-3) is produced in 83 – 85% yield by the oxidation of 1-methanol with potassium dichromate in dilute sulphuric acid.

In case of secondary alcohols highly insoluble in water, a solution of chromic oxide and glacial acetic acid is the ideal reagent.

An elegant method of secondary alcohols consists of refluxing the alcohol in excess acetone solution with 1.2-1.4 molecular equivalent of aluminium tert-butoxide or the more easily prepared aluminium propoxide by the equation.

CHOH + H_3C–CH_2–H_2C–OH $\xrightarrow{Al[OC(CH_3)_3]_3}$ (cyclohexylidene)=O + H_3C–CH_2–CH_2–OH

cyclohexylidenemethanol propan-1-ol ⟶ cyclohexylidenemethanone propan-1-ol

The overall result is reduction of the secondary alcohol by acetone. The equilibrium is forced to the right by the use of a large excess of acetone. The aluminium iso propoxide serves as catalyst for establishment of the equilibrium; the method of oxidation of secondary alcohols is particularly valuable when other is specific for the carbonyl group. As faster reaction may be obtained by using the higher boiling cyclohexanone as hydrogen acceptor, while with very sensitive compounds, the oxidation may be affected with acetone at room temperature during a period of several weeks.

Reactions for the Oxidation of Alcohols

Primary Alcohols

Primary alcohol first give aldehydes and the acids both containing same number of c-atoms

(i) $CH_3OH \xrightarrow{o} HCHO + H_2O$

formaldehyde

↓ o

HCOOH

Formic acid

(ii) $CH_3CH_2OH \xrightarrow{o} CH_3CHO + H_2O$

Ethyl Alcohol → Acetaldehyde

↓ o

CH_3COOH

Acetic acid

Secondary Alcohols

On oxidation first give ketones containing the same number of C-atoms but on further oxidation they give acids with fewer number of C-atoms

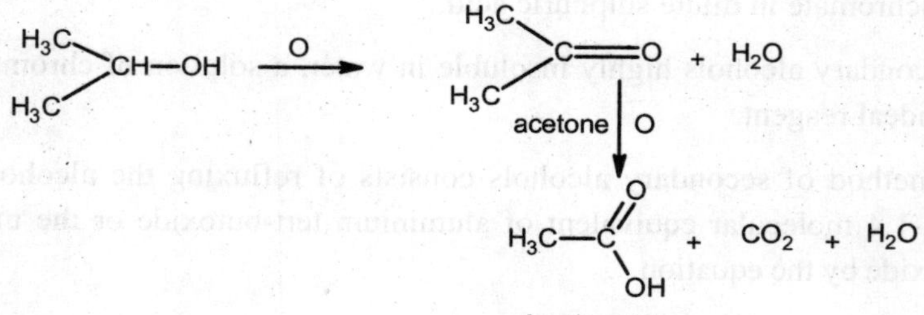

$(H_3C)_2CH{-}OH \xrightarrow{O} (H_3C)_2C{=}O + H_2O$

acetone ↓ O

$H_3C{-}C(=O){-}OH + CO_2 + H_2O$

Acetic acid

Tertiary Alcohols

These are not easily oxidised strong agent, however convert them into ketones and acids each containing smaller number of C-atoms

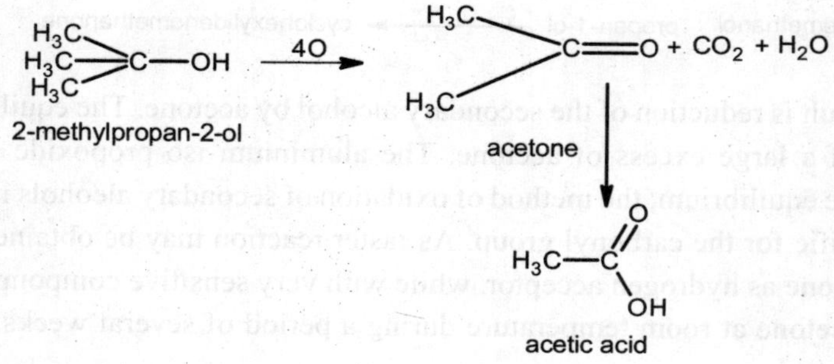

$(H_3C)_3C{-}OH \xrightarrow{4O} (H_3C)_2C{=}O + CO_2 + H_2O$

2-methylpropan-2-ol → acetone

↓

$H_3C{-}C(=O){-}OH$

acetic acid

Thus we can say that above behaviour of alcohols on oxidation may be employed to destroy primary and secondary and tertiary alcohols

(6) Action of hot reduced Copper: The vapours of an alcohol are passed over hot reduced copper at 300°C when

Primary Alcohol: Aldehydes and hydrogen

$$CH_3CH_2OH \xrightarrow[300^\circ C]{Cu} H_3C-CH{=}O + H_2$$

Ethyl Alcohol → acetaldehyde

$$CH_3OH \xrightarrow{Cu} H_2C{=}O + H_2$$

Methyl Alcohol → formaldehyde

Secondary Alcohol: Yields ketones and hydrogen

$$(H_3C)_2CH{-}OH \xrightarrow{Cu^+} (H_3C)_2C{=}O + H_2$$

propan-2-ol → acetone

Tertiary alcohol: They are dehydrated and gives olefines*

$$(H_3C)_3C{-}OH \xrightarrow{Cu^+} (H_3C)_2C{=}CH_2 + H_2$$

2-methylpropan-2-ol, Trimethyl carbinol → 2-methylprop-1-ene, Diethyl Ethylene

(7) Victor Mayor's Reaction: The alcohol is first converted into nitro-paraffin, which is further treated with nitrous acid and finally made alkaline with potassium hydroxide**

Primary alcohol gives red colour

$$H_3C{-}CH_2{-}OH \xrightarrow{\text{red P and } I_2} H_3C{-}CH_2{-}I \xrightarrow{AgNO_3} H_3C{-}CH_2{-}\overset{+}{N}({=}O){-}O^- \xrightarrow{HNO_2} H_3C{-}C(NO_2){=}N{-}OH \xrightarrow{KOH} H_3C{-}C(NO_2){=}N{-}O^-K^+$$

ethanol → iodoethane → nitroethane compound with methane (1:1) → (1Z)-N-hydroxy-1-nitroethanimine → Potassium Nitrolate red

Secondary Alcohols give blue colour

$$(H_3C)_2CH{-}OH \xrightarrow{\text{Red P and } I_2} (H_3C)_2CH{-}I \xrightarrow{AgNO_3} (H_3C)_2CH{-}\overset{+}{N}({=}O){-}O^- \xrightarrow{HNO_2} (H_3C)_2C(NO_2){-}N{=}O$$

propan-2-ol, Isopropyl Alcohol → 2-iodopropane → 2-nitropropane → 2-nitro-2-nitrosopropane blue

---

*This reaction is also used for distinguished primary secondary and tertiary alcohol.

**These reactions are extremely useful for distinguishing the alcohol groups (the primary secondary and tertiary).

Tertiary alcohol does not give any colour

$(CH_3)_3C{-}OH$ (2-methylpropan-2-ol) $\xrightarrow{\text{Red P and } I_2}$ $(CH_3)_3C{-}I$ (2-iodo-2-methylpropane) $\xrightarrow{AgNO_3}$ $(CH_3)_3C{-}N^+(=O)O^-$ (2-methyl-2-nitropropane compound with methane (1:1)) $\xrightarrow{HNO_2}$ no reaction

***(8) Action of Acid Chloride*:** Esters are formed

C_2H_5OH	+	CH_3COCl	→	$CH_3COOC_2H_5$	+	HCl
Alcohol				Acetyl Chloride		Ethyl acetate

***(9) Action of Acid Anhydrides*:** With acetic anhydride

C_2H_5OH	+	$C_4H_6O_3$	→	$CH_3COOC_2H_5$	+	CH_3COOH
Alcohol		Acetic Anhydride		Ethyl acetate		Acetic acid

***(10) Action of Grignard Reagent*:** Paraffins are formed

$CH_3{-}Mg{-}I \xrightarrow{CH_3OH} H_3C{-}CH_3$ (ethane) $+ \; Mg(OH)I$

$CH_2{-}CH_3{-}Mg{-}Br \xrightarrow{C_2H_5OH} H_3C{-}CH_2{-}CH_2{-}CH_3$ (butane) $+ \; Mg(OH)Br$

The reaction may also be used for stepping up the homologous series.

Distinction between Primary Secondary and Tertiary Alcohols

Physical Properties

Properties	Primary Alcohol	Secondary Alcohol	Tertiary Alcohol
1. State, Colour, Odour	Aliphatic Primary alcohols of low molecular weight are generally volatile colourless liquids with sweet smell	Aliphatic Primary alcohols of low molecular weight are generally volatile colourless liquids with characteristic smell	Aliphatic Primary alcohols of low molecular weight are generally volatile colourless liquids with characteristic smell
2. Miscibility with water	Lower primary alcohols are miscible with water in all properties	Lower secondary alcohols are miscible with water in all proportions	Lower secondary alcohols are miscible with water in all proportions
3. Characteristic Functional group	Primary alcohols possess $—CH_2OH$	Secondary alcohols possess CHOH group	Tertiary alcohols possess C– OH
4. Boiling point	Among the isomeric alcohols the boiling point of the primary alcohols have higher boiling points	Among the isomeric alcohols the boiling point of the secondary alcohols have higher boiling points higher than tertiary but lower than primary	The tertiary alcohols are lowest B.PS.

Distinction between Primary Secondary and Tertiary Alcohols
Chemical Properties

Properties	Primary Alcohol	Secondary Alcohol	Tertiary Alcohol
Oxidation	On oxidation the primary alcohols are first converted into aldehydes and then to the corresponding acids having the same number of C-atoms as that of the alcohols	On oxidation, the secondary alcohols are first converted into ketones. On further oxidation, the ketone is broken down to produce acids having C-atoms that of the parent alcohols	Tertiary alcohols are not oxidised under ordinary conditions, but under drastic conditions they are oxidised to ketones and then into an acid having fewer number of C- atoms then the parent alcohols
Reactions with halogenhydracids	Reacts with halogen hydracids under suitable conditions to give alkyl halides. The ease of such reaction is the least as compared with the other two categories of alcohols.	Reacts with halogen hydracids under suitable conditions to give alkyl halides. The ease of such reaction is intermediate between the primary and the tertiary alcohol	Reacts with halogen hydracids under suitable conditions to give alkyl halides. The ease of such reaction is the greatest of all the categories of alcohol.

Distinction in Properties of the Methyl and Ethyl Alcohol
Physical Properties

Properties	Methyl Alcohol	Ethyl Alcohol
1. State	Liquid	Liquid (less volatile than methyl alcohol)
2. Class	Aliphatic Alcohol	Aliphatic alcohol (higher homologue of methyl alcohol)
3. Colour	Colourless	Colourless
4. Odour	Pleasant Smell	Pleasant smell wine like (with a burning taste)
5. Boiling Point	Lower Boiling Point	Higher boiling point
6. Stability	Stable does not change on keeping	Stable does not change on keeping
7. Miscibility	Miscible in all properties	Miscible in all properties
8. Toxicity	Poisonous	Non-poisonous

Distinction in Properties of the Methyl and Ethyl Alcohol
Chemical Properties

Properties	Methyl Alcohol	Ethyl Alcohol
(i) Oxidation	It burns in air (oxygen) with non-luminous flame	It burns in air (oxygen) with non-luminous flame
(ii) Reaction	Forms ester (methyl benzene with characteristic sweet smell)	Forms ester (ethyl benzoate) with characteristic sweet smell
(iii) With iodine	No reaction	Produces Iodoform
(iv) Oxidation may be effected by red copper spiral	Gets oxidised to formaldehyde with characteristic smell with turns ammonical silver nitrate solution (Tollen's reagent) black	Gets oxidised to formaldehyde with characteristic smell with turns ammonical silver nitrate solution (Tollen's reagent) black

METHYL ALCOHOL, METHANOL

Formula: CH_3OH

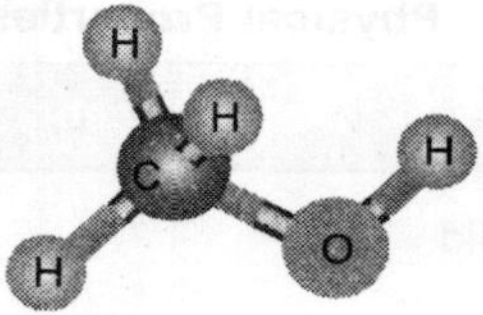

History and Occurrence

In their embalming process, the ancient Egyptians used a mixture of substances, including methanol, which they obtained from the pyrolysis of wood. Pure methanol, however, was first isolated in 1661 by Robert Boyle, who called it *spirit of box*, because he produced it via the distillation of boxwood. It later became known as *pyroxylic spirit*. In 1834, the French chemists Jean-Baptiste Dumas and Eugene Peligot determined its elemental composition. They also introduced the word *methylene* to organic chemistry, The term *methyl* was derived in about 1840 by back-formation from *methylene*, and was then applied to describe *methyl alcohol*. This was shortened to *methanol* in 1892 by the International Conference on Chemical Nomenclature.

In 1923, the German chemist Matthias Pier, working for BASF developed a means to convert *synthesis gas* (a mixture of carbon monoxide and hydrogen derived from coke and used as the source of hydrogen in synthetic ammonia production) into methanol. This process used a zinc chromate catalyst, and required extremely vigorous conditions—pressures ranging from 300–1000 atm, and temperatures of about 400°C. Modern methanol production has been made more efficient through the use of catalysts capable of operating at lower pressures.

Manufacture

For industrial purpose it is manufactured in the following processes.

1. ***Destructive Distillation of Wood***: When wood is heated in a limited supply of air in iron retorts the following important products are obtained (see Fig. 9.1)

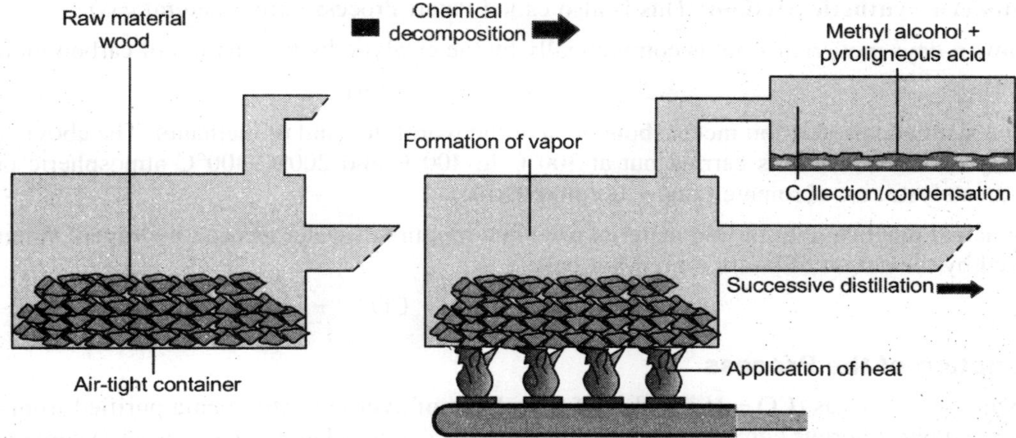

Fig. 9.1. The Industrial Process of Destructive Distillation of Wood

A. ***Wood Gas***: It is mixture of CO, H_2, CH_4, CO_2 etc and is used for fuel.

B. ***Wood tar***: It is thick, black heavy liquid collected at the bottom of the reservoir, i.e. settling tank. It is used for wood preservation.

C. ***Pyroligneous Acid***: It is the brown aqueous upper layer in the settling tank. It contains methyl alcohol (2-4%) acetone (0.5%). Acetic acid (10%) water etc. Methyl alcohol is recovered from it.

Recovery of the Methanol

Pyroligneous acid is separated from the wood tar and distilled. The vapours (containing CH_3OH), acetone acetic acid and water) are passed through hot calcium hydroxide. Acetic acid is neutralized and calcium acetate is formed. The remaining vapours of methanol, acetone and water are condensed in separate receivers. The distillate is dried over lime and then subjected to fractional distillation. More volatile acetone is collected separately at 56°C

Purification

Methyl alcohol obtained in the above process is not pure. It contains 1-2% of acetone as the main impurity. The crude methyl alcohol is treated with fused $CaCl_2$ when a solid compound $CaCl_24CH_3OH$ is formed. The solid product is separated, squeezed to remove acetone, dried decomposed by warming with water and distilled to get methyl alcohol, the distillate still contains traces of acetone, and other major impurities. It is finally purified by treating it with anhydrous oxalic acid, when solid dimethyl oxalate is obtained. This precipitate is washed with a little cold water to remove traces of acetone and then hydrolyzed with boiling potassium hydroxide. Methyl alcohol is recovered from the alkaline liquid by distillation. The product is dried over quick lime and calcium and redistilled. Pure methyl alcohol is thus obtained.

Modern Synthetic Method: This is also called Patart Process (from natural gas)

Now-a-days methyl alcohol is commercially by the catalytic hydrogenation of carbon monoxide

$$2H_2 + CO \rightleftharpoons CH_3OH$$

The synthetic production meets about 90% of the world-demand of methanol. The above reaction is highly exothermic and is carried out at 300°C to 400°C and 200°C-300°C atmospheric pressure using a catalyst, zinc chromate (zinc + chromic oxide).

In actual practice, the starting material used in a mixture of water gas and hydrogen. Water gas is obtained by the action of steam on red-hot coke.

$$CO + H_2O \rightleftharpoons CO + H_2$$

Description of the Process

A mixture of water gas ($CO + H_2$) with half its volume of hydrogen after being purified from arsenic sulphur and phosphorous compound is admitted into the contact chamber (see Fig. 9.2) containing the catalyst and provided with electrically heating coils. The gas mixture coming out of the converter and containing methanol along with unconverted CO_2 and H_2 (conversion is only 20 - 25%) is passed methyl alcohol vapours condense to a liquid, while the unchanged gas mixture is pumped back into the converter.

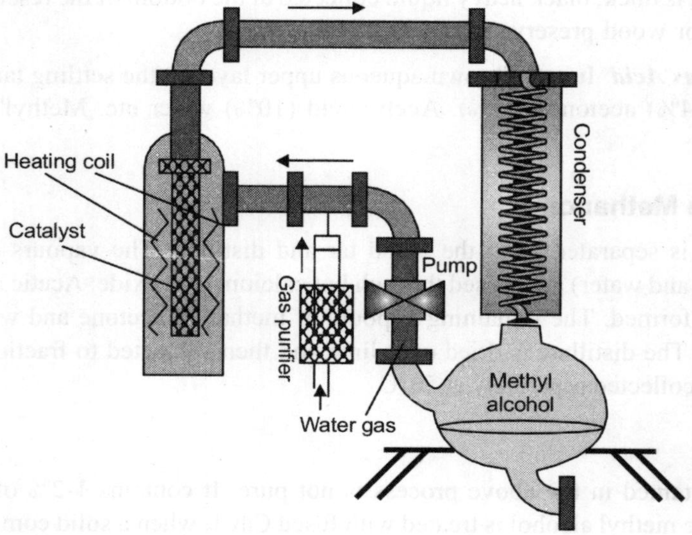

Fig. 9.2. The Industrial Plant for Patart Process

Another Synthetic Process

This process includes in the heating a mixture of carbon dioxide and hydrogen in presence of zinc oxide.

$$CO_2 + 3H_2 \rightarrow CH_3OH + H_2O$$

From methane: Methyl alcohol is also manufactured from methane. On passing a mixture of methane and oxygen (in 9:1 ratio by volume) at 100 atmosphere through a copper tube heated at 200°C methyl alcohol is formed. Copper of the tube acts as a catalyst in the oxidation

$$2CH_4 + O_2 \rightarrow 2CH_3OH$$

Laboratory Process

Although it is very rare that methyl alcohol is prepared in the lab, but still it can be prepared by the distillation of a solution mixture of potassium hydroxide and methyl iodide and the distillate is collected at 60-65°C

$$CH_3I + NaOH \rightarrow CH_3OH + NaI$$

Properties

Physical Properties

Phase behaviour	
Melting point	-97°C
Boiling point	64.7°C
Triple point	175.5 K (-97.7°C)
Critical point	513 K (240°C)78.5 bar
$\Delta_{vap}H$	37.4 kJ/mol
Liquid properties	
$\Delta_f H^0_{liquid}$	-238.4 kJ/mol
S^0_{liquid}	127.2 J/mol°K
C_p	79.5 J/mo°K
Density	0.7918 ×10³ kg/m³
Gas properties	
$\Delta_f H^0_{gas}$	-201 kJ/mol
S^0_{gas}	239.9 J/mol°K
C_p	44.06 J/mol°K
Safety	
Acute effects	Poisonous by ingestion or inhalation, may cause respiratory failure, kidney failure, blindness.
Chronic effects	As acute. Skin contact can cause dermatitis.
Flash point	11°C
Auto ignition temperature	455°C
Explosive limits	7-36%

Chemical Properties and Reactions

The chemical properties and reaction of methyl alcohol have been already discussed. The main characteristics are properties of the alcohol, which are now going to be discussed below:

(1) *Combustion*: Methyl alcohol is highly inflammable liquid, it in air, when ignited with a pale blue flame and the products to combustion (oxidation) are CO_2 and H_2O

$$2CH_3OH + 3O_2 \rightarrow 2CO_2 + 4H_2O$$

(2) *Oxidation*: As a primary alcohol, it is oxidant by common agents [such as acidified $KMnO_4$, $K_2Cr_2O_7$) to form formaldehyde. Then boil the solution (the remaining formaldehyde) is readily oxidised further to formic acid which is ultimately oxidised to CO_2 and H_2O

$$CH_3OH \xrightarrow{\text{oxidation}} HCHO \xrightarrow{\text{oxidation}} HCOOH \xrightarrow{\text{oxidation}} CO_2 + H_2O$$

Methyl alcohol is catalytically oxidised to formaldehyde when in vapour, mixed with requisite amount of air or oxygen, is passed over copper spiral or platinum 250°C-300°C

$$CH_3OH + O_2 \xrightarrow{\text{Cu/Pt } 250^\circ - 350^\circ} 2HCHO + 2H_2O$$

If excess of air (or oxygen is used or when its vapour mixed with requisite amount of air oxygen is passed over copper spiral or platinum catalyst at 250°C-300°C

$$2CH_3OH + O_2 \xrightarrow{\text{Cu/Pt } 250^\circ - 300^\circ C} 2HCHO + 2H_2O$$

If excess of air (or oxygen) used or when platinum black is used as the catalyst in the above process

(3) *Reaction with Sulphuric Acid:* When slightly warmed with conc. H_2SO_4 methyl alcohol forms methyl hydrogen

$$CH_3OH + H_2SO_4 \rightarrow \underset{\text{Methyl Hydrogen Sulphate}}{CH_3HSO_4} + H_2O$$

$$2CH_3HSO_4 \rightarrow \underset{\text{Dimethyl Sulphate}}{(CH_3)_2SO_4} + H_2SO_4$$

On heating excess of methyl alcohol with sulphuric acid at about 140°C or methyl alcohol with methyl hydrogen sulphate dimethyl ether is produced

$$2CH_3OH \xrightarrow{\text{Conc } H_2SO_4/140^\circ C} \underset{\text{Dimethyl Ether}}{CH_3 - O - CH_3} + H_2O$$

$$(CH_3)_2SO_4 + CH_3OH \rightarrow CH_3 - O - CH_3 + H_2SO_4$$

(4) *Reaction with hydrogen Chloride (gas)*[*]: On passing dry HCl gas through hot methyl alcohol in the presence of a dehydrating agent like anhydrous (fused) $ZnCl_2$ methyl chloride is formed

$$CH_3OH + HCl \xrightarrow{ZnCl_2} CH_3Cl + H_2O$$

(5) *Reaction with Grignard reagent:* With Grignard reagent alkenes are formed

$$RMgI + CH_3OH \rightarrow CH_3OMgI + RH$$

(where R is the alkyl group, CH_3)

(6) *Acylation:* An acyl chloride(R.COCl) reacts with methyl alcohol to form the ester CH_3OOCR

$$CH_3OH + RCOCl \rightarrow CH_3OOCR + HCl$$

Thus, with acytyl chloride (CH_3COCl) methyl acetate is produced

$$CH_3OH + CH_3COCl \rightarrow CH_3COOCH_3 + HCl$$

(7) *Reaction with sodium (or potassium)*: At ordinary temperature, sodium (or potassium) readily reacts with methyl alcohol to form the corresponding methoxide (methylate), with the evolution of hydrogen gas. The methoxide is hydrolyzed by water to form methyl alcohol and sodium (or potassium) hydroxide

$$2CH_3OH + 2Na \rightarrow \underset{\text{Sodium ethoxide}}{2CH_3ONa} + H_2$$

(8) *Action of Phosphorous Halides*: Phosphorous Trihalides or PCl_5 react with methyl alcohol to form methyl halides, with PCl_3, the other product is phosphorous acid [in presence of PBr_3 and PI_3 the reaction is generally carried out with red phosphorous and Br_2 or I_2, which produce phosphorous tribromide (PBr_3) and phosphorous triiodide (PBI_3) and methyl bromide (CH_3Br and methyl iodide respectively are formed) with PCl_5 phosphorous oxychloride and HCl are formed along with methyl chloride

$$3CH_3OH + PCl_3 \rightarrow 3CH_3Cl + H_3PO_3$$

$$2P + 3Br_2 \rightarrow 2PBr_3$$

$$3CH_3OH + PBr_3 \rightarrow 3CH_3Br + H_3PO_3$$

$$CH_3OH + PCl_5 \rightarrow CH_3Cl + POCl_3 + HCl$$

(9) *Reduction:* When heated with excess of hydrochloric acid in sealed tube methyl alcohol is reduced to methane.

$$CH_3OH + 2HI \rightarrow CH_4 + I_2 + H_2O$$

The reduction occurs in two steps. In the first step, methyl iodide is formed through the replacement of the alcoholic OH groups by iodide atom, methyl iodide is then reduced to methane.

$$CH_3OH + HI \rightarrow CH_3I + H_2O$$

$$CH_3I + HI \rightarrow CH_4 + I_2$$

***The mixture of anhydrous $ZnCl_2$ and dry HCl to a good reagent for replacing an alcoholic-OH group by Cl. It is known as Lucas Reagent.**

(10) *Esterification With Organic Acids:* In the presence of a small amount of conc. H_2SO_4 which acts both as dehydrating agent and a catalyst, methyl alcohol reacts with organic acids on heating to form methyl ester of the acid.

$$RCOOH + CH_3OH \xrightleftharpoons{\text{(conc. Sulphuric acid)}} RCOOCH_3 + H_2O$$

With acetic acid, methyl acetate is formed in the reaction

$$CH_3COOH + CH_3OH \xrightleftharpoons{\text{(conc. Sulphuric acid)}} \underset{\text{Methyl acetate}}{CH_3COOCH_3} + H_2O$$

A symbol of the important reactions of the methyl alcohol

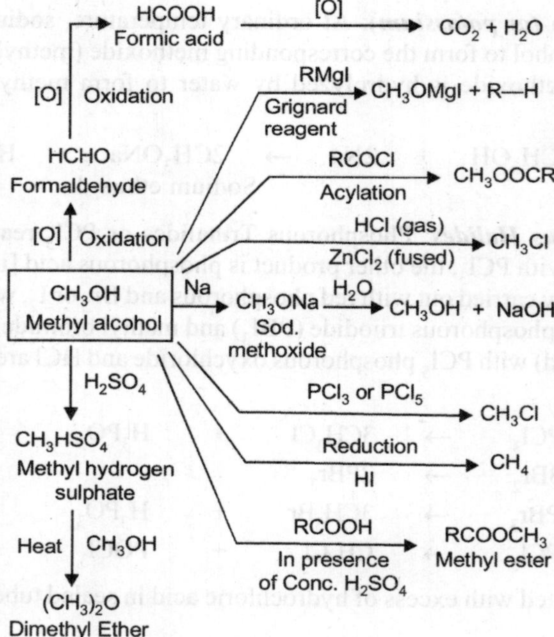

Uses of Methyl Alcohol

Methanol is used on a limited basis to fuel internal combustion engines, mainly by virtue of the fact that it is not nearly as flammable as gasoline. Methanol blends are the fuel of choice in open wheel racing circuits like Champcars, as well as in radio controlled model airplanes.

When produced from wood or other organic materials, the resulting organic methanol (bioalcohol) has been suggested as renewable alternative to petroleum-based hydrocarbons. However, one cannot use BA100 (100% bioalcohol) in modern petroleum cars without modification. Methanol is also used as a solvent and as an anti-freeze in pipelines. The largest use of methanol by far, however, is in making other chemicals. About 40% of methanol is converted to formaldehyde, and from there into products as diverse as plastics, plywood, paints, explosives, and permanent press textiles.

In the 1990s, large amounts of methanol were used in the United States to produce the gasoline additive methyl tert-butyl ether (MTBE). The 1990 Clean Air Act required certain major cities to use MTBE in their gasoline to reduce photochemical smog. However, by the late 1990s, it was found that MTBE had leaked out of gasoline storage tanks and into the groundwater in sufficient amounts to affect the taste of municipal drinking water in many areas. Moreover, MTBE was found to be a carcinogen in animal studies. In the resulting backlash, several states banned the use of MTBE, and its future production remains uncertain. Other chemical derivatives of methanol include dimethyl ether, which has replaced chlorofluorocarbons as the propellant in aerosol sprays, and acetic acid.

Health and Safety

Methanol is toxic, as its metabolites formic acid and formaldehyde cause blindness and death. It enters the body by ingestion, inhalation, or absorption through the skin. Dangerous doses will build up if a person is regularly exposed to fumes or handles liquid without skin protection. If methanol has been ingested, a doctor should be contacted immediately. The usual fatal dose: 100–125 ml (4 oz.). Toxic effects take hours to start, and effective antidotes can often prevent permanent damage. One treatment is the injection of ethanol: this slows down the breakdown of methanol by the liver, so that the toxic metabolites can not build up. Symptoms of methanol ingestion are similar to those of intoxication: headache, dizziness, nausea, lack of coordination, confusion, drowsiness, followed by unconsciousness and death. The ester derivatives of methanol do not share this toxicity. Ethanol is sometimes denatured (adulterated) by the addition of methanol. The result is known as methylated spirits or meths. (The latter should not be confused with *meth*, a common abbreviation for methamphetamine.) Pure methanol has been used in open wheel racing since the mid-1960's. Unlike petroleum fires, methanol fires can be extinguished with plain water. The decision was made shortly after the death of two drivers.

ETHYL ALCOHOL OR ETHANOL

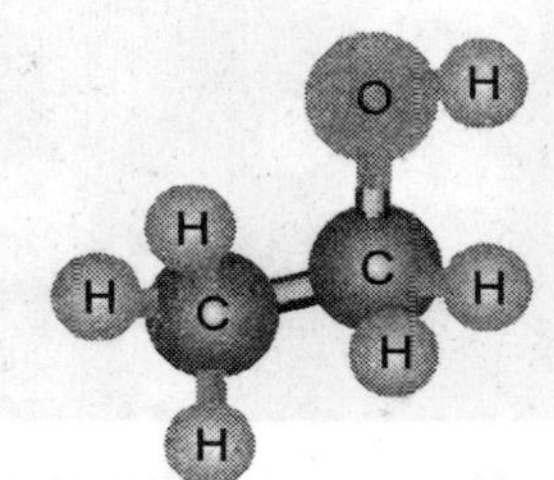

Formula: C_2H_5OH

History and Occurrence

Ethanol has been known to humans since prehistory as the active ingredient of alcoholic beverages. Its isolation as a relatively pure compound was probably achieved first by Islamic alchemists who developed the art of distillation, such as Geber (721-815) and Al-Razi (864-930).

Fermentation

In **fermentation** these pyruvic acid molecules are turned into some "waste" product, and a little bit of energy (only two ATP molecules per molecule of glucose – actually four are produced in glycolysis, but two are used up) is produced. Out of many possible types of fermentation processes, two of the most common types are **lactic acid fermentation** and **alcohol fermentation**.

Alcohol fermentation is done by yeast and some kinds of bacteria. The "waste" products of this process are ethanol and carbon dioxide (CO_2). Humans have long taken advantage of this process in making bread, beer, and wine. In bread making, it is the CO_2 which forms and is trapped between the gluten (a long protein in wheat) molecules that causes the bread to rise, and the ethanol (often abbreviated as EtOH) evaporating that gives it its wonderful smell while baking. The effects of the ethanol in beer and wine are something with which many college students are familiar and it is the CO_2 produced by the process of fermentation that makes these beverages effervescent.

Conditions of Fermentation

(i) ***Temperature:*** The optimum or most favourable temperature for fermentation process lies between 25°C and 35C. Higher temperature destroys the ferment and at lower temperature the reaction is very low

(ii) ***Concentration***: Dilute solution favours fermentation. In highly concentrated solution the action of ferment is either slowed down or stopped.

(iii) ***Air***: The presence of air may improve the activity of the enzyme

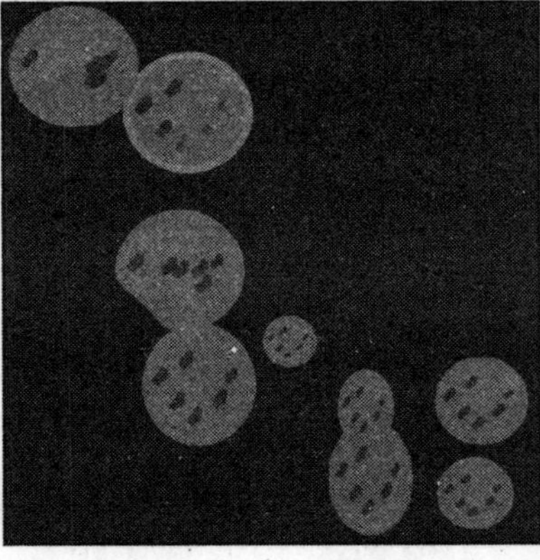

Fig 9.3. The Micrograph of Yeast (x 5000)

Yeast is one of the important ferments. It contains a number of enzymes in it such as Invertase, Maltose, and Zymase etc. Yeast can bring about the fermentation of sugars into alcohol. Cane sugar is hydrolyzed to glucose and fructose in the presence of an enzyme 'Invertase':

$$\underset{\text{cane sugar}}{C_{12}H_{22}O_{11}} + H_2O \xrightarrow{\text{Inveratse}} \underset{\text{glucose}}{C_6H_{12}O_6} + \underset{\text{fructose}}{C_6H_{12}O_6}$$

Zymase another enzyme, present in yeast then converts glucose or fructose into alcohol

$$C_6H_{12}O_6 \xrightarrow{\text{Zymase}} 2C_2H_5OH + 2CO_2$$

The above process in which sugar is changed into alcohol is known as alcoholic fermentation of sugar.

Amylase or diastase in an enzyme, present in germinating barley. It is used to convert starch into maltase sugar

$$\underset{\text{starch}}{2(C_6H_{10}O_5)_n} + {}_nH_2O \xrightarrow{\text{Diastase}} \underset{\text{Maltose}}{{}_nC_2H_{22}O_{11}}$$

Preparation of Ethyl Alcohol

1. **From Sugar**: (Molasses) a dark coloured syrupy liquid is obtained as an important by product in the manufacture of sugar. It contains about 55-60% of sugar and is a valuable source of industrial alcohol. The production of pure alcohol form molasses consists of the following operations

   (a) *Dilution of Molasses*: Molasses are diluted with water so that the sugar concentration in it is reduced to 10-12%

   (b) *Fermentation of Sugar Solution*: Molasses are acidified with dilute sulphuric acid with vapours the growth of yeast (single celled living plant) is added to start fermentation. A little $(NH_4)_3PO_4$ or $(NH_4)_2SO_4$ may also be added to act as a food for the yeast. Fermentation is complete in these four days is accompanied by the evolution of CO_2 and frothing of the liquid. The fermented mass contains about 15% of the alcohol and is technically known as "wash". The following reactions take place:

$$C_{12}H_{22}O_{11} + H_2O \xrightarrow{\text{Yeast(Invertase)}} \underset{\text{glucose}}{2C_6H_{12}O_6} + \underset{\text{fructose}}{C_6H_{12}O_6}$$

$$C_6H_{12}O_6 \xrightarrow{\text{Yeast(Zymase)}} 2C_2H_5OH + 2CO_2\uparrow$$

   CO_2 removed as one of the by products

   (c) *Distillation:* Wash containing about 15% of alcohols is next submitted to distillation in Coffey stills provided with fractionating columns, which work on the counter current principle.

Aeneas Coffey (1780-1852), the inventor of the Coffey or columnar still, was an Irish (born in Dublin) excise man and inventor who worked for the British government in Scotland. During his long tenure as an inspector, he had plenty of time to inspect and study a range of mostly illegal and some legal stills. Over time his astute observation talent and engineering knowledge led him to invent a very efficient and fast still that accomplished three consecutive distilling operations in one step. It is less expensive to operate than any alembic still and much faster. The invention was patented in Dublin (1831).

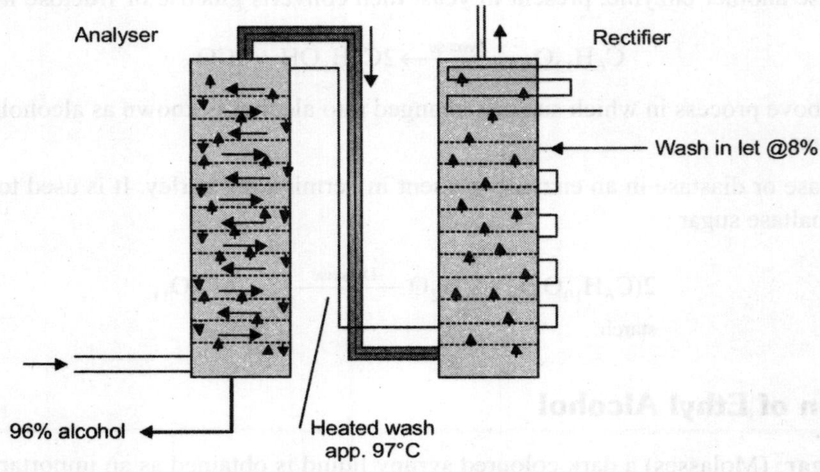

Fig 9.4. The Diagram of the Coffey Stills

The principle of distillation is simple. The important aspect remains the means of separating alcohol from an alcohol containing liquid. A!cohol starts to evaporate at 78.3°C at sea level, hence when an alcohol-containing liquid is heated to that temperature the alcohol starts to vapourize. A simple devise composed of metal container connected to another by means of a swan neck can be employed as a still. In fact such crude devices have been and are still being used illegally in many countries to manufacture potent distillates, many unfit for human consumption due to their high fusel oil and methanol content

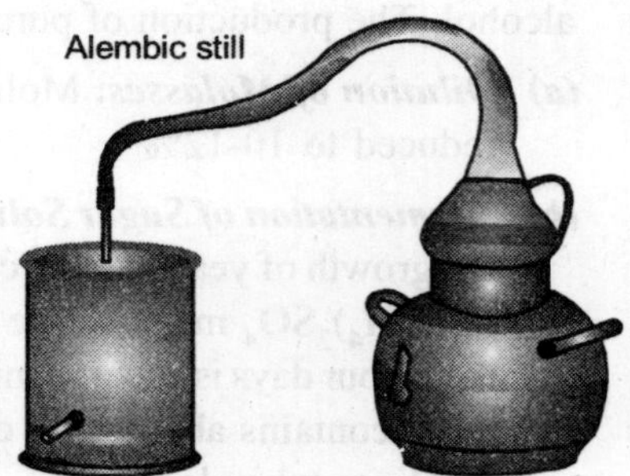

Fig 9.5. The alembic still

(d) ***Purification or Rectification***: The raw spirit is further refined by fractional distillation. The following fractions are obtained: -

I. First Fraction: Products with low boiling point (20-21°C) containing mainly acetaldehyde.

II. Middle Fraction or Rectified Spirit: It is collected at 78-79°C and contains about 96% alcohol.

III. Last Fraction: Products of higher boiling point containing higher alcohols ($C_5H_{11}OH$) – amyl alcohol. The conversion of sugar into alcohol is called alcoholic fermentation of sugar"

Besides molasses other numerous fermentable sugars such as fruit juices beet dry dates etc. can also be employed as raw material for this process.

Absolute Alcohol

The chief impurity present in rectified spirit is water.* It is not possible to remove it completely from rectified spirit by fractional distillation as alcohol and water from constant boiling mixture at 78.15°C

Pure absolute alcohol boiling at 78.3°C and having no trace of water is obtained by distilling rectified spirit over first quick lime [BaO or $CaCl_2$]. Alcohol containing 0.3% of water is thus prepared and is commonly known as "absolute alcohol". The last traces of sodium or calcium and re-distilling it.

Absolute alcohol may also be obtained by distilling alcohol with benzene. At 64.85°C a ternary mixture of benzene, water and alcohol is obtained. At 68.25°C a binary mixture of alcohol and benzene distills over. At 78.3°C pure absolute alcohol is obtained.

From Starch: Potatoes, maize barley etc. are used as the source of starch. The process employing potatoes as the raw material is described below: -

Conversion of Starch into Sugar

(i) ***Malting***: Moist barley is placed in a dark room in 8-12" thick layer at 15°C. It begins to germinate. Germinated barley contains enzyme called diastase capable of converting starch into sugar. Extract of such barley is called Malt.

(ii) ***Saccharification***: The starchy material, i.e. potatoes are heated with steam under pressure at 140-150°C. The starch cells are broken up and brought into milky solution – known as Mash. The process is known as Mashing. This is cooled to maltose. The resulting sweet is known as "wort"

$$\underset{\text{starch}}{2(C_6H_{10}O_5)} + {}_nH_2O \rightarrow \underset{\text{Maltose}}{{}_nC_{12}H_{22}O_{11}}$$

Saccharification is also accomplished on the large scale by using hot dilute sulphuric acid or hydroiodic acid instead of malt.

Alcoholic Fermentation of Sugar

To the solution so obtained yeast is added. Alcoholic fermentation is allowed to proceed at about 30°C. The maltose is first converted into glucose by the enzyme maltose present in yeast.

$$C_{12}H_{22}O_{11} \xrightarrow{\text{Zymase}} 2C_2H_5OH + 2CO_2 \uparrow$$

The liquid so obtained contains 5-10% of alcohol.

Alcohol from Starch

$$\text{STARCH} \xrightarrow{\text{Hot dilute acid}} \text{GLUCOSE} \xrightarrow{\text{zymase of yeast}} \text{ALCOHOL}$$

STARCH —(diastase of malt)→ MALTOSE —(Maltose of Yeast)→ ALCOHOL

*The presence of a small amount of water in alcohol can be detected by adding a solution of ethyl formate and anhydrous sodium ethylate (C2H5ONa) when a white precipitate is produced.

Distillation

The fermented liquor is distilled is Coffey Stills when raw spirit is obtained.

Rectification

The raw spirit is then purified as before and absolute alcohol is prepared.

By Products of Alcoholic Industry

The important by products of alcohol industry are: -

(a) ***Carbon Dioxide:*** It is compressed in iron cylinders and stored for use in aerated water. It is also used for making commercial dry ice for refrigeration purpose.

(b) ***Acetaldehyde:*** It is obtained during the first fractional distillation during purification.

(c) ***Fusel Oil:*** The last fractional distillate obtained during rectification is known as fusel oil. It contains higher alcohols – amyl alcohol ($C_5H_{11}OH$).

(d) ***Spent wash:*** It is liquid left after distillation of wash and is used as a cattle food.

(e) ***Argol or Tarter:*** It is potassium hydrogen tartrate and is obtained as a brown crust during the fermentation of grape juice. It is used in the manufacture of tartaric acid.

(f) ***Glycol:*** The amount of glycerol in the fermentating liquid can be increased from 3% to about 37% by adding sodium tartrate, (Na_2CO_3) sodium sulphite (Na_2SO_3) sodium hydrogen sulphite ($NaHSO_3$)

$$\underset{\text{Glucose}}{C_6H_{12}O_6} \rightarrow \underset{\text{Acetaldehyde}}{CH_3CHO} + \underset{\text{Glycerol}}{C_3H_8O_3} + CO_2\uparrow$$

Alcoholic Beverages

Alcoholic liquors also known as alcoholic beverages divided into two classes (a) distilled beverages and (b) undistilled beverages.

Whisky From barley, **Gin** from Maize, **Brandy** from grape juice and **Rum** from molasses are distilled beverages. Their alcohol content varies from 35% - 40%. They are obtained by the concentration of fermented liquor.

Beer, wine and champagne are undistilled beverages. Beer is made from bariey and contains 3-6% of alcohol besides some contain 8-15% of ethyl alcohol champagne is also obtained from grape juice. Alcohol content of these beverages is raised to 20% by adding pure alcohol and they are known as fortified wines (i.e. sherry and port).

They contain colouring and flavoring material besides a large content of water. When taken in small quantities alcohol causes excitement (but large doses can cause fatal conditions even death). Due to its oxidation in the blood stream it immediately causes warmth in the body and is recommended in cause of exposure to cold.

Alcoholometry

It is the method of determining the quantity of alcohol in the fermented liquors. The strength of alcohol in a sample can be ascertained by determining its specific gravity at a standard temperature and then in refining it directly to weight percentage tables in which the specific gravity of any mixture of alcohol and water is given.

Proof Spirit

This name was originally gives to the weakest spirit capable of firing gun powder when brought into contact with it and ignited. It is a standard and contains 49.28% alcohol by weight or 57.10% by volume and has the specific gravity 0.9197 at 15°C spirits are termed as *under* or *over* proof according as they are weaker or stronger than proof spirit. Thus a spirit at 20°C under proof contains 80 volumes of proof spirit and 20 volumes of water at 15°C while 100 volumes of spirit at 20°C over proof gives 120 volumes of proof spirit.

Methylated Spirit

Methylated spirit or denaturized alcohol i.e. alcohol containing materials which make it unfit for drinking purposes. The common materials used as denaturants are crude methyl alcohol, benzene and pyridine bases. The industrial methylated spirit contains 95% ethyl alcohol, 14% of water and 4% crude methyl alcohol. For general use mineralized methylated spirit is sold. It is a mixture of 80% ethyl alcohol, 11% water and 7.5% methyl alcohol and a small amount of mineral naphtha and colouring water.

Properties

Physical Properties

Phase behaviour	
Melting point	-114.3°C,
Boiling point	78.4°C,
Triple point	-114°C,
Critical point	241°C,63 bar
Heat of Fusion $\Delta_{fus}H$	4.9 kJ/mol
Standard energy of formation $\Delta_{fus}S$	31 J/mol·K
Heat of Formation $\Delta_{vap}H$	38.56 kJ/mol
Solubility	Soluble in water

Acid-base properties	
pKa	15.9 (H^+ from -OH group)
pH	7.0 (Neutral)
Liquid properties	
$\Delta_f H^0_{liquid}$	-277.38 kJ/mol
S^0_{liquid}	159.9 J/mol·K
Heat Capacity C_p	112.4 J/mol·K
Density	789 gram/L
Gas properties	
Heat of Formation $\Delta_f H^0_{gas}$	-235.3 kJ/mol
Standard Heat of Formastion S^0_{gas}	283 J/mol·K
Heat capacity C_p	65.21 J/mol·K

Chemical Properties

(i) ***Combustion:*** Ethyl alcohol is a combustible liquid. It burns in air, when ignited, with a non-luminous light blue flame. The products of combustion (oxidation) are carbon dioxide and water

$$C_2H_5OH + 3O_2 \rightarrow 2CO_2 + 3H_2O$$

(ii) ***Oxidation:*** When heated with oxidizing agent like sulphuric acid potassium dichromate or potassium permanganate sulphuric acid mixture, ethyl alcohol is first oxidised to acetaldehyde. If allowed to remain within the reaction mixture it is further oxidised to acetic acid

It may be noted that, the oxidation of ethanol which is a primary alcohol produces first an aldehyde and then an acid both of which contain the same number of carbon atom in their molecules as their parent alcohol.

$$\underset{\text{ethanol}}{C_2H_5OH} \xrightarrow{K_2Cr_2O_7\ +H_2SO_4} \underset{\text{acetaldehyde}}{CH_3CHO} \xrightarrow{K_2Cr_2O_7\ +H_2SO_4} \underset{\text{sodium formate}}{CH_3COONa}$$

Catalytic Oxidation or Dehydrogenation

When the vapour of the ethyl alcohol is passed over heated catalyst, two hydrogen atoms are liberate form the molecule.

$$C_2H_5OH \xrightarrow{Cu/250-300^\circ C} CH_3CHO + H_2$$

But when a mixture of the alcohol vapour with excess air or an oxygen is passed over heated platinum black catalyst, ethyl alcohol is converted into acetic acid with the production of water.

$$C_2H_5OH + O_2 \xrightarrow{\text{Pt Catalyst/heat}} CH_3COOH + H_2O$$

Oxidation by Chlorine

On passing chlorine gas through ethyl alcohol at ordinary temperature the alcohol is obtained by the halogen to acetaldehyde

$$C_2H_5OH \quad + \quad Cl_2 \quad \rightarrow \quad CH_3CHO \quad + \quad 2HCl$$

If the passage of chlorine is continued the – CH_3 group of acetaldehyde is progressively chlorinated till with prolonged passage of chlorine (i.e. with excess of chlorine) trichloroacetaldehyde or chloral is formed. This reaction though oxidation no doubt is commonly treated to as chlorination of acetaldehyde.

$$CH_3CHO \quad + \quad 3Cl_2 \quad \rightarrow \quad \underset{\text{Chloral}}{CCl_3CHO} \quad + \quad 3HCl$$

Bromine reacts with ethyl alcohol similarly with the formation of bromal or tribromoacetaldehyde

$$C_2H_5OH \quad + \quad Br_2 \quad \rightarrow \quad CH_3CHO \quad + \quad 2HBr$$

$$CH_3CHO \quad + \quad 3Br_2 \quad \rightarrow \quad \underset{\text{Bromal}}{CBr_3CHO} \quad + \quad 3HBr$$

Iodine does not react with ethyl alcohol in these similar conditions.

Aerial Oxidation in the Presence of Certain Bacteria

In the presence of air, bacteria like mycoderma acetii oxidizes ethyl alcohol to certain extent into acetic acid. The souring of a dilute aqueous solution of ethyl alcohol (or wine) on long exposure to air is due to such formation for acetic acid through the agency of the bacteria present in the air. Vinegar, a dilute aqueous solution of acetic acid (4-8%) is prepared by the natural process.

(iii) Reduction: When heated with excess of hydroiodic acid in a sealed tube ethyl alcohol is reduced to ethane.

$$C_2H_5OH \quad + \quad 2HI \xrightarrow{\text{Heat and Pressure}} C_2H_6 + H_2O + I_2$$

Reduction with metallic Sodium

Metallic sodium reacts with ethyl alcohol at ordinary temperatures to form sodium ethoxide or sodium ethylate with the evolution of hydrogen gas. Sodium ethoxide remains in the solution. The reaction often turns vigorous so that cooling of the reaction mixture may be necessary for its smooth running. Metallic potassium reacts with alcohol in the similar method producing potassium ethylate and hydrogen gas

This is the method for the preparation of sodium ethylate in the lab, for cooling perform the reaction in an evaporating dish, and while the reaction keep the dish in a water bath.

$$2C_2H_5OH + 2Na \rightarrow \underset{\text{sodium ethoxide}}{2C_2H_5ONa} + H_2\uparrow$$

$$C_2H_5OH + 2K \rightarrow \underset{\text{Potassium ethoxide}}{2C_2H_5OK} + H_2\uparrow$$

***(iv) Reaction with Phosphorous Halides*:** Phosphorous trihalides (chloride, bromide and iodide) and phosphorous pentachloride react separately with ethyl alcohol to form the corresponding ethyl halide. With the tri-halides phosphorous acid is formed along with ethyl halides, while with PCl_5, phosphorus oxychloride and HCl are produced along with the ethyl chloride.

PCl_5 and PCl_3 are used in the reaction as such. But the reactions with PBr_3 and PI_3 is generally carried out by heating the alcohol with a mixture of red phosphorous and bromine or iodine. Red phosphorous and iodine or bromine first react to form PBr_3 or PI_3 which reacts with ethyl alcohol

$$3C_2H_5OH + PCl_5 \rightarrow 3C_2H_5Cl + H_3PO_3$$

$$[2P + 3BR_2 = 2PBr_3 \; ; \; 2P + 3I_2 = 2PI_3]$$

$$3C_2H_5OH + PBr_3 \rightarrow 3C_2H_5Br + H_3PO_3$$

(v) Reactions with Concentrated Sulphuric Acid: Depending upon the condition ethyl alcohol results with conc. Sulphuric acid to give different products

(a) at about 100°C equimolecular proportion of ethyl alcohol and sulphuric acid react together to form ethyl hydrogen sulphate (an ester of the alcohol with the inorganic acid H_2SO_4) and water

$$C_2H_5OH + H_2SO_4 \xrightarrow{100^\circ} C_2H_5SO_4 + H_2O$$

(b) When excess of ethyl alcohol is heated at about 140°C with conc. Sulphuric acid diethyl ether is produced, the elimination of the molecule of water from 2 molecules of ethyl alcohol

$$2C_2H_5OH \xrightarrow{\text{conc } H_2SO_4\ (-H_2O)} C_2H_5OC_2H_5 + H_2O$$

The reaction proceeds in two steps. At the first step ethyl hydrogen sulphate is produced at the lower temperatures. In the second step the diethyl ether is produced (at 140°C) through the elimination of a molecule of H_2SO_4 from ethyl hydrogen sulphate by the action of another molecule of ethyl alcohol.

$$C_2H_5OH + H_2SO_4 \rightarrow C_2H_5HSO_4 + H_2O$$

$$C_2H_5HSO_4 + C_2H_5OH \rightarrow C_2H_5OC_2H_5 + H_2O$$

When ethyl alcohol is heated with excess of concentrated sulphuric acid at 168°-170°C molecule of the alcohol is dehydrated (through the molecule of H_2O) to form ethylene.

$$C_2H_5OH \xrightarrow[\text{(- }H_2O)]{\text{Conc } H_2SO_4 / 165^\circ - 170^\circ C} C_2H_4$$

This reaction also proceeds in two steps, in the first step hydrogen sulphate is formed at lower temperatures. In the second step, a molecule of H_2SO_4 is eliminated from a molecule of ethyl hydrogen sulphate at the higher temperatures to form ethylene

$$C_2H_5OH \xrightarrow{\text{Conc } H_2SO_4 / 165^\circ - 170^\circ C} C_2H_4 + H_2SO_4$$

(vi) ***Dehydration:*** When the vapour of ethyl alcohol is passed over Al_2O_3 heated at 350°C, the alcohol is dry dehydrated (through the elimination of a molecule of H_2O from a molecule of the alcohol) to ethylene

$$\underset{\text{vapour}}{C_2H_5OH} \xrightarrow{Al_2O_3 / 350^\circ C} C_2H_4 + H_2O$$

(vii) ***Acetylation:*** When heated with acetyl chloride (CH_3COCl) or acetic anhydride $(CH_3CO)_2O$ ethyl alcohol is acetylated to form ethyl acetate (an ester). In these reactions the hydrogen atom of the –OH group of the alcohol is substituted by alcohol group (%CO.CH_3)

$$C_2H_5OH + \underset{\text{acetyl chloride}}{CH_3COCl} \rightarrow \underset{\text{ethyl acetate}}{CH_3COOC_2H_5} + HCl$$

$$C_2H_5OH + \underset{\text{acetic anhydride}}{(CH_3CO)_2O} \rightarrow \underset{\text{ethyl acetate}}{CH_3COOC_2H_5} + HCl$$

(viii) ***Esterification:*** Ethyl alcohol reacts with acetic acid (organic or inorganic) generally in the presence of a dehydrating agent to form an ester and water. The reaction is reversible and the dehydrating agent is used to absorb the water that is formed as one of the products and thus to shift reaction towards completion. This reaction is known as esterification.

Conc. Sulphuric acid used in the reaction serves both as the dehydrating agent and a catalyst.

Ethyl alcohol reacts with glacial acetic acid when heated in the presence of a little conc. Sulphuric acid to form ethyl acetate (ester) and water.

$$C_2H_5OH + CH_3COOH \underset{}{\overset{\text{Conc. } H_2SO_4}{\rightleftharpoons}} \underset{\text{Ethyl acetate}}{CH_3COOC_2H_5} + H_2O$$

When dry hydrogen chloride gas is passed through absolute ethyl alcohol in the presence of anhydrous zinc chloride (dehydrating agent) ethyl chloride is formed. This reaction may be look upon as the esterification of ethyl alcohol with an inorganic acid (HCl) The formation of ethyl alcohol is also the esterification of ethyl alcohol.

$$C_2H_5OH + HCl \overset{ZnCl_2}{\rightleftharpoons} C_2H_5Cl + H_2O$$

(ix) ***Reaction with Halogen in the presence of alkali:*** The ethyl alcohol reacts with halogens (Cl_2, Br_2, I_2) in the presence of an alkaline to form a trihalogenated methane (or trihalomethane) which is commonly known as chloroform, $CHBr_3$ bromoform and CHI_3 Iodoform. This reaction is known as haloform reaction. Thus as solution of chlorine in caustic soda or sodium hypochlorite reacts with ethyl alcohol to form chloroform. The reaction involves three steps (as already discussed). The oxidation of ethyl alcohol to acetaldehyde (by chlorine) the chlorination of acetaldehyde to trichloroacetaldehyde or chloral (by chlorine) and the hydrolysis of chloral to chlroform (by caustic soda solution).

$$C_2H_5OH \xrightarrow{Cl_2\,(oxidation)} CH_3CHO \xrightarrow{Cl_2\,(oxidation)} CCl_3CHO \xrightarrow{Cl_2\,(oxidation)/NaOH} CHCl_3 + HCOONa$$

ethanol acetaldehyde chloral chloroform

or

$$C_2H_5OH + 4Cl_2 + 6NaOH \rightarrow CHCl_3 + HCOONa + 5NaCl + 6H_2O$$

Chloroform is also formed when ethyl alcohol is distilled with a thin aqueous paste of bleaching powder.

$$C_2H_5OH + Ca(OCl)Cl + H_2O \rightarrow CHCl_3 + \underset{\text{Calcium formate}}{(HCOO)_2Ca}$$

When ethyl alcohol is warmed with alkaline iodine solution iodoform is produced

$$C_2H_5OH + 4I_2 + 6NaOH \rightarrow \underset{\text{Iodoform}}{CHI_3} + HCOONa + 5NaI + 5H_2O$$

$$C_2H_5OH + 4Br_2 + 6NaOH \rightarrow \underset{\text{Bromoform}}{CHBr_3} + HCOONa + 5NaBr + 5H_2O$$

Ethanol is not the only compound that under goes the haloform reaction. Isopropyl alcohol (C_3H_8O) and acetone (C_3H_6O), methyl ethyl ketone (C_4H_8O) also respond to this reaction (methyl alcohol does not take part in this reaction). In fact an organic compound which contain in its molecule a CH_3CO group or any such group then can be easily oxidised to CH_3CO group

(eg. $H_3C-\underset{OH}{\underset{|}{C}}=$ group) under goes this reaction.

The summery of reaction of ethyl alcohol is given below:

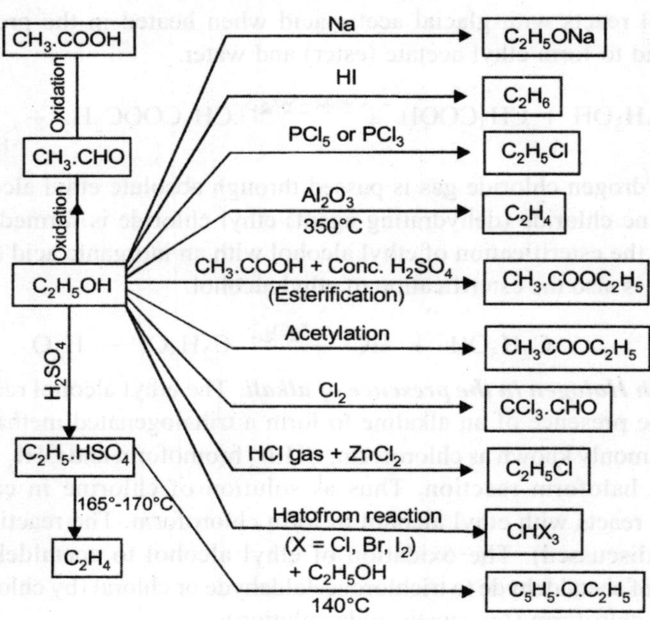

Test of Alcohol

(i) Ethyl alcohol has a sweet smell, on warming ethyl alcohol with iodine and sodium hydroxide a light yellow crystals of iodoform having characteristic smell are obtained.

(ii) Ethyl alcohol is heated with fused sodium acetate and conc. H_2SO_4. A fruity smell of ethyl acetate (ester) is obtained.

(iii) About 5 ml of ethyl alcohol are taken in a test tube and is converted into acetaldehyde by repeatedly plunging a red hot copper spiral into it, then paper soaked into ammonical silver nitrate is held in the mouth of the test tube which turns black due to the reduction of silver nitrate into metallic silver by acetaldehyde.

Presence of hydroxyl group in the molecule of ethyl alcohol: proof

The –OH group of a compound is replaced by a chlorine atom when it reacts with PCl_5; the number of HCl molecules liberated in this reaction indicates the number of %OH group present. The reaction of ethyl alcohol with PCl_5 produces C_2H_5Cl with the liberation of 1 molecule of HCl per molecule of the alcohol

$$C_2H_5OH + PCl_5 \rightarrow C_2H_5Cl + POCl_3 + HCl$$

The result of this reaction prove that ethyl alcohol contains one –OH group in its molecule, orit is a monohydric alcohol. The formation of mono-acetyl derivative, when acetyl chloride or acetic anhydride reacts with ethyl alcohol

Use

Ethanol is used as a fuel (often mixed with gasoline) and in a wide variety of industrial processes. Ethanol is also used in antifreeze products for its low melting point. The state in the United States that produces the most ethanol used in automobiles is Iowa. It is easily soluble in water in all proportions with a slight overall decrease in volume when the two are mixed. Absolute ethanol and 95% ethanol are themselves good solvents, somewhat less polar than water and used in perfumes, paints and tinctures. Other proportions of ethanol with water or other solvents can also be used as a solvent. Alcoholic drinks have a large variety of tastes because various flavor compounds are dissolved during brewing.

A solution of 70-85% of ethanol is commonly used as a disinfectant. It kills organisms by denaturing their proteins and dissolving their lipids and is effective against most bacteria and fungi, and many viruses, but is ineffective against bacterial spores. Because of this disinfectant property, alcoholic beverages can be stored for a long time.

The hydroxyl group on the ethanol molecule is an extremely weak acid, but upon treatment alkali metal or a very strong base, an H^+ can be removed to form an **ethoxide ion**, $C_2H_5O^-$.

Hazards

- Ethanol and mixtures with water greater than about 50% ethanol are flammable and easily ignited, although there are some solvents and organic compounds which are even more flammable.
- Although ethanol is not highly toxic, death from ethyl alcohol consumption is possible when blood alcohol level reaches 0.4%. A blood level of 0.5% or more is commonly fatal. Levels of even less than 0.1% can cause intoxication, with unconsciousness often occurring at 0.3-0.4%. The legal limits for driving are about 0.08-0.10% in most states, with a trend toward lowering the limit in recent years.

CONVERSION OF ALCOHOLS

(a) Primary Alcohol to Secondary Alcohol

A primary alcohol may be converted into secondary alcohol according to the following steps:

$$\underset{\text{n-propyl alcohol}}{CH_3CH_2CH_2OH} \xrightarrow{\text{HI / Heat}} \underset{\text{n - propyl iodide}}{CH_3CH_2CH_2I} \xrightarrow{\text{alcoholicKOH}} \underset{\text{propane}}{CH_3CH}$$

$$= CH_2 \xrightarrow{\text{HI}} \underset{\text{Isopropyl Iodide}}{CH_3CHICH_3} \xrightarrow{\text{aqueousKOH}} \underset{\text{Iso-propyl Alcohol (Sec alcohol)}}{CH_3CHOHCH}$$

(b) Secondary Alcohol to Tertiary Alcohol

(i) $$\underset{\text{Isopropyl Alcohol}}{CH_3CHOHCH_3} \xrightarrow{\text{Oxidation Potassium Dichromate + Sulphuric Acid}} \underset{\text{Acetone}}{CH_3COCH_3}$$

(ii) $$\underset{\text{acetone}}{H_3C{-}C({=}O){-}CH_3} \xrightarrow[\text{Dry ether}]{\text{Grignard Reagent}} \underset{\text{Acetyl magnesium Iodide}}{H_3C{-}C(CH_3)_2{-}MgI} \xrightarrow{H_2O} \underset{\text{2-methylpropan-2-ol (Tertiary alcohol)}}{H_3C{-}C(CH_3)_2{-}OH}$$

(c) Primary Alcohol to Tertiary Alcohol

It may be done by the methods outlined above by first converting the primary alcohol into the secondary alcohol which is then converted to the tertiary alcohol or by the method schematically given below:

$$\underset{\text{2-methylpropan-1-ol}}{(CH_3)_2CH{-}CH_2{-}OH} \xrightarrow{HI} \underset{\text{1-iodo-2-methylpropane}}{(CH_3)_2CH{-}CH_2{-}I} \xrightarrow[-HI]{Al\ KOH} \underset{\text{2-methylprop-1-ene}}{(CH_3)_2C{=}CH_2} \xrightarrow{HI} \underset{\text{isobutane}}{(CH_3)_2CH{-}CH_3} \xrightarrow[KOH]{H_2O} \underset{\text{2-methylpropan-2-ol}}{(CH_3)_3C{-}OH}$$

(d) Ethyl Alcohol to Methane (CH_4)

$$\underset{\text{ethanol}}{C_2H_5OH} \xrightarrow{\text{Oxidation}} \underset{\text{acetaldehyde}}{CH_3CHO} \xrightarrow{\text{Oxidation}} \underset{\text{acetic acid}}{CH_3COOH}$$

$$+Na_2CO_3 \rightarrow \underset{\text{soidum acetate}}{CH_3COONa} \xrightarrow{\text{Dry salt + soda lime}} \underset{\text{methane}}{CH_4}$$

(e) Ethyl Alcohol To (i) Ethylene (C_2H_4) (ii) Diethyl Ether ($C_2H_5O\ C_2H_5$) (iii) Ethylene Glycol ($C_2H_6O_2$)

(i) $$C_2H_5OH \xrightarrow{\text{Conc } H_2SO_4\text{(excess)165-170°C}} C_2H_4$$

$$C_2H_5OH \xrightarrow{\text{Red P + } Br_2 \text{ Heat}} \underset{\text{ethyl bromide}}{C_2H_5Br} \xrightarrow{\text{Alcoholic KOH/heat}} \underset{\text{ethylene}}{C_2H_4} + KBr + H_2O$$

(ii) $$\underset{\text{(ethanol excess)}}{2C_2H_5OH} + \text{Conc } H_2SO_4 \xrightarrow{140°\,C} \underset{\text{diethyl ether}}{C_2H_5OC_2H_5} + H_2O$$

$$\underset{\text{Ethanol}}{C_2H_5OH} \xrightarrow{\text{Conc } H_2SO_4\text{(excess)165-170°C}} \underset{\text{ethylene}}{C_2H_4} \xrightarrow{\text{Alkaline and dilute } KMnO_4} \underset{\text{diethyl ether}}{C_2H_6O_2}$$

(f) Ethyl Alcohol to Acetylene (C_2H_2)

$$\underset{\text{ethanol}}{C_2H_5OH} \xrightarrow{\text{Conc } H_2SO_4\text{ / excess(165°–170°)}} \underset{\text{ethylene}}{C_2H_4} \xrightarrow{Br} \underset{\text{ethylene bromide}}{C_2H_4Br_2} \xrightarrow{\text{AlcoholicKOH / heat}} \underset{\text{acetylene}}{C_2H_2}$$

(g) Ethyl Alcohol to Ethane (C_2H_6)

$$C_2H_5OH + 2HI \xrightarrow[\text{ethane}]{\text{Heated / under pressure}} C_2H_6 + H_2O + I_2$$

(h) Ethyl Alcohol to Propane

$$\underset{\text{ethanol}}{C_2H_5OH} \xrightarrow{\text{oxidation}} \underset{\text{acetic acid}}{CH_3COOH} \xrightarrow{Ca(OH)_2} \underset{\text{calcium acitate}}{(CH_3COO)_2Ca} \xrightarrow{\text{dry salt}} \underset{\text{acetone}}{CH_3COCH_3}$$

$$\underset{\text{acetone}}{CH_3COCH_3} \xrightarrow{\text{Zn / Hg com HCl(clemmensen's reaction)}} \underset{\text{propane}}{CH_3CH_2CH_3}$$

(i) Ethyl Alcohol to Butane (C_4H_{10})

$$C_2H_5OH \xrightarrow{Red P+I_2} C_2H_5I + Na \xrightarrow{Dry\ ether(wurtz\ reaction)} C_4H_{10} + 2NaI$$

(j) Ethyl Alcohol to acetaldehyde (CH_3CHO), Acetic Acid (CH_3COOH) Acetone (CH_3COCH_3)

$$\underset{ethanol}{C_2H_5OH} \xrightarrow[K_2Cr_2O_7+H_2SO_4]{Oxidation} \underset{acetaldehyde}{CH_3CHO} \xrightarrow[K_2Cr_2O_7+H_2SO_4]{Oxidation} \underset{acetic\ acid}{CH_3COOH} \xrightarrow[salt\ distilled]{Dry\ calcium} \underset{acetone}{CH_3CHOCH_3}$$

(k) Ethyl Alcohol to Benzene (C_6H_6)

$$\underset{ethanol}{C_2H_5OH} \xrightarrow[(160^\circ-170^\circ)]{Conc\ H_2SO_4} \underset{ethylene}{C_2H_4} \xrightarrow{Br_2} \underset{ethylene\ bromide}{C_2H_4Br_2} \xrightarrow{Alcoholic\ KOH} \underset{acetylene}{C_2H_2} \xrightarrow[red\ hot\ iron]{passed\ through} \underset{benzene}{C_6H_6}$$

(l) Ethyl Alcohol to Methyl Alcohol (CH_3OH)

$$\underset{ethanol}{C_2H_5OH} \xrightarrow{Oxidation\ K_2Cr_2O_7+H_2SO_4} \underset{acetic\ acid}{CH_3COOH} \xrightarrow{NH_4OH} \underset{Ammonium\ acetate}{CH_3COONH_4} \xrightarrow{Heat\ 200^\circ C} \underset{Acetamide}{CH_3CONH_2}$$

$$\underset{Acetamide}{CH_3CONH_2} \xrightarrow{Br+KOH\ Sol} \underset{methanamine}{CH_3NH_2} \xrightarrow{HNO_2\ (NaNO_2+dilHCl)} \underset{methanol}{CH_3OH} + N_2 + H_2O$$

(m) Methyl Alcohol to Ethyl Alcohol (C_2H_5OH)

$$\underset{methanol}{CH_3OH} \xrightarrow{Red P+I_2} \underset{methyl\ iodide}{CH_3I} \xrightarrow{KCN} \underset{acetonitrile}{CH_3CN} \xrightarrow{Reduction(Na+C_2H_5OH)} \underset{ethyl\ amide}{CH_3CH_2NH_2} \xrightarrow{HNO_2} \underset{ethanol}{C_2H_5OH}$$

Highly Poisonous

(n) Methyl Alcohol to Acetic Acid (CH_3COOH)

$$\underset{methanol}{CH_3OH} \xrightarrow{Red\ P+I_2} \underset{Methyl\ Iodide}{CH_3I} \xrightarrow{KCN} \underset{Acetonitrile}{CH_3CN} \xrightarrow{Hydrolysis(H^+\ H_2O)} \underset{acetic\ acid}{CH_3COOH}$$

(o) Methanol to Ethane

$$\underset{methanol}{CH_3OH} \xrightarrow{Red\ P+I_2} \underset{Methyl\ Iodide}{CH_3I} \xrightarrow{Dry\ ether/wurtz\ reaction} \underset{ethane}{CH_3CH_3} \text{ or } C_2H_6 + 2NaI$$

DIHYDRIC ALCOHOLS (GLYCOLS)

Glycols or Dihydroxy derivatives of paraffins were derived by Wurtz in 1856. They from a homogeneous series of general formula $C_nH_{2n}(OH)_2$ and are closely related to the monohydric alcohol. Glycols are usually named after the corresponding paraffins by adding the termination of "diol".

Ethylene Glycol, Ethane-diol

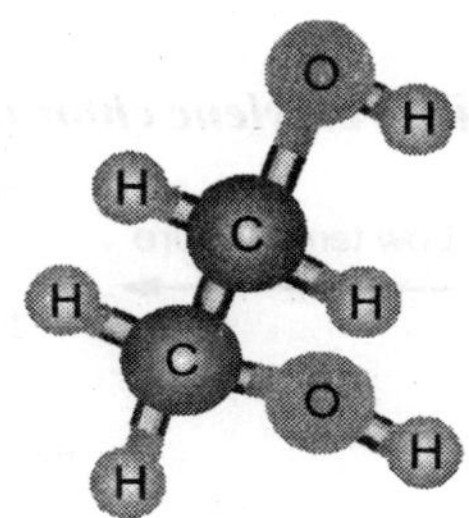

Formula: $C_2H_6O_2$

Preparation (Laboratory)

1. ***By the hydrolysis of ethylene bromide with potassium carbonate solution***

$$\underset{\text{1,2-dibromoethane}}{BrCH_2-CH_2Br} + \underset{\text{water}}{H_2O} \longrightarrow \underset{\text{ethylene glycol}}{HOCH_2-CH_2OH} + \underset{\text{hydrogen bromide}}{HBr}$$

$$K_2CO_3 + 2HBr \rightarrow 2KBr + H_2O + CO_2$$

Ethylene bromide is heated with 14% K_2CO_3 solution under a reflux condenser, when the reflux is complete the solution is slowly evaporated and extracted with a mixture of alcohol and ether in which only glycol is soluble. The liquid id filtered and subjected to fractional distillates. Glycols is collected at 197°C - 198°C.

2. ***By Hydrolysis of ethylene glycol acetate***

$$\underset{\text{1,2-dibromoethane}}{BrCH_2-CH_2Br} + \underset{\text{Silver acetate}}{2CH_3COOAg} \rightarrow \underset{\text{2-(acetyloxy)ethyl acetate}}{CH_3COOCH_2-CH_2OOCCH_3} + 2AgBr \xrightarrow{2H_2O} \underset{\text{ethylene glycol}}{HOCH_2-CH_2OH} + \underset{\text{acetic acid}}{H_3C-COOH}$$

3. ***By oxidation of Ethylene with dilute alkaline*** $KMnO_4$ ***(Bayer's Reagent)***

$$2KMnO_4 + 2KOH \longrightarrow 2K_2MnO_4 + H_2O + O$$

Bayer's Reagent

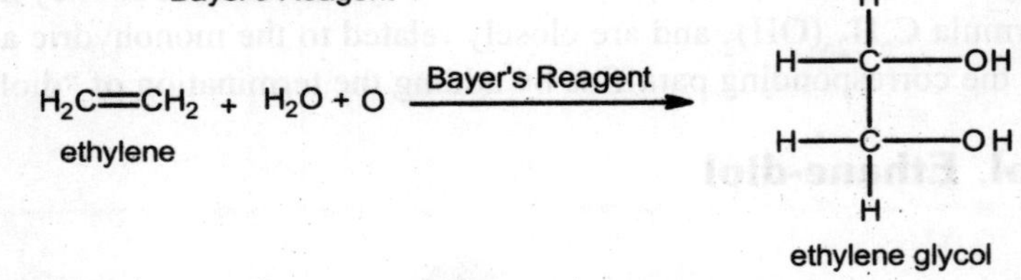

$$H_2C{=}CH_2 + H_2O + O \xrightarrow{\text{Bayer's Reagent}} \text{H–C(H)(OH)–C(H)(OH)–H}$$

ethylene → ethylene glycol

5. MANUFACTURE: - ***By the hydrolysis of ethylene chlorohydrin***

$$H_2C{=}CH_2 + HClO \xrightarrow{\text{Low temperature}} H_2C(OH){-}H_2C(Cl)$$

ethylene → 2-chloroethanol (ethylene chlorohydrin)

$$H_2C(OH){-}H_2C(Cl) + NaHCO_3 \xrightarrow{\text{heat}} H_2C(OH){-}H_2C(OH) + NaCl + CO_2$$

2-chloroethanol (ethylene chlorohydrin)

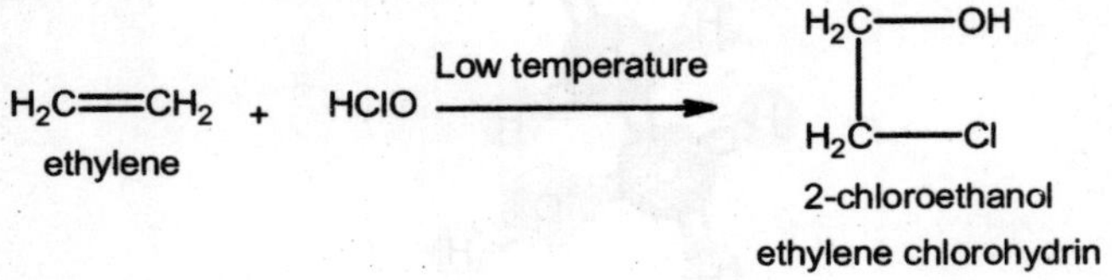

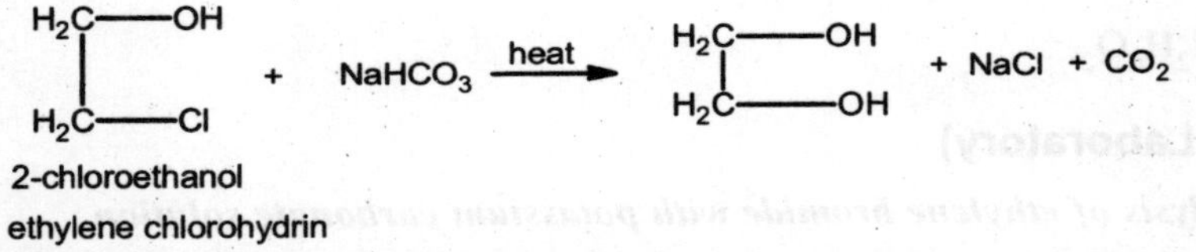

Ethylene obtained from alcohol or from natural gas or by cracking of petroleum is treated with hypochlorous acid. Ethylene chlorohydrin is thus formed. This is separated by steam distillation and then hydrolyzed with either hot sodium bicarbonate solution or milk of lime.

6. ***Hydrolysis of Ethylene oxide by dilute acid***

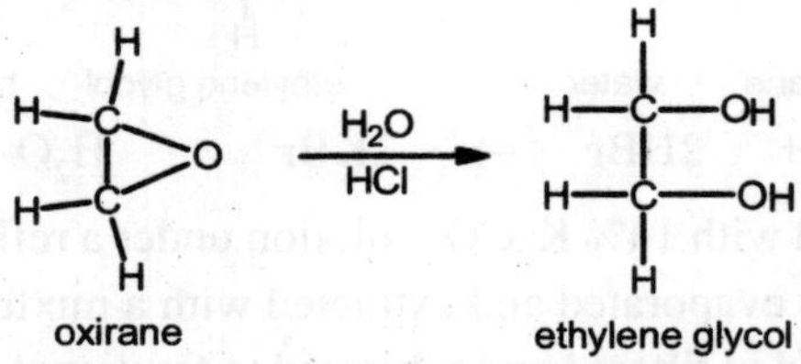

$$\text{oxirane} \xrightarrow[HCl]{H_2O} \text{H–C(H)(OH)–C(H)(OH)–H (ethylene glycol)}$$

oxirane → ethylene glycol

(Ethylene oxide is conveniently prepared by the direct combination of ethylene and oxygen under high pressure and temperature in the pressure of silver catalyst)

Synthesis: Glycol may be synthesized.

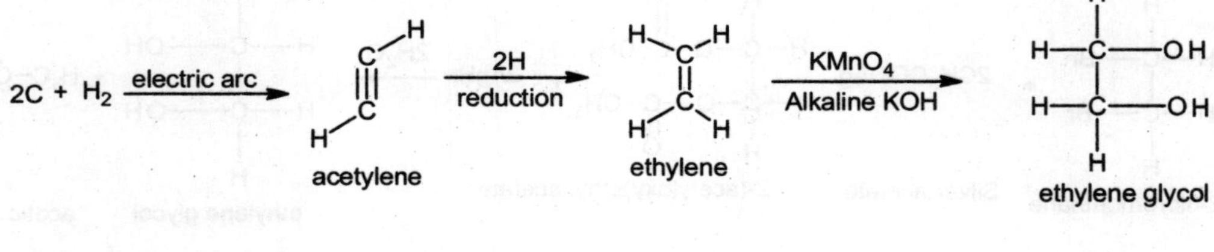

$$2C + H_2 \xrightarrow{\text{electric arc}} HC{\equiv}CH \xrightarrow[\text{reduction}]{2H} H_2C{=}CH_2 \xrightarrow[\text{Alkaline KOH}]{KMnO_4} \text{H–C(H)(OH)–C(H)(OH)–H}$$

acetylene → ethylene → ethylene glycol

$$H_2C{=}CH_2 \xrightarrow{Br_2} BrH_2C{-}CH_2Br \xrightarrow{2KOH} HOH_2C{-}CH_2OH$$

ethylene → 1,2-dibromoethane → Ethylene glycol

Properties

Physical Properties

Phase behavior	
Melting point	260.2 K (–12.9 °C)
Boiling point	470.4 K (197.3 °C)
Triple point	256 K (-17 °C)
Critical point	720 K (447°C) 8.2 MPa
$\Delta_{fus}H$	9.9 kJ/mol
$\Delta_{fus}S$	38.2 J/(mol·K)
$\Delta_{vap}H$	65.6 kJ/mol
Solubility	Miscible with water
Liquid Properties	
$\Delta_f H^0_{liquid}$	-460 kJ/mol
S^0_{liquid}	166.9 J/(mol·K)
C_p	149.5 J/(mol·K)
Density	1.1132 g/cm^3
Gas properties	
$\Delta_f H^0_{gas}$	-394.4 kJ/mol
S^0_{gas}	311.8 J/(mol·K)
C_p	78 J/(mol·K)
Safety	
Acute effects	Nausea, vomiting. CNS paralysis. Kidney damage.
Chronic effects	Kidney damage
Flash point	111 °C
Autoignition temperature	410 °C
Explosive limits	1.8–12.8%

Chemical Properties

It is a diprimary alcohol and gives all the reactions of monohydric alcohols.

1. ***Action of Sodium***: Mono and di sodium derivatives are produced

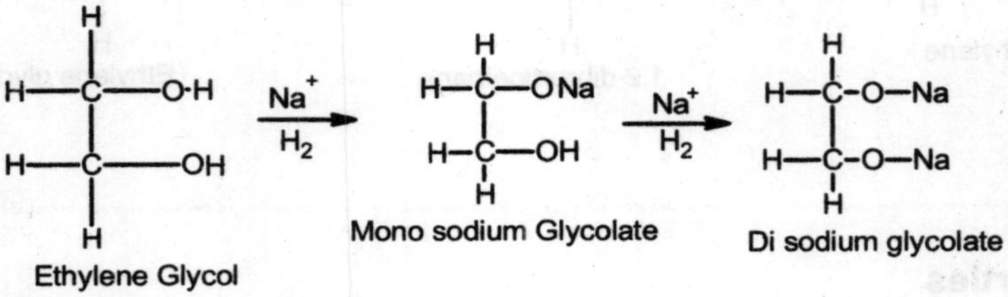

$$\underset{\text{Ethylene Glycol}}{H_2C(OH)-CH_2OH} \xrightarrow[H_2]{Na^+} \underset{\text{Mono sodium Glycolate}}{H_2C(ONa)-CH_2OH} \xrightarrow[H_2]{Na^+} \underset{\text{Di sodium glycolate}}{H_2C(O-Na)-CH_2(O-Na)}$$

2. ***Action of Phosphorus Pentachloride***: Ethylene chloride is formed

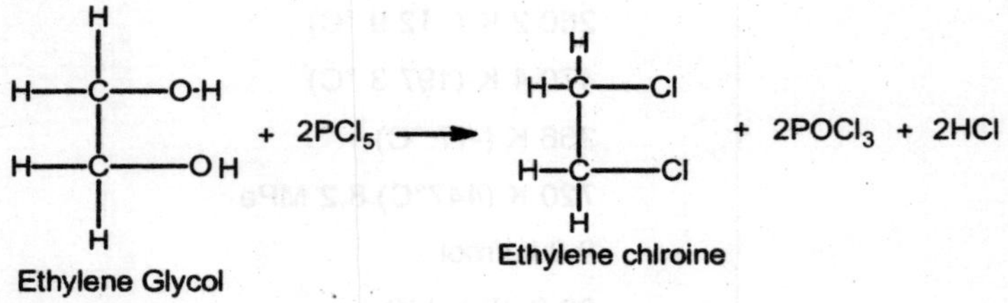

$$\underset{\text{Ethylene Glycol}}{H_2C(OH)-CH_2OH} + 2PCl_5 \longrightarrow \underset{\text{Ethylene chlroine}}{H_2C(Cl)-CH_2Cl} + 2POCl_3 + 2HCl$$

with P and I_2 it gives ethylene

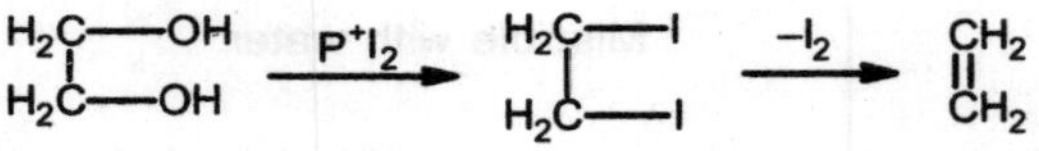

$$H_2C(OH)-H_2C(OH) \xrightarrow{P^+I_2} H_2C(I)-H_2C(I) \xrightarrow{-I_2} CH_2{=}CH_2$$

3. ***Acetylation***: Action of acetyl chloride or acetic acid, anhydride esters are produced.

(a)

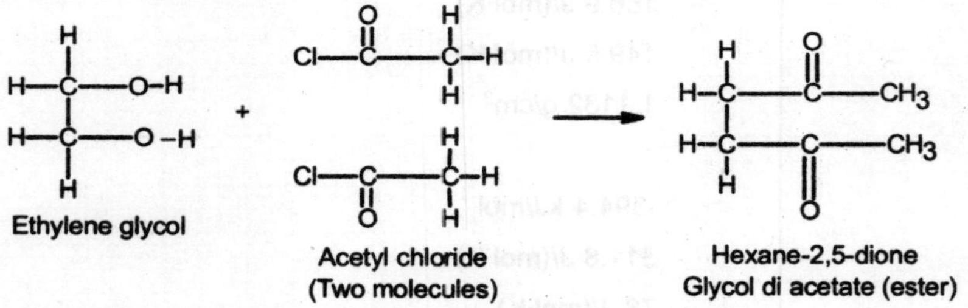

$$\underset{\text{Ethylene glycol}}{H_2C(O-H)-CH_2(O-H)} + \underset{\text{Acetyl chloride (Two molecules)}}{2\,Cl-CO-CH_3} \longrightarrow \underset{\text{Hexane-2,5-dione Glycol di acetate (ester)}}{H_2C(-CO-CH_3)-CH_2(-CO-CH_3)}$$

The reaction can also take place into stages:

(b)

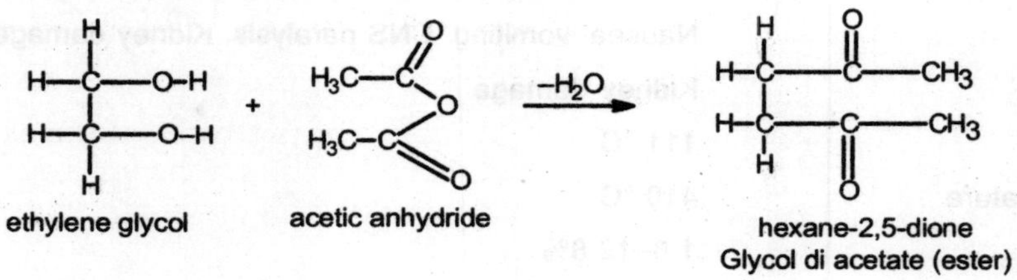

$$\underset{\text{ethylene glycol}}{H_2C(O-H)-CH_2(O-H)} + \underset{\text{acetic anhydride}}{(H_3C-CO)_2O} \xrightarrow{-H_2O} \underset{\text{hexane-2,5-dione Glycol di acetate (ester)}}{H_2C(-CO-CH_3)-CH_2(-CO-CH_3)}$$

4. Esterification: (action of acids) esters are formed

(a) with acetic acid

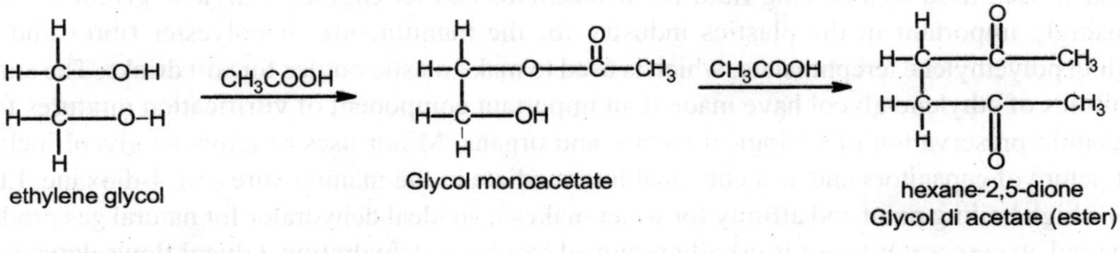

ethylene glycol $\xrightarrow{CH_3COOH}$ Glycol monoacetate $\xrightarrow{CH_3COOH}$ hexane-2,5-dione Glycol di acetate (ester)

(b) with Hydrochloric acid

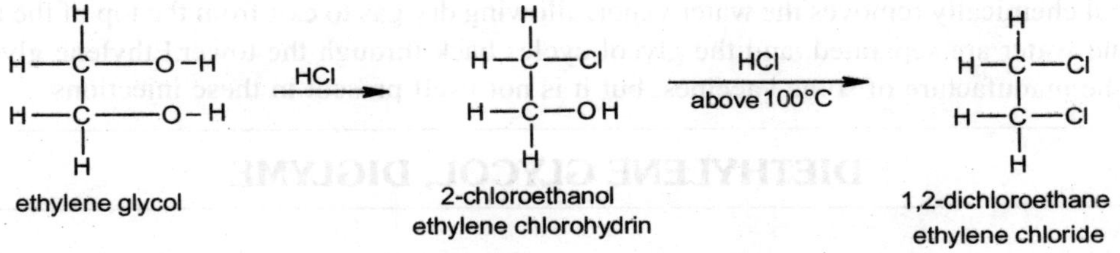

ethylene glycol $\xrightarrow{HCl}$ 2-chloroethanol ethylene chlorohydrin $\xrightarrow[\text{above } 100°C]{HCl}$ 1,2-dichloroethane ethylene chloride

(c) With conc. nitric acid in the presence of conc. sulphuric acid

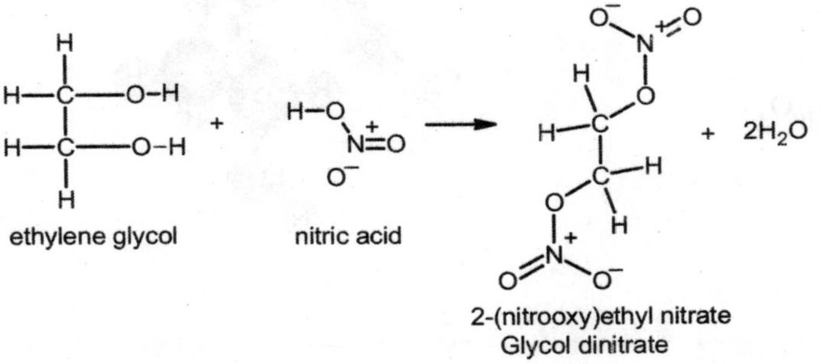

ethylene glycol + nitric acid $\longrightarrow$ 2-(nitrooxy)ethyl nitrate Glycol dinitrate + $2H_2O$

6. Oxidation: **Oxidation of glycol gives rise to a number of products, depending upon the nature of the oxidizing agent used and also upon the nature of the temperature and concentration.**

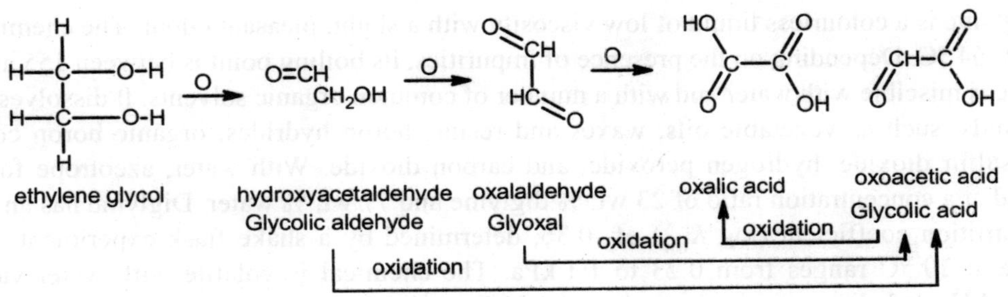

ethylene glycol $\xrightarrow{O}$ hydroxyacetaldehyde Glycolic aldehyde $\xrightarrow{O}$ oxalaldehyde Glyoxal $\xrightarrow{O}$ oxalic acid

oxoacetic acid Glycolic acid

(oxidation of hydroxyacetaldehyde → oxoacetic acid; oxidation of oxalaldehyde → oxoacetic acid; oxidation of oxoacetic acid → oxalic acid)

Uses

The major use of ethylene glycol is as an engine coolant and antifreeze. Due to its low freezing point, it has also been used as a deicing fluid for windshields and jet engines. Ethylene glycol has become increasingly important in the plastics industry for the manufacture of polyester fibres and resins, including polyethylene terephthalate, which is used to make plastic bottles for soft drinks. The antifreeze capabilities of ethylene glycol have made it an important component of vitrification mixtures for low-temperature preservation of biological tissues and organs. Minor uses of ethylene glycol include the manufacture of capacitors and as a chemical intermediate in the manufacture of 1,4-dioxane. Ethylene glycol's high boiling point and affinity for water makes it an ideal dehydrator for natural gas production. In the field, excess water vapor is usually removed by glycol dehydration. Glycol flows down from the top of a tower and meets a rising mixture of water vapour and hydrocarbon gases from the bottom. The glycol chemically removes the water vapor, allowing dry gas to exit from the top of the tower. The glycol and water are separated, and the glycol cycles back through the tower.Ethylene glycol is also used in the manufacture of some vaccines, but it is not itself present in these injections.

DIETHYLENE GLYCOL, DIGLYME

Formula: $C_4H_{10}O_3$

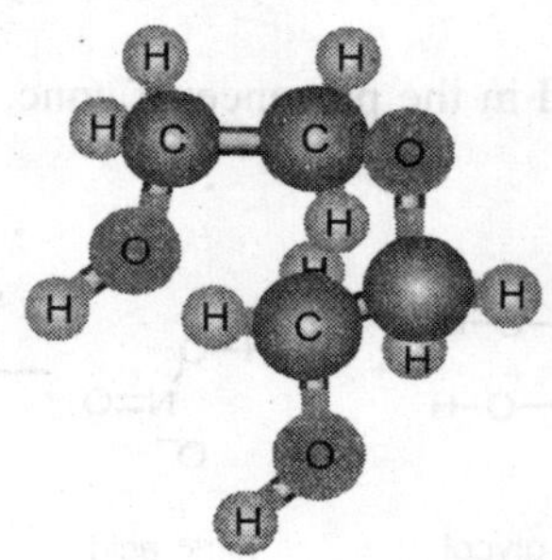

Introduction

Diglyme Sis also known as bis(2-methoxyethyl)ether (IUPAC name), diethylene glycol dimethyl ether, DEGDM(E), dimethyl carbitol, and 2,5,8-trioxynonane. It belongs to the group of ethylene glycol ethers (EGEs). The molecular structure of diglyme ($C_4H_{10}O_3$)

Diglyme is a colourless liquid of low viscosity with a slight, pleasant odour. The chemical freezes at about -64 °C. Depending on the presence of impurities, its boiling point is between 155 and 165 °C. Diglyme is miscible with water and with a number of common organic solvents. It dissolves numerous compounds, such as vegetable oils, waxes and resins, boron hydrides, organic boron compounds, sulfur, sulfur dioxide, hydrogen peroxide, and carbon dioxide. With water, azeotrope formation is observed at a concentration ratio of 23 wt. % diglyme and 77 wt. % water. Diglyme has an *n*-octanol/water partition coefficient (log K_{ow}) of -0.36, determined by a shake flask experiment. Its vapour pressure at 20 °C ranges from 0.23 to 1.1 kPa. The chemical is volatile with water vapour. The calculated Henry's law constant is given as 0.041 Pa·m³/mol

The conversion factors for diglyme for the gas phase (101.3 kPa, 20 °C) are as follows:

1 mg/m³ = 0.18 ppm

1 ppm = 5.58 mg/m³

Diglyme is chemically stable. In the presence of strong oxidation agents, peroxide may form. Commercial products typically contain peroxides at a concentration of 5 mg/kg. To avoid the further formation of peroxides, commercial products may contain antioxidants, such as 2,6-di-*tert*-butyl-4-methylphenol

Analytical Methods

Two general methods for the determination of glycol derivatives in ambient and workplace air are described:

- adsorption onto modified silica containing cyanopropyl groups, synthetic polymers such as XAD 2 or XAD 7, or modified activated charcoal, with subsequent solvent elution (e.g., acetone, dichloromethane, dichloromethane/methanol); and
- adsorption onto TENAX TA with subsequent thermal desorption.

In either case, detection is carried out via gas chromatography/flame ionization detection (GC/FID) or gas chromatography/mass spectrometric detection (GC/MSD). For the determination of diglyme in indoor air, the chemical was adsorbed onto activated charcoal, eluted with dichloromethane/methanol, and determined by capillary GC/MSD (internal standard toluene-d8 and 1,2,3-trichloropropane). The detection limit was 3 μg/m³; data on recovery rate and standard deviation are not available.

The enrichment of diglyme from water samples is also in general carried out by adsorption onto XAD 4 or XAD 8 material with subsequent solvent elution (e.g., diethylether, dichloromethane) and determination by capillary GC/MSD (Morra et al., 1979; Lauret et al., 1989). Recovery rates and standard deviations are not available. A detection limit of 0.01 μg/litre is reported.

Analytical methods for the determination of diglyme in soil or sediment are not available.

Diglyme was determined together with other glycol ethers in human urine by enrichment on diatomaceous earth, extraction with dichloromethane/acetone (90:10), and detection by capillary GC/FID. The validation results of the method are given only as ranges for total glycol ethers: detection limits 0.25–1 mg/litre, standard deviation 1.5–17.1% (at 5 mg/litre), and recovery rates 92.0–125.2% (at 2, 5, and 10 mg/litre). [^{14}C]Diglyme was determined in rat urine for metabolism studies by high-performance liquid chromatography/scintillation detection on a reversed-phase C18 column (gradient elution with methanol/acetic acid) after acidification of the sample. Information on detection methods for other biological materials is not available.

The metabolite 2-methoxyacetic acid is assumed to play a major role in the toxic effects of diglyme. Therefore, common detection methods for this compound in urine after inhalation exposure to related EGEs are described briefly here. The basis of these methods is an esterification of 2-methoxyacetic acid with diazomethane after lyophilization of the alkaline urine solution and uptake in hydrochloric acid/dichloromethane (Groeseneken et al., 1986) or with trimethylsilyldiazomethane after extraction of the acid urine solution with dichloromethane/isopropyl alcohol. The determination was carried out in both cases with a GC/FID using a capillary column. The recovery rates were reported to be 31% (Groeseneken et al., 1986) and 98%. The detection limits were 0.15 mg/litre and 0.05 mg/litre.

Sources of Human and Environmental Exposure

Natural sources

There are no known natural sources of diglyme.

Anthropogenic sources

Diglyme is manufactured in a closed system by the catalytic conversion of dimethyl ether and ethylene oxide under elevated pressure (1000–1500 kPa) and temperatures (50–60 °C) with a maximum yield of 60%. The by-products tri- and tetraethylene glycol dimethyl ether and small amounts of a high-molecular-mass ethylene glycol dimethyl ether are separated by fractional distillation. This process is based on the classic Williamson ether synthesis.

In 1982, about 47,200 tonnes of diglyme were produced in the USA (HSDB, 1983). In 1990, about 400 tonnes of the chemical were manufactured in Germany, of which 200 tonnes were exported. More recent data or data from other countries are not available. Diglyme is registered as a high-production-volume chemical by the Organisation for Economic Co-operation and Development (OECD) (i.e., its production volume in at least one OECD member state is >1000 tonnes/year).

Properties

Physical Properties

Physical State	Clear liquid
Melting Point	-10°C
Boiling Point	244 - 245°C
Specific Gravity	1.117 - 1.12
Solubility in Water	Soluble
pH	8.2
Vapour Density	3.66
Autoignition	700°C
NFPA Ratings	Health: 1; Flammability: 1; Reactivity: 0
Flash Point	93°C
Stability	Stable under ordinary conditions.

Chemical Properties

It is characterised by two hydroxyl groups along with two ether linkages, which contribute to its high water solubility, hygroscopicity, solvent properties and reactivity with many organic compounds. DEG is used in the synthesis of morpholine and 1,4-dioxane. The TEG is displacing diethylene glycol in many of these applications on account of its lower toxicity. The TEG finds use as a vinyl plasticizer, as an intermediate in the manufacture of polyester resins and polyols, and as a solvent in many miscellaneous applications. Triethylene glycol (TEG) is derived as a coproduct in the manufacture of ethylene glycol from ethylene oxide, and from "on-purpose" TEG production using diethylene glycol. Some capacities are based on total capacity for ethylene glycols. The main uses for TEG depend upon its hygroscopic properties. Air conditioning systems use TEG as dehumidifiers and, when volatilized, as an air disinfectant for bacteria and virus control. Glycols, having high boiling point and affinity for water, are employed as liquid desiccant for the dehydration of natural gas. The dehydration means the removal of water vapor in refinery tower so that dry hydrocarbon gases can exit from the top of the tower. There are wide range of glycol ethers which have bifunctional nature of ether and alcohol. cellosolves are monoether derivatives of ethylene glycol. They are excellent solvents, having solvent properties of both ethers and alcohols. Glycol family products are versatile compounds used in the fields include;

Uses

Because of its dipolar aprotic properties and its chemical stability, diglyme is used mainly as a solvent, as an inert reaction medium in chemical synthesis, and as a separating agent in distillations. These uses include industrial applications, such as polymerization reactions (e.g., of isoprene, styrene), the manufacture of perfluorinated organic compounds (BUA, 1993a), reactions in boron chemistry and its application as a solvent for, for example, textile dyes, lacquers, and cosmetics.

Diglyme is also used in the manufacture of integrated circuit boards, primarily as a solvent for the photoresists. These are used as photosensitive materials for the coating of the wafer during microlithographic patterning in the photo/apply process and in the production of semiconductors.

Diglyme is included in the European Inventory of Cosmetics Ingredients in the solvent category. Its use in cosmetics in Germany and Canada was not reported, personal communication, 2000; R. Gomes, Health Canada, personal communication, 2001). Data for other countries are also not available.

The EGEs in general are also used as auxiliary solvents in water-based paints that are industrially applied (e.g., in the spraying of automobiles, metal furniture, household appliances, and machines). It is not possible with the available data to estimate the annual amount of diglyme in this field of use or its application in water-based paints for consumer use.

TRIHYDRICALCOHOLS

GLYCEROL, GLYCERINE

The most important member of this family is Glycerol, glycerine, 123 tryhydroxy propane

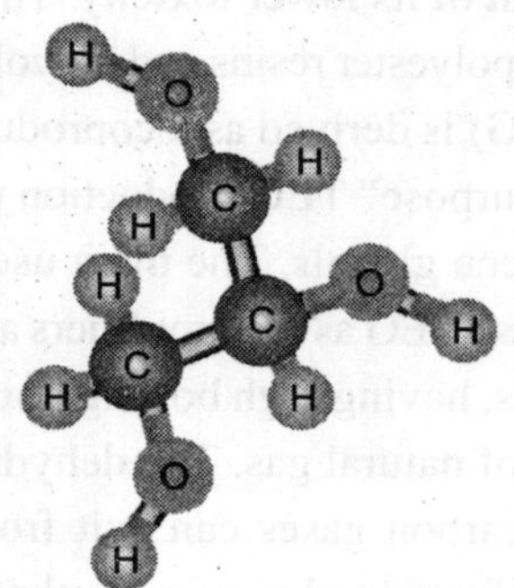

Formula: $C_3H_6O_3$

Occurrence

It occurs in fats oils which are the esters of glecerol and higher acids, such as steric acid $C_{17}H_{35}COOH$, palmatic acid $C_{15}H_{31}COOH$ and oleic acid $C_{17}H_{33}COOH$. It is also present in small amounts in the blood. Glecerol is produced as an intermediate product in aloholic fermentation.

Manufacture

Glycerol is prepared technically in large quantities as a by product in

1. Soap Industry
2. Candle Industry
3. Alcohol Industry

(1) From Soap Industry: During the manufacture of soap fact and oils by soponification with hot caustic alkalies an important by product known as spent lye is obtained. The chief component of spent lye are alkali common salt and 4 – 5 % of Glycerol.

(a) ***Recovery of Glycerol from Spent Lye:*** The lye si taken into iron "settling tank" where it is cooled and hevier matter settles down. The clear supernatent liquid is thenbrought into a "treating tank". Here alkali present in the spent lye id mostly neutralized by means of hydrochloric acid. The remaining alkalies however neutralized by adding basic sulphate

$$Fe_2(SO_4)_3 + 6NaOH \rightarrow 2Fe(OH)_3 + 4Na_2SO_4$$

This treatment with also precipitate any acids present as iron soaps.

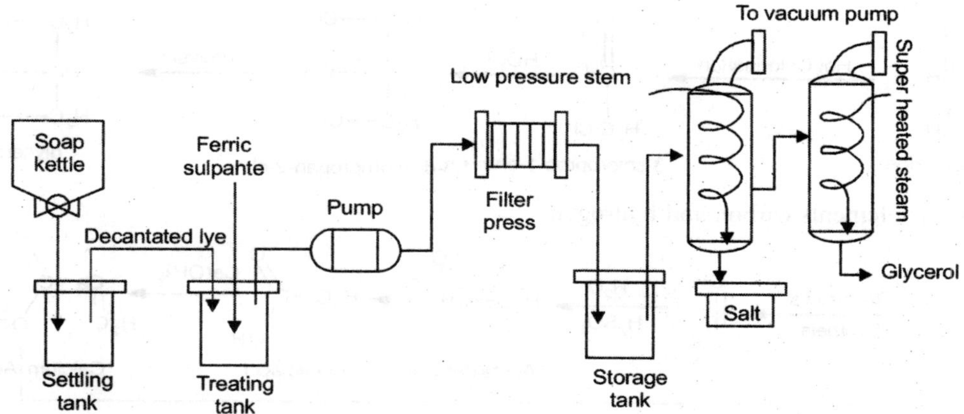

Fig. 9.6. The Glycerol Recovery from Spent Lye

The precipitate of $Fe(OH)_2$ and iron soap is now allowed to settle when it also carries down with it the albumenous impurities.

The lye is then filtered through the filter press and the clear is brought into thje "storage tank". From here the lye is transferred into "steam heated vacuum pans" where it is concetrated under reduced pressure, most of the liquid evaporates and common salt being insoluble separates out and is removed, Glycerol is left behind

Glycerol so obtained is allowed to pass through a bed of hot animal charcoal to decolorage it. It is further purified by redistillation with supper heated steam under reduced pressure. The ditillate is again distilled in vacuum when colourless glycerol is obtained.

2. **From Candle Industry**: During the manufacture of candles from fats oils by hydrolysis with superheated steam and a little dilute sulphuric acid (catalyst) a by product known as "sweet water" is obtained. Sweet water contains glycerol volatiled fally acids and water.

   (a) ***Recovery of Glycerol From Sweet Water***: It is treated with calculated quantity of sodium carbonate to neutralize the free fatty acids. The liquid is then steam distilled under pressure. Glycerol distilled over while non-voltile sodium salts of the fatty acids are left behind. Glycerol is purified as sated before.

3. **From Alochol Industry**: During the alcoholic fermentation of sugar, about 4% of glycerol is also produced. The yiled of gycerol is greatly imposed in the presence of alkaline substance such is Na_2CO_3 $NaHCO_3$ or Na_2SO_3(sodium Sulphite), however gives the best results.

$$\underset{\text{Glucose}}{C_6H_{12}O_6} \rightarrow \underset{\text{aldehyde}}{CH_3CHO} + \underset{\text{glycerol}}{C_3H_8O_3} + CO_2$$

Synethsis of Glycerol: Glycerol has been synthesized technically from propylene which is manufactured by the cracking the petroleum

prop-1-ene —Hot Chlorination→ 3-chloroprop-1-ene —HOCl→ 1,3-dichloropropan-2-ol —hydrolysis→ glycerol

From its elements carbon and hydrogen

$2C + H_2$ —Berthelot's Syntheis→ $CH \equiv CH$ —H_2O, H_2SO_4→ H_3C-CHO (Acetaldehyde) —O→ $H_3C-COOH$ (Acetic acid) —$Ca(OH)_2$→ Calcium Acetate

Calcium Acetate —Reduction, 2H→ acetone —Distil→ propan-2-ol (Isopropyl Alcohol) —$-H_2O$, H_2SO_4→ prop-1-ene (Propylene)

prop-1-ene —Cl_2→ 1,2-dichloropropane —ICl→ 1,2,3-trichloropropane —$3H_2O$, 170°C→ glycerol

Properties

Physical Properties

Boiling Point	290°C
Freezing Point	-18°C
Density at 20°C	1.26108
Dielectric constant of 9.88% Glycerine	75.98
Refractive Index	1.35749
Visocity at 80°C	32.18 Centipoises/mPa s
Vapour pressure at 182°C	20 mm Hg
Heat of Vapourization of Glycerine at 55°C	21060 cal/mol
Molar heat of solution of Glycerine	2960 cal/joules
Thermal expansion at 50°C	10240

Chemical Properties

It is trihydric alcohol and contains two primary alcohol and are secondary alcohol groups. It shows the following important reactions.

1. With sodium: Glycolates are produced

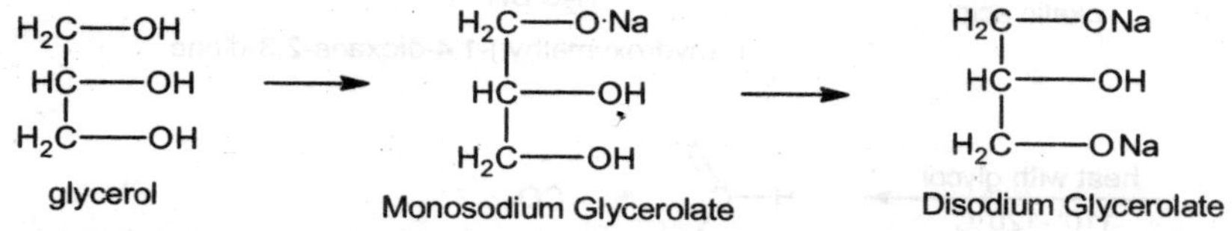

H₂C—OH / HC—OH / H₂C—OH (glycerol) → H₂C—ONa / HC—OH / H₂C—OH (Monosodium Glycerolate) → H₂C—ONa / HC—OH / H₂C—ONa (Disodium Glycerolate)

The H-atoms of the secondary alcoholic group in not replaced

2. With Phosphorous Pentachloride: Glycerol trichloride or 1,2,3 trichloride propane is formed

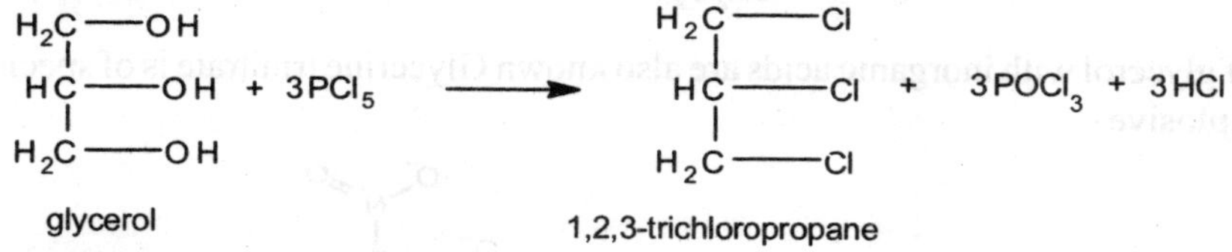

$$\underset{\text{glycerol}}{C_3H_5(OH)_3} + 3PCl_5 \longrightarrow \underset{\text{1,2,3-trichloropropane}}{C_3H_5Cl_3} + 3POCl_3 + 3HCl$$

3. With Acids: Esters are formed

(a) With acetic acid glycerol monoacetate di acetate and triacetate are produced

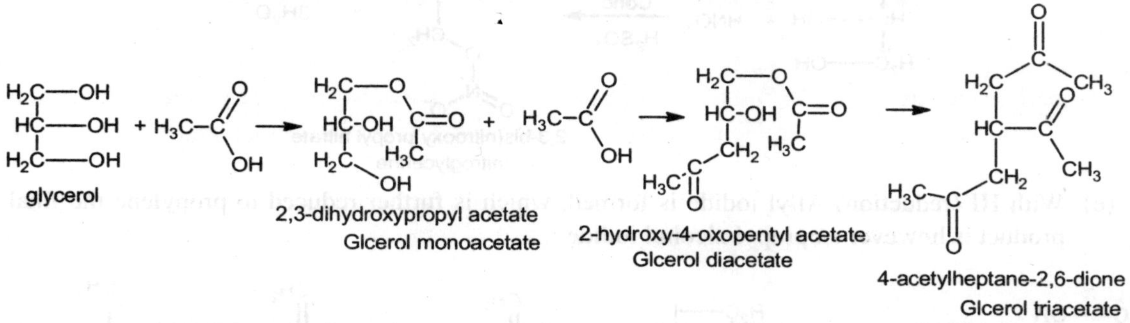

glycerol + $H_3C{-}COOH$ → 2,3-dihydroxypropyl acetate (Glcerol monoacetate) + $H_3C{-}COOH$ → 2-hydroxy-4-oxopentyl acetate (Glcerol diacetate) → 4-acetylheptane-2,6-dione (Glcerol triacetate)

(b) higher fatty acids such as Stearic Acid, palmitic acid and oleic acid forms esters are known as fats and oils.(this is formed in the human metabolism)

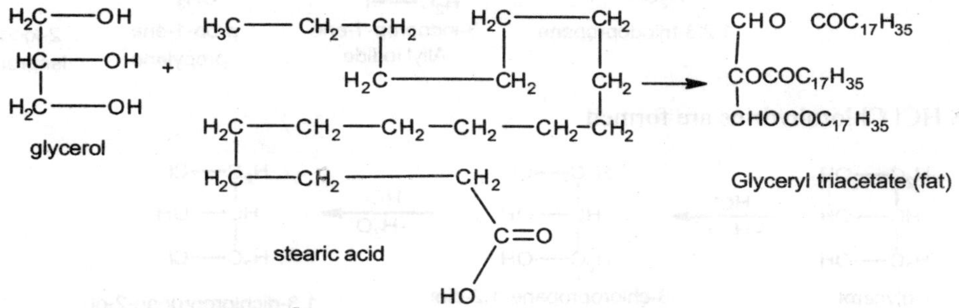

glycerol + stearic acid → CHO COC₁₇H₃₅ / COCOC₁₇H₃₅ / CHOCOC₁₇H₃₅ (Glyceryl triacetate (fat))

$$\underset{\text{Glycerol}}{C_3H_8O_3} + \underset{\text{stearic acid}}{3C_{17}H_{35}COOH} \rightarrow \underset{\text{Glyceryl triacetate (fat)}}{C_{52}H_{107}O_6}$$

(c) With oxalic acid either allyl alcohol or formic acid according to the experimental conditions.

(i) glycerol + oxalic acid $\xrightarrow{230\text{-}250°C}$ 5-(hydroxymethyl)-1,4-dioxane-2,3-dione $\xrightarrow{heat/2\text{-}CO_2}$ prop-2-en-1-ol allyl alcohol

H₂C—OH, HC—OH, H₂C—OH (glycerol) + HO—C=O, HO—C=O (oxalic acid) → $CH_2=CH-CH_2OH$

(ii) oxalic acid $C_2H_2O_4$ $\xrightarrow[110°\text{-}120°C]{\text{heat with glycol}}$ formic acid CH_2O_2 + CO_2

(d) Esters of glycerol with inorganic acids are also known Glycerine trinitrate is of special importance as an explosive.

glycerol + HNO_3 $\xrightarrow[H_2SO_4]{Conc.}$ 2,3-bis(nitrooxy)propyl nitrate (nitroglycerine) + $3H_2O$

(e) With HI (reduction) Allyl iodide is formed, which is further reduced to propylene the final product is however isopropyl alcohol iodide

glycerol + 3HI $\xrightarrow{-3H_2O}$ 1,2,3-triiodopropane $\xrightarrow{-I_2}$ 3-iodoprop-1-ene (Allyl iodide) $\xrightarrow[reduction]{HI}$ prop-1-ene (propylene) $\xrightarrow{HI}$ 2-iodopropane (Isopropyl iodide)

(f) With HCl Chlorohydrins are formed

glycerol $\xrightarrow[-H_2O]{HCl}$ 3-chloropropane-1,2-diol (glyceryl-α-chlorohydrin) $\xrightarrow[-H_2O]{HCl}$ 1,3-dichloropropan-2-ol (glyceryl-α-α-chlorohydrin)

4. ***Dehydration***: with $KHSO_4$ or P_2O_5 Acrolein (an unsaturated aldehyde) is formed.

$$\begin{array}{l} H_2C-OH \\ HC-OH \\ H_2C-OH \end{array} \xrightarrow{-H_2O} \begin{array}{l} CH_2 \\ \| \\ CH \\ | \\ HC=O \end{array}$$

glycerol — acrylaldehyde (acrolein)

Acrolein posses a disagreeable odour and therefore dehydration may be used as a test for glycerol.

5. ***Oxidation***: With $KMnO_4$ or dil HNO_3 Glycol on oxidation gives a variety of products according to the conditions under which it is carried out.

glycerol (H_2C-OH, $HC-OH$, H_2C-OH) $\xrightarrow{O}$ 2,3-dihydroxypropanal, glyceric aldehyde (H_2C-OH, $CH-OH$, $HC=O$) $\xrightarrow{2O}$ 2,3-dihydroxypropanoic acid, Glyceric acid (H_2C-OH, $CH-OH$, $HO-C=O$) $\xrightarrow{2O}$ hydroxymalonic acid, tartaric acid ($O=C-OH$, $CH-OH$, $HO-C=O$) $\xrightarrow{O}$ oxomalonic acid, mesooxalic acid ($O=C-OH$, $C=O$, $HO-C=O$)

glycerol $\xrightarrow{O}$ 1,3-dihydroxyacetone ($H_2C\text{-}OH$, $C=O$, H_2COH) → oxomalonic acid, mesooxalic acid $\xrightarrow{O}$ oxalic acid ($HO-C(=O)-C(=O)-OH$)

If glycerol is poured in drops on the potassium permanganate, the mass catches fire due to oxidation as the temperature increases the flash point of the glycerol (104°C)

Uses

Glycerin has lots of uses besides being used to make nitroglycerin (note: glycerin is not an explosive substance by itself. It has to be turned into nitroglycerin before it becomes explosive, so it's safe to work with in your kitchen). Some uses for glycerin include: conserving preserved fruit, as a base for lotions, to prevent freezing in hydraulic jacks, to lubricate molds, in some printing inks, in cake and candy making, and (because it has an antiseptic quality) sometimes to preserve scientific specimens in jars in your high school biology lab.

Glycerin is also used to make clear soaps. Highly glycerinated clear soaps contain about 15%-20% pure glycerin. Known as "Melt and Pour" soaps, these soaps are very easy for the hobbyist to work with. They melt at about 71°C, and solidify fairly rapidly. Because of their high glycerin content, the soaps are very moisturizing to the skin. Unfortunately, this high glycerin content also means that the soaps will dissolve more rapidly in water than soaps with less glycerin, and that if the bar of soap is left exposed to air, it will attract moisture and "glisten" with beads of ambient moisture.

These downsides, however are more than compensated by the emollient, skin loving and gentle nature of this soap which is especially good for tender skin and children.

The pure chemical product is called Glycerol (which shows that it is an alcohol), while the impure commercial product is called Glycerin.

NITROGYCERINE OR NOBEL'S OIL

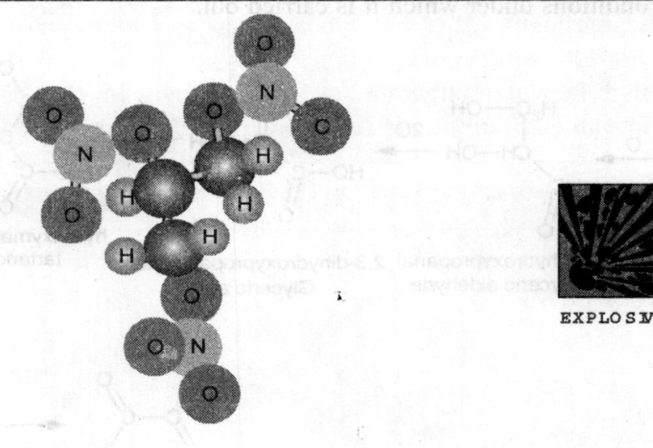

Formula: $C_3H_5N_3O_9$

Preparation

It is prepared by adding anhydrous glycerol to a mixture of conc. Sulphuric acid and fuming nitric acid between 10-20°C in lead vessels. The product is passed in water when nitroglycerine separates out as heavy oil. It is washed with a solution of sodium carbonate and finally filtered.

$$\underset{\text{glycerol}}{\begin{matrix} H_2C-OH \\ HC-OH \\ H_2C-OH \end{matrix}} + 3HNO_3 \xrightarrow[H_2SO_4]{\text{conc.}} \underset{\text{nitroglycerine}}{\begin{matrix} H_2C-O-N^+(=O)O^- \\ CH-O-N(=O) \\ CH_2-O-N^+(=O)O^- \end{matrix}}$$

Students should never try this reaction because it is explosive and risky.

What is detonation?

Nitroglycerin and any or all of the diluents mentioned above can certainly deflagrate, or burn. However, the explosive power of nitroglycerin is derived from detonation: a shock propagates through the fuel-rich medium at a supersonic speed. In other words, the initial burn sets up a pressure gradient that pre-ignites unshocked material, creating a fast-moving transition zone, which (due to the nature of the material) can detonate any appropriate material it encounters.

This generates a self-sustained cascade of hyper-instantaneous pressure-induced combustion that grows upon itself exponentially. This is quite unlike deflagration, which depends solely upon available fuel, regardless of pressure or shock. An explosion is essentially a very fast combustion, and combustion requires fuel and an oxidant. Nitroglycerin, as can be seen from its composition and structure (below), essentially contains both of these components. If it is detonated under pressure, it explodes to form thousands of times its original volume in hot gas. One of these gases is nitrogen gas. N_2 is very stable so its production is highly exothermic, which is why nitrogen is a main constituent of most explosives.

Manufacturing

The industrial manufacturing process uses a 50 : 50 mixture of fuming sulfuric acid (fuming means it is very concentrated) and red fuming nitric acid. This produces nitronium ions in situ, which are attacked by glycerin's nucleophilic oxygen atoms. The nitro group NO_2 is thus added.

The use of strong acids almost always results in an exothermic reaction (i.e., heat is produced), and this reaction is no exception. However, if the mixture becomes too hot, it explodes. Thus, the acid mixture is added slowly to the reaction vessel containing the glycerin. The reaction vessel itself is cooled with ice-cold water or some other coolant mixture at about zero °C. The vessel itself has an emergency trap door at its bottom, which hangs over a large pool of very cold water. If sensors in the mixture detect the temperature rising too rapidly, then the whole mixture can be dumped into the ice-cold water, which prevents an explosion if done in time.

Properties

Physical

Boiling point	explosion°C
Melting point	13°C
Relative density (water = 1)	1.6
Solubility in water	poor–
Vapour pressure, Pa at 20°C	0.03
Relative vapour densitys (air = 1)	7.8
Auto-ignition temperature	270°C

Nitroglycerine is a colourless odour less heavy oily liquid, it is less soluble in water but more so in alcohol, benzene etc.

If it is struck with force it detonates violently giving CO, O_2, N_2

The sudden liberations of large volume of gas results in violent explosion

It is hydrolyzed boiling caustic alkalies

$$\text{Nitroglycerine} + 3KOH \longrightarrow \underset{\text{glycerol}}{H_2C(OH)\text{–}HC(OH)\text{–}H_2C(OH)} + 3KNO_3$$

It is also attacked by Hydrogen sulphide

$$\text{Nitroglycerine} + 12H_2S \longrightarrow \underset{\text{glycerol}}{H_2C(OH)\text{–}HC(OH)\text{–}H_2C(OH)} + 6H_2O + 3NH_3 + 12S$$

Use

In medicine, nitroglycerin (sometimes called Glyceryl trinitrate) is used as a heart medication (under the trade names Nitrospan and Nitrostat). It is used as a medicine for angina pectoris (ischaemic heart disease) in tablets, ointment, or solution for intravenous use. A recent medical development will include a small amount of nitroglycerin in the tip of a new durex condom to stimulate erection during intercouse. "The CSD500 condom contains a chemical in its teat, called glyceryl trinitrate (GTN), which is absorbed into the skin and causes blood vessels to dilate."

The principal action of nitroglycerin is vasodilation, that is, widening of the blood vessels. The main effects of nitroglycerin in episodes of angina pectoris are

- **chest pain subsides**
- **blood pressure decreases**
- **heart rate increases**

These effects arise because nitroglycerin is converted to nitric oxide in the body (by a mechanism that is not completely understood), and nitric oxide in turn is a well-known natural vasodilator. Recently, it has also become popular in an off-label use at reduced (0.2%) concentration in ointment form as an effective treatment for anal fissure.

HIGHER ALCOHOLS

Propyl Alcohol

Also known is propenol, normal and Iso , $CH_3 - CH_2 - CH_2OH$ and $CH_3 - CHOH - CH_3$ are made in a commercial scale, the iso is more important their the normal

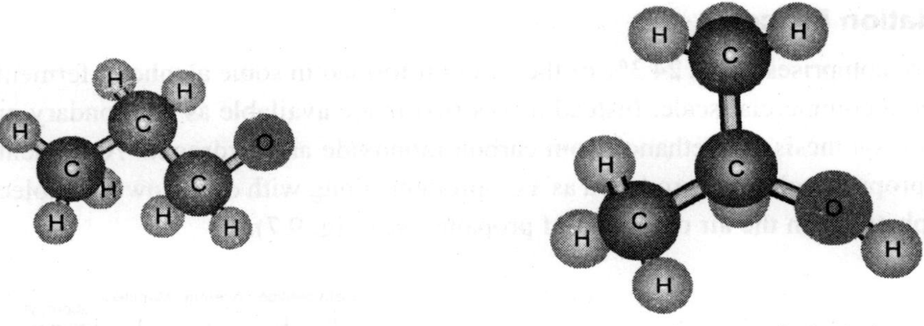

n – Propyl Alcohol *Iso* propyl alcohol

***n*-Propyl Alcohol** or 1-propenol occurs in fusel oil and was formerly isolated there form fractional distillation; the process is no longer economic. It is manufactured from ethylene co and H_2 in presence of iron hydrocarbonyl $Fe(CO)_4H_2$ catalyst and a promoter e.g. sodium salt of dimethylglycine $(CH_3)_2 N - CH_2 - COONa$ by the Oxo process.

The Oxo process

The oxo process is the reaction of a terminal alkene with CO and H_2 (1:1) under pressure (100-300 atmos) at 90°-150°C in presence of a soluble cobalt carbonyl complex to give an aldehyde. Catalytic carbonyl complex to give an aldehyde. Catalytic reduction of the aldehyde yields primary alcohol.

$$\underset{\text{alkene}}{R - CH = CH_2} + CO + H_2 \xrightarrow{CO(CO)_2} \underset{\text{aldehyde}}{R - CH_2 - CH_2 - CHO}$$

The active catalyst, cobalt Tricarbonyl hydride $HCO(CO)_3$ formed from cobalt carbonyl and H_2 co-ordinate with p and d bonds of the alkene and inserts CO – H and CO groups into the C = C bond n and iso – butyl alcohols are thus manufactures from propylene.

$$\underset{\substack{\text{prop-1-ene} \\ \text{propylene}}}{H_3C\text{-}CH{=}CH_2} \xrightarrow[H(CO)(CO)_3]{CO^+H_2} \underset{\text{butyraldehyde}}{H_3C{-}CH_2{-}CH_2{-}CHO} + \underset{\substack{\text{2-methylpropanal} \\ \text{isobutyraldehyde}}}{(CH_3)_2CH{-}CHO} \xrightarrow{H_2/Ni} \underset{\substack{\text{2-methylpropan-1-ol} \\ \text{isobutyl alcohol} \\ \text{b.p 108°C}}}{(CH_3)_2CH{-}CH_2{-}OH}$$

(butyraldehyde ↑) $H_3C{-}CH_2{-}CH{=}CH{-}OH$ (1Z)-but-1-en-1-ol major product 117°C

These are separated by fractional distillation.

$$\underset{\text{ethylene}}{H_2C{=}CH_2} + CO + H_2 \xrightarrow{180 - 200°C} \underset{\text{propionaldehyde}}{H_3C{-}CH_2{-}CH{=}O} \xrightarrow{H_2Ni} \underset{\text{propan-1-ol}}{H_3C{-}CH_2{-}CH_2{-}OH}$$

The Air Oxidation Process

n- propyl alcohol comprises about 24.3% of the fusel oil formed in some alcoholic fermentation but is not separated on a commercial scale. Instead it was first made available as a secondary product from the high pressure synthesis of methanol from carbon monoxide and hydrogen. At present the largest quantities of *n*-propyl alcohol are produced as a co-product, along with other lower – molecular weight oxygenated aliphatic from the air oxidation of propane. (see Fig. 9.7).

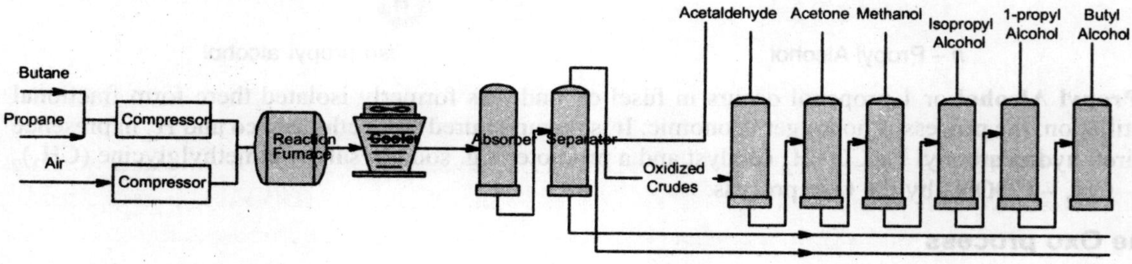

Fig. 9.7. The Manufacturing of the 1-propanol

Properties

Physical Properties

Boiling point	97.15°C
Melting point	-127°C
Relative density (water = 1)	0.8
Solubility in water	miscible
Critical temperature	264°C
Critical pressure	49.9 atmos
Vapour pressure, kPa at 20°C	2.0
Viscosity at 20°C	2.256 centipoise
Relative vapour density (air=1)	2.1

Relative density of the vapour/air-mixture at 20°C (air=1)	1.02
Flash point: 15°C c.c.Auto-ignition temperature	371°C
Explosive limits, vol% in air	2.1-13.5
Octanol/water partition coefficient as log Pow	0.25
Lower explosive limit	2.6 % by vol in air
Upper explosive limit	13.5% by vol in air

Chemical Properties

It forms several azeotropes with water (e.g. one with 29:3) water boils at 87.7°C) so cannot be dehydrated by fractional distillation. A cold saturated solution of $CaCl_2$ dissolves methanol and ethanol by complex formation but not 1-propenol by heating with H_2SO_4 with H_2SO_4 which dehydrates it to propylene and hydrolyses the later one again following the Markowinkoff's rule.

$$H_3C-CH_2-CH_2-OH \xrightarrow[-H_2O]{H_2SO_4} [H_3C-CH=CH_2] \xrightarrow{H_2O} H_3C-CH(CH_3)OH$$

propan-1-ol → prop-1-ene → propan-2-ol

Uses

1-Propenol is a solvent for printing ink, nail polish for spinning acrylonible fibres in wool dyeing and for making carbomethyl cellulose propionaldehyde Propionic acid and its esters are made from 1-propanol.

ISOPROPYL ALCOHOL

Isopropanol is a clear and flammable liquid at room temperature with odour resembles that of a mixture of ethanol and acetone; completely miscible with water, ethanol, acetone, chloroform, and benzene; melting at -89 C and boiling at 82 C. It undergoes all chemical reactions typical of secondary alcohols. It reacts violently with strong oxidizing agents. In a fire, it may decompose to form toxic gases, such as carbon monoxide. It is produced from propene by indirect hydration (strong-acid process) or direct catalytic reduction of acetone. It is a low cost solvent in many applications as Isopropanol is similar to ethyl alcohol in solvent properties and evaporation rate. Its high latent solvent power for cellulose nitrate, cellulose acetate butyrate and cellulose acetate propionate, along with its moderate evaporation rate and its complete miscibility with most solvents, make it useful in lacquers, inks and thinners. Isopropanol is a mature commodity which is projected to grow at a global rate of 1 to 3 percent per year. IPA consumption for the production of monoisopropylamine for herbicides (primarily glyphosate) is expected to be the fastest growing segment, while ketone derivatives used as solvents in coatings and inks will remain flat or increase only slightly. Government regulations covering volatile organic compounds have been, and will continue to be, a major consideration in future planning by IPA producers. IPA is used also in the production of acetone (oxidation of isopropanol is now the major source of acetone) and its derivatives and other chemicals (such as isopropyl acetate, isopropylamine, diisopropyl ether, isopropyl xanthate, fatty acid esters, herbicidal esters, and aluminium isopropoxide). Other uses include the application as a coolant in beer manufacture, a coupling agent, a dehydrating agent, a polymerization modifier in the production of polyvinyl fluoride, a foam inhibitor, a de-icing agent, a preservative, a heat-exchange medium, and in windscreen wiper concentrates. It is also used as a flavouring agent and in household and personal care products, pharmaceuticals.

Manufacture

The Elli's process, with improvements, is the conventional process used today

$$C_3H_6 + H_2SO_4 \rightarrow C_3H_7HSO_4$$

$$C_3H_6 + C_3H_7HSO_4 \rightarrow (C_3H_7)_2SO_4$$

$$C_3H_7HSO_4 + H_2O \rightarrow H_2SO_4 + C_3H_7OH$$

$$(C_3H_7)_2SO_4 + H_2O \rightarrow C_3H_7HSO_4 + C_3H_7OH$$

$$(C_3H_7)_2SO_4 + C_3H_7OH \rightarrow (C_3H_7)_2O + C_3H_7HSO_4$$

Isopropyl alcohol is produced by the cracking of petroleum hydrocarbons by thermal catalytic process. Gases containing propylene are scrubbed to remove hydrogen sulphide and mercaptans. Higher molecular weight hydrocarbons are removed by distillation. The concentrated propylene (50% or more) is then absorbed under suitable conditions of pressure and temperature in concentrated sulphuric acid to produce mixed isopropyl sulphates. The mole ratio of propylene absorbed per mole of sulphuric acid runs from 1.0 to 1.5.

In the flow plant shown in the (fig. 9.8) propylene and concentrated sulphuric acid are fed to an absorber operated under a pressure of several atmospheres. The major portion of the propylene is removed in the absorption process. The spent gas after scrubbing with caustic soda, may be used for fuel or it may be processed further. The heart of reaction in the absorber in removed by recycling through a water cooler a sufficient quantity of extract (mono and di-iso propyl sulphates) to maintain the desired temperature level. The extract is withdrawn from the absorber at a uniform rate through an extract mixer, where it is diluted with water for hydrolysis, from the extract mixer, the dilute solution flows to the generator, where, by means of live steam under low pressure, the alcohol is removed from the weak acid solution. This weak acid free of alcohol and isopropyl alcohol passes from the bottom of the generator and is concentrated to the proper strength for feed to the absorber. The alcohol vapours pass through a scrubbing tower, where sulphur dioxide and extrained acid are removed by caustic soda. Subsequently, the neutralized vapours condensed and the distillate is pumped to the storage tank.

The crude alcohol, containing 45 to 55 volume % of the isopropyl alcohol, some isopropyl ether, water and a small percentage of hydrocarbon is pumped from storage tanks to a finishing system

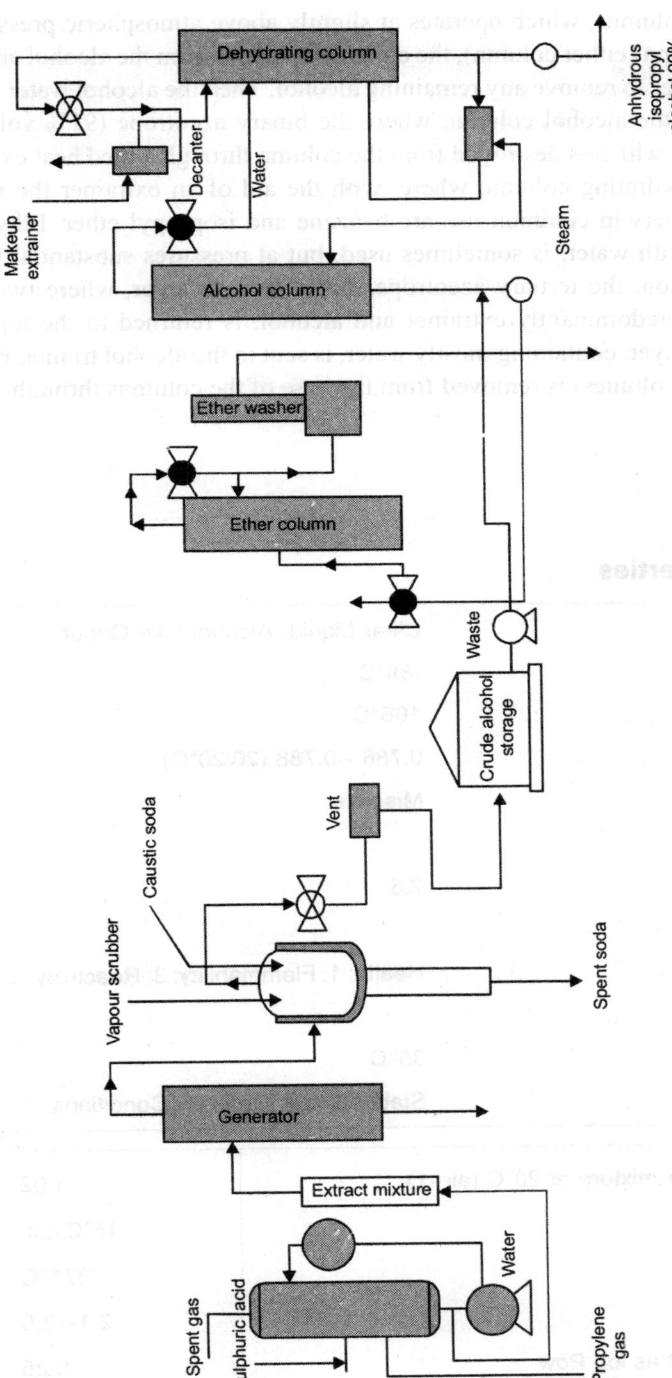

Fig. 9.8. Showing the Industrial Process of Isopropyl alcohol preparation.

consisting of three fractionating columns, which operates at slightly above atmospheric pressure and auxiliary facilities. In the first column (either column), the ether is separated from the alcohol and water and is passed through a water washer to remove any remaining alcohol. Then the alcohol water mixture passes from the either column to the alcohol column, where the binary azeotrope (91% volumetric alcohol) is distilled from the water, which is destroyed from the column through a feed heat exchange. The 91% alcohol is sent to a dehydrating column, where, with the aid of an extrainer the water is removed from the alcohol. Extrainers in common use are benzene and isopropyl ether. Ethyl ether, which forms a binary azeotrope with water, is sometimes used, but at pressures substantially above atmospheric. Upon the condensation, the tertiary azeotrope, flows to a decanter, where two layers from. The upper layer, which is predominantly extrainer and alcohol, is returned to the top of the column as reflux, while the lower layer, containing mostly water, is sent to the alcohol trainer. Finished anhydrous isopropyl alcohol (+99 volumes) is removed from the base of the columns through a cooler to storage tank.

Properties

Physical and Chemical Properties

Physical State	Clear Liquid, Alcohol Like Odour
Melting Point	-88°C
Boiling Point	108°C
Specific Gravity	0.786 - 0.788 (20/20°C)
Solubility in Water	Miscible
pH	
Vapour Density	2.6
Autoignition	
NFPA Ratings	Health: 1; Flammability: 3; Reactivity: 0
Refractive Index	
Flash Point	35°C
Stability	Stable Under Ordinary Conditions

Relative density of the vapour/air-mixture at 20°C (air=1)	1.02
Flash point	15°C c.c.
Auto-ignition temperature	371°C
Explosive limits, vol% in air	2.1-13.5
Octanol/water partition coefficient as log Pow	0.25

Uses

Process solvent

- Extraction and purification of natural products, such as vegetable and animal oils and fats, gums resins, waxes, colours, flavorings, alkaloids, vitamins, kelp and alginates.
- Carrier in the manufacture of food products
- Purification, crystallization and precipitation of organic chemicals

Coating and dye solvent

- Synthetic polymers, such as phenolic varnishes and nitrocellulose lacquers
- Cements, primers, paints, and inks

Cleaning and drying agent

- Manufacture of electronic parts, for metals and photographic films and papers, in glass cleaners, liquid soaps and detergents, and in aerosols

Solvent in topically applied preparations

- Pharmaceutical products: embrocations, massage solutions, such topically as rubbing alcohol (70% 2-propanol aerosols
- Cosmetics: hair tonics, perfumes, skin lotions, hair dye rinses and permanent wave lotions, skin cleaners and deodorants, nail polishes, shampoos

Aerosol solvent

- Cleaners, waxes, polishes, paints, de-icers, shoe and sock sprays, insect repellants, hair sprays, deodorants, air-fresheners
- Medical and veterinary products: antiseptics, foot fungicides, first aid and medical vapour sprays, skin soothers, veterinary pink eye, wound, and dehorning sprays, house and garden type insecticides

BUTYL ALCOHOL

Formula: C_4H_9OH

There are four structurally isomeric butyl chloride, noted below: Of these (i) and (ii) are primary alcohols and (iii) and (iv) are secondary alcohols and tertiary alcohols respectively. Only (iii) contains an asymmetric carbon atom (shown in bold). It exists optional rotation in plane-polarized light as

dextro and lævo forms: - (i) and (iii) are called 1-butanol-2-butanol (iii) and (iv) 2-methyl 1-propanol and 2-methyl-2-propanal respectively but it is incorrect to call isobutanol alcohol (IUPAC rule).

(i) $H_3C—CH_2—CH_2—CH_2OH$
butan-1-ol
bp 117°C fp -90°C

(ii) $H_3C—C(OH)=CH_2$
isobutyl alcohol
bp 108°C fp -108°C

(iii) $H_3C—CH_2—CH(CH_3)—OH$
butan-2-ol
bp 99.5°C fp -115°C

(iv) $H_3C—C(CH_3)_2OH$
2-methylpropan-2-ol
bp 82.5°C mp -25°C

Of these isobutyl alcohol occurs appreciably in fusel oil alcoholic fermentation; it is recovered there form by fractional distillation.

The b.p. and solubility in water of butyl alcohols differ considerably corresponding to this molecular substances. The straight chain *n*-alcohol has the highest boiling point of 117°C as it can associate by hydrogen bonding without much stearic hindrance; further more, the Van der Waals force operate on a large molecular surface. The higher branched trimethyl carbinol has the lowest boiling point 82.5°C as stearic hindrance to association is maximum and the molecular surface available for Van der Waals force minimum (ii) and (iii) have intermediate b.p. consisted with the structure solubility in water, dependant on forming hydrogen bond. Where as only 7.08 gm of 1-butanol dissolves in 100 g of water at 30°C 1-butyl alcohol is infinitely soluble in water. Alkane character due to Van der Waals forces predominates in 1-butanol to a much greater extent than 1-butyl alcohol, which least associated can water molecular more easily and get dissolved. At expected 1-butanol and isobutanol and common feature of their physical property in the formation of binary azeotropes with water. Those of *n* - *iso* - sec and *t*-butyl alcohols with 42.4 33.0 and 118% water boil respectively at 92.6°C 90°, 88.5°C and 79.9°C hence none can be dehydrated by fractional distillation of an aqueous solution.

n-BUTYL ALCOHOL

Formula: C_4H_9OH

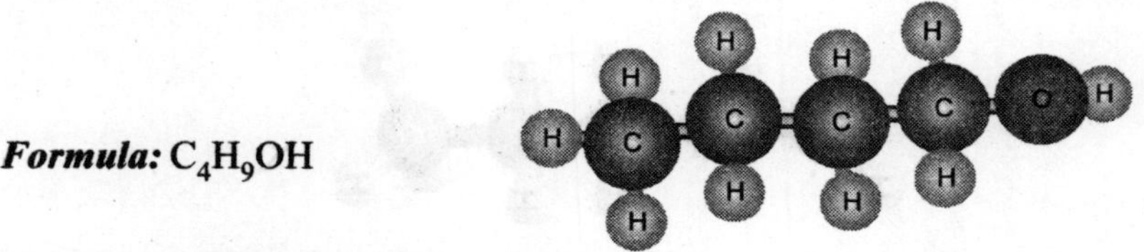

This was mainly obtained up to 1945 by fermenting sterilized molasses (5.71 sugar content) at about 37°C and pH 5-7 for 36-48 hrs with Weizmann Bacteria Clostridium acetobutylicum in absence of air. A mixture of approximately 60% butanol 302 acetylene and 10% ethanol thus results CO_2 and H_2

are liberated during the process Inorganic phosphate and ammonium salt are added as sources of phosphorous and nitrogen for bacteria. The fermented liquor is fractionally distilled wet butanol is dehydrated by redistillation; 1-butanol water azeotrope (bp) 92.6°C with 42.4% water by weight) goes off first 1-butanol (b.p 117°C) next 100 litres of molasses yield about 15 Kg of 1-butanol.

$$\underset{\text{sucrose}}{C_{12}H_{22}O_{11}} \xrightarrow{H_2O} \underset{\text{Glucose}}{C_6H_{12}O_6} \longrightarrow \underset{\text{butan-1-ol}}{H_3C-CH_2-CH_2-CH_2OH} + \underset{\text{acetone}}{H_3C-\overset{O}{\overset{\|}{C}}-CH_3} + \underset{\text{ethanol}}{H_3C-CH_2-OH} + CO_2 + H_2$$

Laboratory Methods

(i) ***From Alkyl Halides by Hydrolysis:*** A primary alkyl halide yields a primary alcohol on hydrolysis with aqueous NaOH or AgOH. The method is practical for alkyl halides obtained from sources other than primary alcohols, e.g. from alkanes or alkenes.

$$\underset{\text{1,2-dichloroethane}}{\begin{matrix} H_2C-Cl \\ | \\ H_2C-Cl \end{matrix}} \xrightarrow[+H_2O]{\text{hydrolysis}} \underset{\text{butan-1-ol}}{H_3C-CH_2-CH_2-CH_2OH} + Cl_2$$

A secondary alkyl halide forms a secondary alcohol and some alkane on alkline hydrolysis.

$$\underset{\text{2-bromopropane}}{H_3C-\overset{CH_3}{\overset{|}{C}H}-Br} + NaOH \longrightarrow \underset{\text{propan-2-ol}}{H_3C-\overset{CH_3}{\overset{|}{C}H}OH}$$

synthetic methods from acetaldehyde or propylene now provide a product at economic cost in countries where acetylene and propylene are readily available.

(ii) ***From acetaldehyde***: Aldol condensation of acetaldehyde (obtained from direct hydration of acetylene) in presence of a dilute base is the first step. Distillation of aldol with acetic acid yields crotonaldehyde which is catalytically reduced to 1-butanol.

$$\underset{\text{acetylene}}{\begin{matrix} CH \\ \| \\ CH \end{matrix}} \xrightarrow[Hg^{++}]{\text{Dil } H_2SO_4} \underset{\text{ethylenol vinyl alcohol}}{H_2C=CHOH} \longrightarrow \underset{\text{acetaldehyde}}{H_3C-\overset{O}{\overset{//}{C}H}} \xrightarrow[C_2H_5OH]{Ni\backslash H_2} \xrightarrow{OH^-} \underset{\text{ethanol}}{H_3C-CH_2OH}$$

$$H_3C-CHO \xrightarrow{25°C,\ OH^-} \underset{\text{isopropyl formate}}{H_3C-\underset{CH_3}{\underset{|}{C}H}-CH=O} \xrightarrow[CH_3COOH]{\text{distil}} \underset{\text{but-2-enal isobutyl alcohol}}{H_3C-CH=CH-HC=O} \xrightarrow[200°C\text{-}240°C]{H_2\backslash Cu - SiO_2} \underset{\text{butan-1-ol}}{H_3C-CH_2-CH_2-CH_2OH}$$

(iii) ***From Propylene***: By the oxo process, 1- butanol (with some isopropyl alcohol) is manufactured form propylene (of petroleum refineries) CO and H_2.

1-Butanol is a water which substance, mobile liquid (b.p. 117°C sp. Gr. 0.813) with the smell of fusel oil. In industry it is most important of the butyl alcohols; the nitrocellulose lacquer industry consumes. The major portion of the alcohol and the acetate ester 1-butanol is also a commercial solvent. The acetate is a fruit essence. *Di-n*-butyl phthalate is a plasticizer in the plastic industry. The alcohol is also a component of the hydraulic break fluids.

sec-BUTYL ALCOHOL

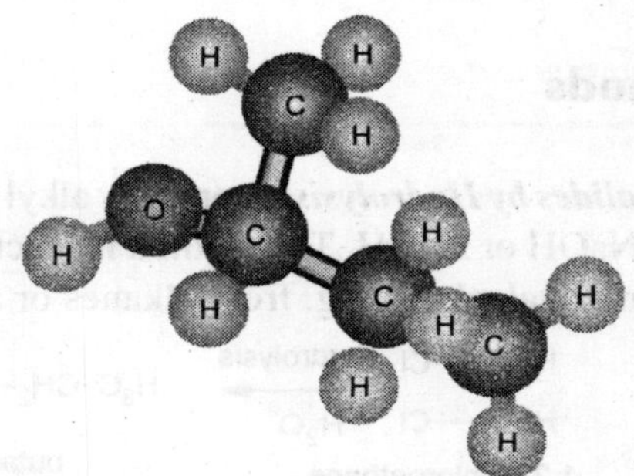

Formula: C_4H_9OH

Also called Iso-butyl Alcohol it is pleasant smelling liquid is mostly dehydrogenated to methyl ethyl ketone, $CH_3 - CO - CH_2 - CH_3$ b.p 80°C available low-boiling solvent for cellulose acetate, butyrate nitrate etc. and polyurethanes. The alcohol is also soluble in shellac. It is made by indirect acid catalyzed hydration of 1 - and 2 - butane mixture of petroleum cracking gases. Since the addition of water takes place according to Markowinkoff's rule both yield sec - butyl alcohol on hydration after absorption in concentrated H_2SO_4 followed by hydrolysis.

$$\underset{\text{1 - butane}}{H_3C{-}CH_2{-}CH{=}CH_2} \xrightarrow[H_2SO_4]{H_2O} \underset{\substack{\text{butan-2-ol} \\ \text{sec-butyl alcohol}}}{H_3C{-}CH_2{-}\overset{\displaystyle OH}{\overset{|}{C}H}{-}CH_3}$$

The synthetic alcohol as usual is optically inactive i.e. racemic* being an equimolecular mixture of dextro and leavo forms they can be resolved into d and l - forms.

t-BUTYL ALCOHOL

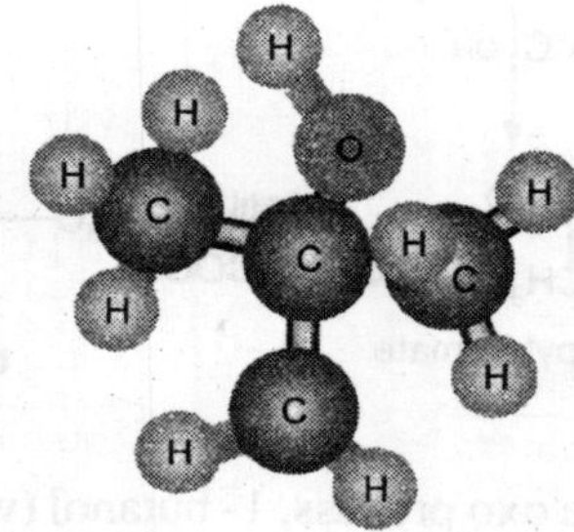

Formula: $(CH_3)_3COH$

*systematic parallel branching like grapes, from the Latin resemus.

A low melting solid (mp 25°C sp. Gr 776) is the least important of the four. The crystals are hydroscopic and smell like camphor. It is used in artificial mask and denaturizing alcohol. It is make by hydration of isobutylene under mild conditions (50-65%) H_2SO_4 at 10°-30° to avoid polymerization and diluted considerably (to avoid dehydration to isobutylene) the alcohol is extracted with solvent e.g. cresols.

AMYL ALCOHOL

Fusel Oil

It is the by-product of the alcoholic fermentation of starches and sugar and is obtained by the redistillation of the oxide ethyl alcohol. The preparation of fusel oil is about 0.2 to 1.1% of the ethyl alcohol and is highest when potatoes are fermented.

Fusel oil is mainly composed of 3 - methyl - 1 - butanol (isoamyl alcohol) and 2 - methyl - 1 - butanol (levorotatory) together with ethyl, *n*- propyl (about 3 - 5%) and isobutyl (about 20%) alcohols water and traces of *n*- butyl alcohol, 1 - pentanol and hexyl and heptyl alcohols. Only primary alcohols are believed to be present. Traces of aldehydes, acids esters pyridine and alkaloids have been reported. The strong residual odour of crude fusel oil is due to some of those non-alcoholic impurities.

Refined fusel oil is obtained by a combination of washing, chemical treatment and distillation. The composition is bout 85% 3 - methyl - 1 - butanol and 15% 2 - methyl - 1 - butanol.

Types of Amyl Alcohol

There are eight (A compound that exists in forms having different arrangements of atoms but the same molecular weight) isomers of **amyl (A liquor or brew containing alcohol as the active agent) alcohol** ($C_5H_{11}OH$) (Students are requested to Create the 3-d Model of these Alochols in Chem-sketch):

Name	Formula	Alcohol structure	IUPAC Name
1. normal amyl alcohol	$CH_3{\cdot}(CH_2)_4{\cdot}OH$	primary	1-pentanol
2. isobutyl carbinol or isoamyl alcohol	$(CH_3)_2{\cdot}CH{\cdot}CH_2{\cdot}CH_2OH$	primary	3-methyl-1-butanol
3. active amyl alcohol	$(CH_3)(C_2H_5){:}CH{\cdot}CH_2OH$	primary	2-methyl-1-butanol
4. tertiary butyl carbinolor neopentyl alcohol	$(CH_3)_3C{\cdot}CH_2OH$	primary	2,2-dimethyl-1-propanol
5. diethyl carbinol	$(C_2H_5)_2CH{\cdot}OH$	secondary	3-pentanol
6. methyl (n) propyl carbinol	$(CH_3.CH_2{\cdot}CH_2)(CH_3){:}CH{:}OH$	secondary	2-pentanol
7. methyl isopropyl carbinol	$(CH_3)_2{:}CH(CH_3){:}CHOH$	secondary	3-methyl-2-butanol
8. dimethyl ethyl carbinolor tertiary amyl alcohol	$(CH_3)_2{\cdot}(C_2H_5){\cdot}{:}C{\cdot}OH$	tertiary	2-methyl-2-butanol

Three of these alcohols, active amyl alcohol, methyl(n) propyl carbinol, and methyl isopropyl carbinol, contain an asymmetric carbon atom and can consequently each exist in two optically active, and one optically inactive form. The most important is isobutyl carbinol, this being the chief constituent of fermentation amyl alcohol, and consequently a constituent of (A mixture of amyl alcohols and propanol and butanol formed from distillation of fermented liquors) fusel oil. It may be separated from fusel oil by shaking with strong (A strong solution of salt and water used for pickling) brine solution, separating the oily layer from the brine layer and distilling it, the portion boiling between 125°C and 140°C being collected. For further purification it may be shaken with hot lime water, the oily layer separated, dried with (A deliquescent salt; used in de-icing and as a drying agent) calcium chloride and fractionated, the fraction boiling between 128 °C and 132 °C only being collected. It may be synthetically prepared from isobutyl alcohol by conversion into isovaleraldehyde, which is subsequently reduced to isobutyl carbinol by means of sodium amalgam.

It is a colourless liquid of density 0.8247 g/cm³ (0°C), boiling at 131.6°C, slightly soluble in water, easily soluble in alcohol, (a colourless volatile highly inflammable liquid formerly used as an inhalation anesthetic) ether, (a volatile liquid haloform ($CHCl_3$); formerly used as an anesthetic) chloroform and (a colourless liquid hydrocarbon; highly inflammable; carcinogenic; the simplest of the aromatic compounds) benzene. It possesses a characteristic strong smell and a sharp burning taste. When perfectly pure, it is not a poison, although the impure product is. On passing its vapour through a red-hot tube, it undergoes decomposition with production of (A colourless flammable gas used chiefly in welding and in organic synthesis) acetylene, (a flammable colourless gaseous alkene; obtained from petroleum and natural gas and used in manufacturing many other chemicals; sometimes used as an anesthetic) ethylene, propylene, etc. It is oxidized by (An unstable acid known only in solution and as chromate salts) chromic acid mixture to isovaleraldehyde; and it forms crystalline addition compounds with (A deliquescent salt; used in de-icing and as a drying agent) calcium chloride and tin(IV) chloride.

The other amyl alcohols may be obtained synthetically. Of these, tertiary butyl carbinol has been the most difficult to obtain, its synthesis having first been accomplished in 1891 by the reduction of a mixture of trimethyl acetic acid and trimethylacetyl chloride with sodium amalgam. It is a solid which melts at 48 to 50 °C and boils at 112.3°C.

A Monohydric Alcohol into Dihydric Alcohol

$$C_2H_5OH \xrightarrow{H_2SO_4} C_2H_5HSO_4 \xrightarrow[160^\circ C]{heat} CH_2{=}CH_2 \xrightarrow{Br_2} H_2C(Br){-}CH_2(Br) \xrightarrow[aqueous]{K_2CO_3} H_2C(OH){-}CH_2(OH)$$

Ethyl Alcohol → Ethyl Hydrogen Sulphate → ethylene → 1,2-dibromoethane → ethylene glycol

Physical Properties of different amyl alcohols

Constant	1-pentanol	2-pentanol	3-pentanol	2-methyl 1-butanol	3-methyl 1-butanol	2-methyl 2-butanol	3-methyl 2-butanol	2-2 dimethyl
Melting point °C	–78.2	—	–74	–69	–117.2	–11.9	—	53
Boiling point °C	137.8	119.0	114.8	–127.5	130.6	101.8	101.8	113.5
Density	0.8144	0.8303	0.8218	0.8193	0.812	0.809	0.819	—
Refractive Index	1.4099	1.4037	1.4104	1.4107	1.4053	1.4052	1.4095	—
Flash Point								
Open cup (°C)	57.78	42.3	38.9	50	56	22	—	—
Viscosity at 25°C (centipoises)	3.31	—	4.12	—	—	3.70	—	—
Heat of Vapourization cal/g	120.6	—	96.8	100	105.4	106.2	—	—
sp heat cal/g	0.712	—	—	—	0.686	0.753	—	—
Co-efficient of exp /°C	0.00092	—	0.00149	0.00078	0.00089	0.00133	—	—

A Monohydric Alcohol into Dihydric Tri Hydric Alcohol

$$\underset{\text{Ethyl Alcohol}}{C_2H_5OH} \xrightarrow{O} \underset{\text{acetaldehyde}}{H_3C\text{–}CHO} \xrightarrow{O} \underset{\text{acetic acid}}{H_3C\text{–}COOH} \xrightarrow{Ca(OH)_2} Ca(CH_3COO)_2 \xrightarrow{\text{heat}} \underset{\text{acetone}}{H_3C\text{–}CO\text{–}CH_3}$$

$$\longrightarrow \underset{\text{propane}}{CH_3\text{–}CH_2\text{–}CH_3} \xrightarrow{-H_2O} \underset{\text{prop-1-ene}}{CH_3\text{–}CH{=}CH_2} \xrightarrow[500^{\circ}C]{Cl_2} \underset{\text{3-chloroprop-1-ene}}{CH_2Cl\text{–}CH{=}CH_2} \xrightarrow{HOCl} \underset{\text{1,3-dichloropropan-2-ol}}{CH_2Cl\text{–}CH(OH)\text{–}CH_2Cl} \xrightarrow{2NaOH} \underset{\text{glycerol}}{CH_2(OH)\text{–}CH(OH)\text{–}CH_2(OH)}$$

Reactions of all the Alcohols

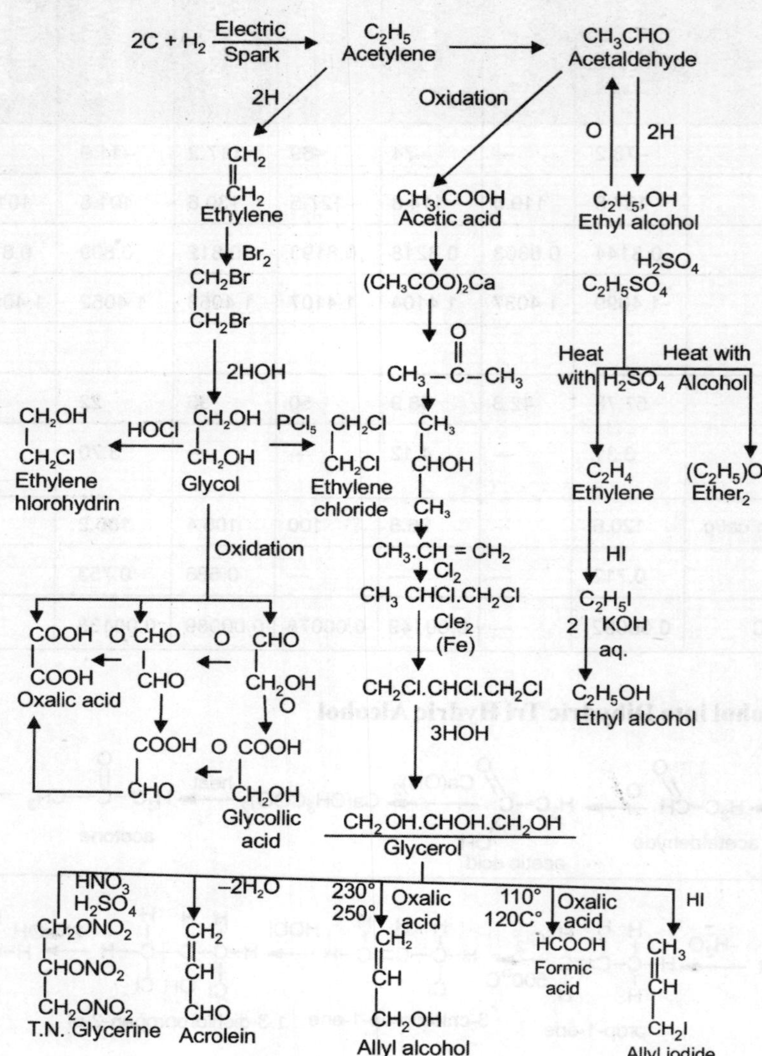

CHAPTER

10

Ether

History and Invention

Ether was discovered in 1275 by Spanish Alchemist **Raymundus Lullius**. This new discovery was given the name "sweet vitriol" or rather "Dulcum Vitrium" In 1540, the synthesis of ether was described by German scientist **Valerius Cordus**. At about the same time, Swiss physician and alchemist Paracelsus discovered the hypnotic effects of ether. Later, in 1730, German scientist **W.G. Frobenius** changed the name of sweet vitriol to ether.

General Information

Ethers form a homologous series of the general formula $C_nH_{2n+1} - O - C_nH_{2n+1}$ or $R - O - R$ where R is the alkyl group. Now if the two alkyl groups are same, the ether is called *simple ether*, and if the ether have two different alkyl group it is called *mixed ether* e.g.

Diethyl Ether	$C_2H_5 - O - C_2H_5$	Simple ether (both the alkyl group are same)
Methyl Ethyl Ether	$CH_3 - O - C_2H_5$	Mixed ether (both the alkyl group are same)

Diethyl ether or simple ether is a typical member of this family.

General Methods of Preparation

Ethyl ethers are prepared by the following general methods:

1. **From Alkyl Halides**: (i) By heating with sodium alkyl oxides (Williamson's synthesis)

$$C_2H_5\,\boxed{I + Na}\text{-}O\text{-}C_2H_5 \longrightarrow C_2H_5\text{-}O\text{-}C_2H_5$$

List of Ethers

Ether	Formula	Formula wt.	B.p. (°C.)	d_4^{10}	n_D^{20}
Aliphatic Ethers					
	Saturated				
Methyl *n*-propyl	$CH_3OCH_2CH_2CH_3$	74.12	38.9	0.738	1.3579
Methyl *tert*-butyl	$CH_3OC(CH_2)_2$	88.15	55.1	0.7406	1.3690
Ethyl *n*-butyl	$C_2H_5O(CH_2)_3CH_3$	102.17	91.4 (90-93)	0.752	—
n-Propyl	$(CH_3CH_2CH_2)_2O$	102.17	91	0.7360	1.3807
tert-Butyl	$(CH_3)_3C)_2O$	130.23	106.5-107	0.6758	1.3949
Isoamyl	$(CH_3)_2CHCH_2CH_2)_2O$	158.28	172.5-173.0	0.78073_{15}^{15}	1.408
Bis(2-ethylhexyl)	$(CH_3(CH_2)_3CH(C_2H_6)CH_2)_2O$	242.43	269.4	0.812_{20}^{20}	—
	Substituted				
Bis(chloromethyl)	$(ClCH_2)_2O$	114.97	0.05	1.315	1.4346
1-Chloroethyl ethyl	$CH_3CHClOC_2H_5$	108.57	28_{57}	0.9495	$1.4021^{28.5}$
2-Chloroethyl ethyl	$ClCH_2CH_2OC_2H_5$	108.57	105-107	0.9945	1.412
	Unsaturated				
Vinyl	$(CH_2{:}CH)_2O$	70.09	39 (28.3)	0.774_{20}^{20}	—
Vinyl methyl	$CH_2{:}CHOCH_3$	58.08	5.0-6.0	0.7511	1.3947^{-25}
Vinyl ethyl	$CH_2{:}CHOC_2H_5$	72.10	35.0-35.5	0.7533	1.3739^{25}
Vinyl *n*-butyl	$CH_2{:}CHO(CH_2)3CH_3$	100.16	93.3-94.1	0.7735^{25}	1.3997^{25}
Vinyl isobutyl	$CH_2{:}CHOCH_2CH(CH_2)_2$	100.16	82.9-83.3	0.768^{25}	1.3938^{20}
Allyl	$(CH_2{:}CHCH_2)_2O$	98.14	94.84	0.8053	1.4165
Bis (2-methylally)	$(CH_2{:}C(CH_3)CH_2)_2O$	126.19	105.44	0.8627	1.4206
Allyl ethyl	$CH_2{:}CHCH_2OC_2H_5$	86.13	67.6	0.765	1.38810
1-Methyleneallyl methyl (methoxy-prene)	$CH_2{:}CHC({:}CH_2)OCH_3$	84.11	74-74.5	0.8272	1.4432
Ethynyl ethyl	$CH{:}COC_2H_5$	70.09	49	0.8001	1.3736
Ethynyl n-butyl	$CH{:}CO(CH_2)_3CH_3$	98.14	102-104	0.8078^{25}	1.4033^{25}
Alicyclic Ethers					
Cyclopropyl methyl	$CH_2.CH_2.CHOCH_3$	73.11	44.73	0.8100	1.3802
Aromatic Ethers					
Benzyl methyl	$C_6H_5CH_2OCH_3$	122.16	$170.4_{749.3}$	0.9745^{25}	1.5031
Benzyl ethyl	$C_6H_5CH_2OC_2H_5$	136.19	185	0.949	—

Both simple and mixed ether can be obtained by this method:

$$\underset{\text{ethyl iodide}}{C_2H_5I} + Ag_2O \longrightarrow 2AgI + \underset{\text{diethyl ether}}{H_3CCH_2—O—CH_2CH_3}$$

2. **From Monohydric Alcohols**: (i) By catalytic dehydration vapours of primary alcohol (ROH) are passed over Al_2O_3 or thorium oxide at 240°C - 260°C [Sabatier and Mailhe's Method]

$$\underset{\text{ethyl iodide}}{C_2H_5OH} \xrightarrow{Al_2O_3} \underset{\text{diethyl ether}}{H_3CCH_2—O—CH_2CH_3} + H_2O$$

The method is used for the preparation of diethyl ether in large scale.

(ii) By heating with conc. sulphuric acid at 140°C (continuous fraction process)

(a) $$C_2H_5OH + H_2SO_4 \rightarrow \underset{\text{ethyl hydrogen sulphate}}{C_2H_5HSO_4} + H_2O$$

(b) $$C_2H_5HSO_4 + C_2H_5OH \rightarrow \underset{\text{diethyl ether}}{C_2H_5-O-C_2H_5}$$

This preparation is employed for the preparation diethyl ether in the lab.

Properties

Physical Properties

1. Diethyl ether and methyl ethyl ether are gases at ordinary temperatures, all other common ethers are colourless, volatile, pleasant smelling liquids.
2. They are lighter than water.
3. They are freely miscible with alcohol but less so in water.
4. They are highly inflammable.

Chemical Properties

Chemically, ethers are relatively inert. They do not react with alkali metal; (thus sodium is used to dry ethyl ether), and they are not attacked by alkalies. They do not undergo hydrogenolysis. On long standing in air, ethers autoxidize to peroxides, which are dangerously explosive. Strong oxidizing conditions split ether molecules. However, an ether group can remain unattacked during the oxidation of a double bond or a free primary or secondary alcoholic group in the same molecule. Hence hydroxyl groups are often "protected" by methylation during synthetic reactions.

Cleavage of one or both of the O-linkages can be affected by inorganic acids, most readily by hydroiodic or nitric acid, by acyl halides, or by phosphorus pentachloride:

$$ROR + HI \xrightarrow{cold} RI + ROH$$

$$ROR + 2HI \xrightarrow{heat} 2RI + H_2O$$

$$ROR + \underset{\text{methyl carbinyl iodide}}{CH_3COI} \rightarrow RI + CH_3COOR$$

$$ROR + PCl_4 \rightarrow 2RCl + POCl_3$$

With hydroiodic acid mixed ethers yield a mixture of alkyl iodides (and alcohols in the cold); if one of the R's is methyl and the other is a higher radical other than a tertiary radical, methyl iodide is produced:

$$CH_3OCH_2CH_2CH_3 + HI \rightarrow CH_3I + CH_3CH_2CH_2OH$$

This reaction of mixed ethers with hydroiodic acid constitutes a useful method *(Zeisel method)* for determining the methoxyl content of organic compounds. The relative stability of various alkoxy and aryloxy groups has been studied by cleavage of ethers with acetyl bromide or with hydrogen bromide in glacial acetic acid. Cleavage of ethers with hydrobromic acid requires somewhat higher temperatures than hydri-odic acid, but refluxing with constant-boiling hydrobromic acid in acetic acid solution is a convenient method for demethylating nonvolatile methoxy compounds, as for removing "protecting" methyl groups. Ethers are also decomposed by heat, especially readily in the presence of alumina, to yield olefins, water, and sometimes various other products

A characteristic property of ethers is their ability to form "addition compounds." With hydrochloric acid and other strong inorganic acids ethers form oxonium com-pounds having the general formula $[R_2OH]^+X^-$. The ability to dissolve in cold concentrated sulphuricacid to form oxonium sulfates distinguishes ethers from paraffin hydrocarbons and alkyl halides and enables ethers to be separated from these other types, of compounds. Ethers can be recovered from sulphuricacid solutions by dilution with water; hydrolysis can be prevented by pouring the solution slowly on ice. Ethers also form soluble complexes with inorganic halides such as mercuric bromide and boron trifluoride and with Grignard reagents.

The hydrocarbon groups of ethels can be chlorinated or brominated somewhat more readily than paraffin hydrocarbons.

The inertness of ethers toward most reagents used to test for the presence of characteristic groups in other compounds serves as a guide in the identification of an unknown substance as ether.

Reactions of Ethers

Although ethyl ether is relatively unreactive, like ethers in general, a few of its reactions are worth noting. The spontaneous formation of nonvolatile peroxides on standing in air, especially in the light, is important from the standpoint of the resulting hazards. The peroxides remain as a residue when ether evaporates and may explode violently in a distillation carried out to dryness. Furthermore, peroxides are pulmonary irritants and must therefore be absent in anesthetic ether. The structure and even the composition of ether peroxides are subject to uncertainty. Among the structures that have been proposed are $CH_3CH(OH)OOCH(OH)CH_3$ (dihydroxydiethyl peroxide, bis(l-hydroxyethyl) peroxide) by Wieland and CH_3CH-$(OOH)OC_2H_6$ (ethyl ether hydroperoxide, 1-ethoxyethylhydroperoxide). Still other formulas suggested represent cyclic and oxonium structures. Weak acids or water bring about peroxide decomposition to hydrogen peroxide and ethyl alcohol.

Under vigorous oxidation ethyl ether undergoes cleavage. In the presence of a copper catalyst and platinum black, ether vapor and air at 100°C. form acetaldehyde and formaldehyde. Chromic anhydride (CrO_3) or nitric acid oxidizes ethyl ether to acetic acid. The reaction with absolute nitric acid is so violent that it constitutes an explosion hazard; some ethyl nitrate is formed at -15°C. and nitrogen oxides at room temperature. Because of the volatility of ether there is danger of concentrating a solution of nitric acid until an explosion takes place. In the presence of concen-trated sulphuricacid at room temperature ether and nitric acid explode violently.

DIETHYL ETHER

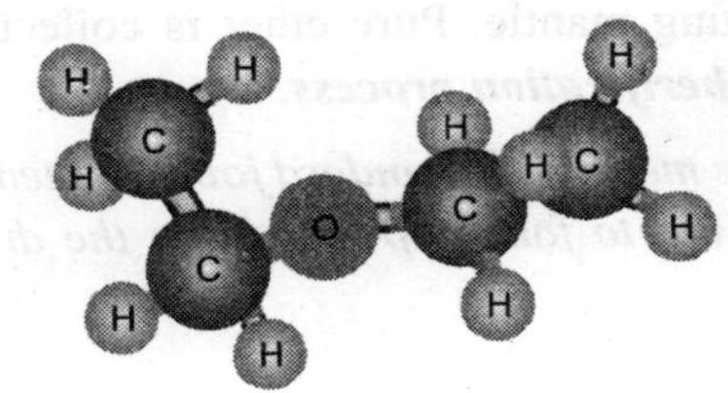

Formula: $C_4H_{10}O$

Laboratory Preparation

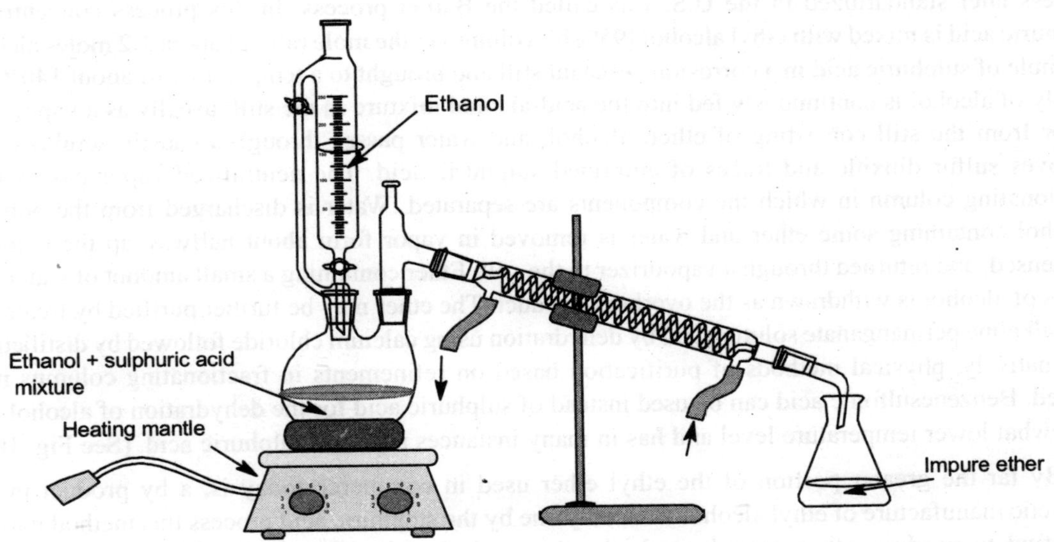

Fig. 10.1. The Lab preparation of Diethyl Ether

Theory: Ether is prepared in the lab by heating a mixture of ethyl alcohol and conc sulphuric acid.

Apparatus	Reagents	Temperature	Catalyst	pressure
Reaction flask, heating mantle, condenser, thermometer Reservoir flask Dropping funnel	Ethyl alcohol, conc sulphuric acid	140°C	Aluminium sulphate	normal

Reaction equation

$$C_2H_5OH + H_2SO_4 \rightarrow C_6H_5HSO_4 + H_2O$$

$$C_6H_5HSO_4 + C_2H_5OH \rightarrow C_2H_5-O-C_2H_5 + H_2SO_4$$

Drying the ether*:** The distillate consist of alcohol and some sulphur dioxide. It is shaking with dilute NaOH solution. The ethereal layer is separated by means of separating funnel. It is dried over fused $CaCl_2$ and redistilled on heating mantle. Pure ether is collected at 34-35°C. The process is continuous and called ***continuous etherification process.

***Precaution*:** *All the set up will be made with standard joint, the temperature should be maintained at 140°C. The ethyl alcohol is allowed to fall drop wise from the dropping funnel. Heating mantle should be used.*

Manufacture

The continuous dehydration of ethyl alcohol by sulphuric acid was first described by Boullay, but the process later standardized in the U.S. was called the Barbet process. In this process concentrated sulphuric acid is mixed with ethyl alcohol (95% by volume) in the mole ratio of about 1.2 moles alcohol per mole of sulphuric acid in a corrosion-resistant still and brought to a temperature of about 140°C. A supply of alcohol is continuously fed into the acid-alcohol mixture in the still, usually as a vapor. The vapor from the still consisting of ether, alcohol, and water passes through a caustic scrubber that removes sulfur dioxide and traces of entrained sulphuric acid. The neutralized vapor passes to a fractionating column in which the components are separated. Water is discharged from the bottom. Alcohol containing some ether and water is removed in vapor form about halfway up the column, condensed, and returned through a vapourizer to the still. Ether containing a small amount of water and traces of alcohol is withdrawn as the overhead product. The ether may be further purified by treatment with alkaline permanganate solution and by dehydration using calcium chloride followed by distillation. Alternatively, physical methods of purification based on refinements in fractionating columns may beused. Benzenesulfonic acid can be used instead of sulphuric acid for the dehydration of alcohol at a somewhat lower temperature level and has in many instances replaced sulphuric acid. (See Fig. 10.2)

By far the greater portion of the ethyl ether used in commerce today is, a by product in the synthetic manufacture of ethyl alcohol from ethylene by the sulphuric acid process this method may be modified to produce ether entirely or both ether and alcohol. The principal reactions involved are:

$$C_2H_4 + H_2SO_4 \rightarrow C_2H_5HSO_4$$

$$2\,C_2H_4 + H_2SO_4 \rightarrow (C_2H_5)_2HSO_4$$

$$C_2H_5HSO_4 + H_2O \rightarrow C_2H_5OH$$

$$(C_2H_5)_2HSO_4 + 2H_2O \rightarrow 2C_2H_5OH + H_2SO_4$$

$$C_2H_5OH + C_2H_5HSO_4 \rightarrow C_2H_5OC_2H_5 + H_2SO_4$$

$$C_2H_5OH + (C_2H_5)_2HSO_4 \rightarrow C_2H_5OC_2H_5 + C_2H_5HSO_4$$

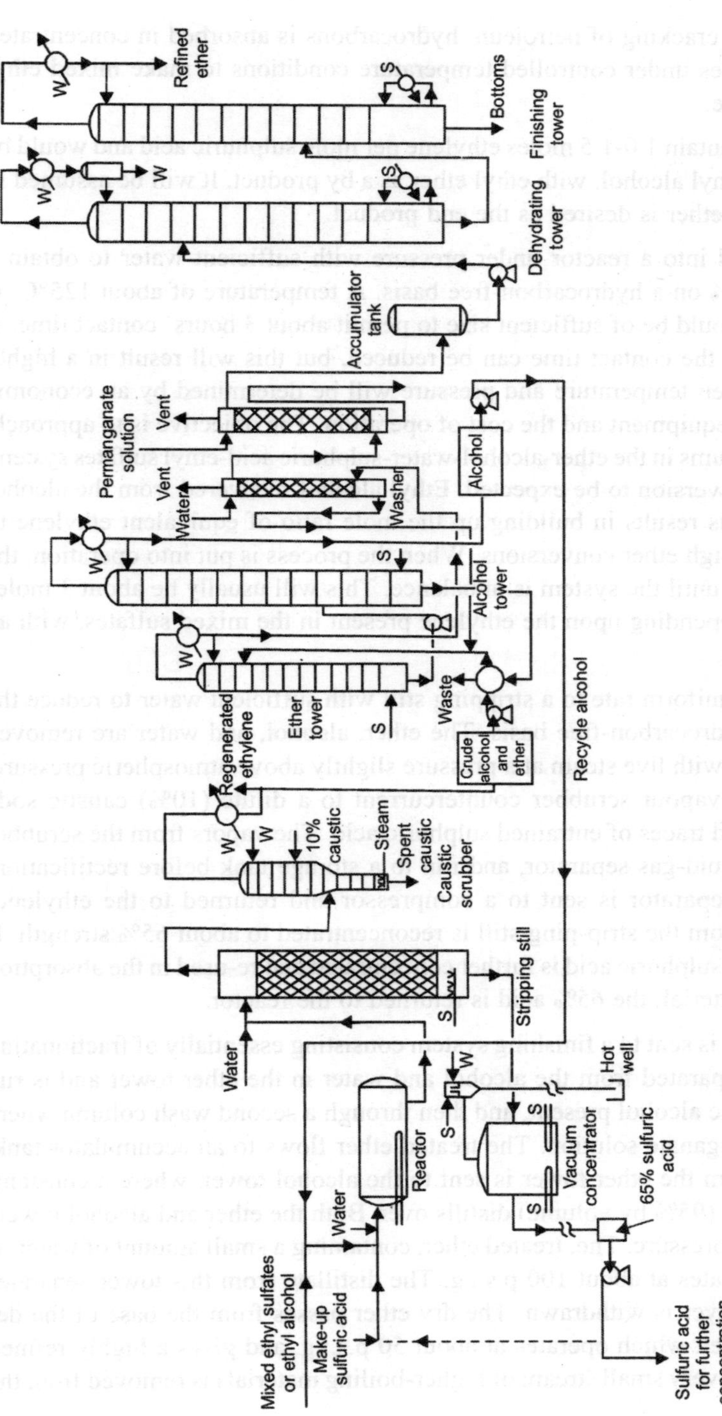

Fig. 10.2. The Manufacturing Process of Diethy Ether

Ethylene *(q.v.)* produced by the cracking of petroleum hydrocarbons is absorbed in concentrated sulphuric acid at fairly high pressures under controlled temperature conditions to make mixed ethyl hydrogen sulfate and diethyl sulphate.

Ordinarily this mixture would contain 1.0-1.5 moles ethylene per mole sulphuric acid and would be hydrolyzed with water to produce ethyl alcohol, with ethyl ether as a by product. It will be assumed in this description, however, that only ether is desired as the end product.

The mixed sulfates are pumped into a reactor under pressure with sufficient water to obtain a sulphuric acid strength of about 60% on a hydrocarbon-free basis. A temperature of about 125°C. is maintained in the re-actor, which should be of sufficient size to permit about 3 hours' contact time. If a higher temperature is maintained, the contact time can be reduced, but this will result in a higher pressure. The proper balance between temperature and pressure will be determined by an economic balance between the expense of the equipment and the cost of operation. The objective is to approach, within reasonable limits, the equilibriums in the ether-alcohol-water-sulphuric acid-ethyl sulfates system, which determines the maximum conversion to be expected. Ethyl alcohol recovered from the alcohol tower is recycled to the reactor. This results in building up the mole ratio of equivalent ethylene to sulphuric acid that is favourable to high ether conversions. When the process is put into operation, the amount of recycle alcohol increases until the system is in balance. This will usually be about 3 moles ethylene per mole sulphuric acid, depending upon the ethylene present in the mixed sulfates, with an ether con-version above 40 per cent.

The reactor mixture is fed at a uniform rate to a stripping still with sufficient water to reduce the acid strength to about 50% on a hydrocarbon-free basis. The ether, alcohol, and water are removed from the sulphuric acid by stripping with live steam at a pressure slightly above atmospheric pressure. The vapours are passed through a vapour scrubber countercurrent to a dilute (10%) caustic soda solution to remove sulfur dioxide and traces of entrained sulphuric acid. The vapors from the scrubber are condensed, passed through a liquid-gas separator, and run to a storage tank before rectification. The ethylene recovered from the separator is sent to a compressor and returned to the ethylene-absorption system. Sulphuric acid from the strip-ping still is reconcentrated to about 65% strength. If ethylene is the raw material, the 65% sulphuric acid is further concentrated and re-used in the absorption of ethylene. If alcohol is the raw material, the 65% acid is returned to the reactor.

The ether-alcohol-water mixture is sent to a finishing system consisting essentially of fractionating towers and washers. The ether is separated from the alcohol and water in the ether tower and is run through a water washer to remove the alcohol present, and then through a second wash column where it is treated with an alkaline permanganate solution. The treated ether flows to an accumulator tank. The ethyl alcohol-water mixture from the ether tower is sent to the alcohol tower, where a constant-boiling mixture of alcohol and water (95% by volume) distills over. Both the ether and alcohol towers operate at substantially atmospheric pressure. The, treated ether, containing a small amount of water, is fed to a dehydrating tower that operates at about 100 p.s.i.g. The distillate from this tower separates into two layers the lower aqueous layer is withdrawn. The dry ether passes from the base of the de-hydrating tower to the finishing tower, which operates at about 50 p.s.i.g. and gives a highly refined ether as an overhead product, while a very small stream of higher-boiling materials is removed from the base. It is desirable to handle the ethyl ether product in a closed system and store it in pressure drums or spheres. An inhibitor such as 1-naphthol is sometimes added to prevent peroxide formation.The proper choice of the materials of construction is particularly important for the part of the process involving the production and recovery of the crude ether. Because of the presence of sulphuric acid under very corrosive conditions, the reactor, stripping still, and concentrator must be carefully constructed. Lead-lined steel vessels with an inner ceramic lining are satisfactory. The stripping still may be packed with porcelain or carbon rings to secure intimate liquid-vapour contact. The vapour scrubber should be made of Monel or some material resistant to both acid and alkali conditions. The equipment for the remainder of the process may be made of steel, copper, or brass. Neither ethyl alcohol nor ethyl ether are corrosive and these materials are usually handled in steel tanks, lines, and other equipment.

Properties

Physical Properties

Freezing point	-116°C
Boiling point	34.5°C
Density	0.7138
Refractive Index	1.3526
Solubility	H_2O, C_2H_5OH
Viscosity at 25°C	0.224 mPa
Thermal conductivity	-0.130 W/mK
Vapour pressure at 34.1°C	100kPa-
Heat of formation	279.5 kJ/mol
Gibb's energy of formation	177.4 J/mol
Standard Molar Entropy	K175.6 J/mol
Heat capacity	K7.19 kJ/mol^{-1}
AZEOTROPIC DATA	
Diethyl ether + pentane	
T_{AZ} (azeotropic b.p)	306.85°K or 33.85°C
Y_{1Az} (refractive index)	0.5500
$P/_{AZ}$ (pa) azeotropic pressure	101.33
type	OX

Uses

Ethyl ether has a wide range of uses. It is a good solvent for many organic substances, among which are oils, fats, gums, resins, nitrocellulose, perfumes, and alkaloids. A mixture of ethyl ether and ethyl alcohol has excellent solvent properties for nitrocellulose and is used in large quantities for this purpose.

Another important use for ethyl ether is as an extracting medium for the concentration of acetic and other organic acids. It is used as a reaction medium in Wurtz and Grignard syntheses, as a starting fuel for diesel engines, and as an entrainer for dehydration of ethyl alcohol and isopropyl alcohol. Ethyl ether is used as a general anesthetic in surgery

H-E Conbimation for Anæsthasia

Halothane and diethyl ether, the two anaesthetics most commonly used in developing countries, have the unusual property that they form an azeotrope when mixed in the ratio of about 2 parts halothane to 1 part ether. An azeotrope is, according to the *Encyclopaedia Britannica*, "... a mixture of liquids that has a constant boiling point because the vapour has the same composition as the liquid mixture . . . The components of the solution cannot be separated by simple distillation." Thus, when the halothane ether azeotrope (**HE**) is placed in an anaesthetic vapourizer, the proportion of halothane and ether that emerges will remain 2:1 for all dial settings and all rates of carrier gas flow. In effect, the azeotrope behaves physically as though it were a single compound. The nature of the chemical bond between the two substances has not been determined with certainty. However, the facts that the volume of the mixture is slightly less than the sum of the volumes of its constituents, that there is a slight exothermic reaction when the components are mixed, and that the boiling point of the mixture is higher than either of its components all support the proposition that chemical bonding does occur. Pharmacologically, **HE** shares the properties of its component parts. Its organ system effects are described briefly below:

Circulation

Patients under **HE** anaesthesia exhibit exceptional hemodynamic stability, probably because the ether moiety stimulates sympatho-adrenal activity and decreases vagal tone, which together tend to offset the direct myocardial depression of ether and halothane. Blood pressure tends to be stable, unless the level of anaesthesia is very deep or the patient is severely hypovolemic. Cardiac rate is well maintained, and arrhythmias are rare. Crossover experiments in dogs at 2% concentration of halothane found that cardiac output, blood pressure, and heart rate were better preserved when the dogs also received ether from the azeotrope. Based upon a clinical series of over 6000 administrations, it suggested that **HE** offers a wider margin of safety than halothane alone that "makes the azeotrope a more desirable agent in the hands of those less skilled than the specialist anesthesiologist." This was confirmed in ventilated pigs, where the median ratio of the lethal concentration to the effective anaesthetic concentration (MAC) was 3.0.8

Respiration

The respiratory stimulating effect of the ether component of **HE** tends to offset the depressant effect of the halothane component, so that spontaneous respiration is well maintained at clinical levels of anaesthesia, with rather increased respiratory frequency. In our experience the use of a moderate dose of an opioid (e.g., pethidine 1mg/kg) slows-down the frequency, somewhat improving respiration. Any tendency for hypoxia due to reduced alveolar ventilation can easily be overcome with a small amount of supplement of oxygen. Spontaneous ventilation under **HE** without supplemental oxygen cannot be recommended. Furthermore, it is important to note that **HE**, like other volatile anaesthetics, abolishes the ventilatory response to hypoxia and can cause intrapulmonary shunt if atelactesis is present. **HE** gas induction is faster and more pleasant than with ether, behaving much like halothane alone. It is not irritating to the airway, and it relaxes bronchial muscles, as do its parent substances. Bronchial secretion is not increased. However, salivation is increased, but not as much as with ether or ketamine. From the point of view of respiration, recovery from **HE** anaesthesia may be safer than with modern anaesthetics. The azeotrope appears to break down in the body, and halothane, having lower blood solubility than ether, is excreted more rapidly during the early phase of recovery. The remaining ether tends to preserve respiratory drive and give analgesia into the early postoperative period, eliminating the need for opioids and their concomitant respiratory depression.

Hepatic Function

No change in liver function has been observed in the early postoperative period after **HE**.

Anaesthetic Potency

The minimum alveolar concentration (MAC) for **HE** in man was found to be 0.71 vol.% (which has about 0.47% halothane), versus 0.65 % for halothane alone, suggesting that the ether component acts synergistically with halothane. Thus, at a given MAC fraction, **HE** is less expensive to use than halothane.

Flammability

Flammability of a vapour is not simply a physical property of the material, but rather is dependent upon the source of ignition used, the carrier gas(es), the geometric configuration of the test chamber, temperature, and other factors. An electric filament used and reported that **HE** was non-explosive in concentrations of 10.7% or less in oxygen. Later a higher-energy electric spark used in a recognized standard test apparatus. He concluded that **HE** is nonflammable in air for all practical purposes, and that its lower limit of flammability in oxygen is 7.25%. Raventos, using a similar apparatus found a lower limit of 8.0% in oxygen. The fact that ether alone has a lower limit of flammability of in oxygen of 2.1% suggests that the presence of halothane in the mixture does little to moderate the flammability of the ether moiety (e.g., 7.25% **HE** contains about 2.4% ether).The flammability tests indicate that the OMV and plenum-type vapourizers should not produce flammable concentrations of the mixture at any dial setting. However, caution should be used with non-calibrated vapourizers; they have not been formally tested as to the output concentration obtained at maximal settings. Like other anaesthetics, **HE** is more flammable in nitrous oxide mixtures than in oxygen/air mixtures Comparative properties of the azeotrope and its components are summarized in Table 10.1.

Tips for Use of HE

The only additional equipment required is a graduated measuring cup or cylinder. Although the azeotrope is non-flammable in air, the ether component is flammable, so the azeotrope should be prepared in a well-ventilated area, away from sources of ignition. A large quantity can be prepared in advance, because the azeotrope is stable for at least 4 months when kept out of bright light.It can be administered with the OMV, any plenum-type vaporizer designed for halothane, or a non-calibrated Boyle's Bottle-type of vapourizer.

Table 10.1: The Properties of Haloethane – Diethyl Ether Mixture

	Halothane	Diethyl Ether	Azeotropic Mixture
Molecular weight	197.4	74.1	158 (calculated)
Boiling point @ 760 mmHg or 100 kPa (°C)	50.2	34.6	52
Vapour pressure @ 20°C (kPa [mmHg])	32.1 [243]	49.1 [373]	28.4 [216]
Liquid density	1.86	0.72	1.48
MAC (% v/v)	0.65 - 0.75	2.0	0.71
Working concentration (% v/v)	0.85 - 1.5	3 - 7	1.0 - 1.5
Flammability	No	2% to 80% in O_2 2% to 36% in Air	7% to 67% in O_2 not in Air
Mask induction	Acceptable, quick	Unpleasant	Acceptable, not long
Clinical signs of depth	No	Yes	Similar to ether
Induction/recovery time	Short	Long	Short
Awake analgesia	+/–	+++	++
Muscular tone	-	Decreased	Decreased
Cardiovascular function	Decreased	Stimulated	Stimulated
Respiratory function	Decreased	Maintained	Maintained

Because the OMV lacks temperature compensation, the concentration emerging from it at any given dial setting will diminish by about half over two hours. **HE** should not be placed in an EMO ether vapourizer, as the thymol from the halothane will ruin it. Inhalation induction with **HE** is neither unpleasant nor unduly long and is similar to that of halothane. With any calibrated vapourizer, the dial setting should be increased progressively to a dial setting of 4-5 and kept there for some minutes after loss of consciousness, carefully attempting from time to time to insert the laryngoscope if intubation is to be done without a muscle relaxant. Surgery can commence after 5-10 minutes.The typical dial setting for maintenance is 1 to 2 (corresponding roughly to 0.9 to 1.5% or 1.3 to 2 MAC). However, owing to the wide margin of safety, higher concentrations can be used, when necessary. Muscle relaxation is fairly good, probably due to the relaxing properties of ether. Non-depolarizing muscle relaxants are potentiated, and can be used sparingly for abdominal surgery and not at all for extra-abdominal interventions. The azeotrope, like its component parts, will relax uterine muscle and should be used only in low concentrations during Caesarean section. In the absence of opioids, muscular tone, blood pressure, pupil size, eyelid reflex, spontaneous respiration depth and rate (if allowed) all help in assessing the depth of anaesthesia.

Awakening time, even after a long procedure, is about 10-15 minutes. Postoperative nausea and vomiting are said to occur in about 10% of patients. The azeotrope of halothane and ether appears to share many of the virtues of its constituent parts, while ameliorating their undesirable properties. It is less expensive than halothane alone and may be a safer in the hands of less experienced anaesthetists. It should be considered for routine use in developing countries.

Other Ethers

Methyl Ether

Methyl ether, (CH_3OCH_3), formula weight 46.07, is a colorless gas; m.p., -138.5°C.; b.p., -23.65°C.; d., 2.091 grams/I.; sp.gr. of the liquid at 20°C., 0.661; flash point (closed cup), -42°F. It is used in refrigeration, as a solvent and extracting agent, as a propellant for sprays, and as a reaction medium. It is available in 50-lb. pressure cylinders and requires a red label for railroad shipment.

Bis (2-chloroethyl) Ether

Bis (2-chloroethyl) ether (3,3-dichlorodiethyl ether) ($ClCH_2CH_2OCH_2CH_2Cl$), formula weight 143.02, is the oxygen analog of mustard gas but is not a vesicant. It is a solvent with a chloroform-like odour; b752, 176-178°C.; b23, 82-83°C.; density at 20°C, 1.22 (10.17 lb./U.S. gal. at 20°C.); refractive index 1.4568; flash point (open cup), 185°F.

Bis(2-chloroethyl) ether is prepared by heating ethylene chlorohydrin at 90-100°C. With concentrated sulphuric acid:

$$2\ ClCH_2CH_2OH \xrightarrow{H_2SO_4} (ClCH_2CH_2)_2O + H_2O$$

It may also be obtained by the reaction of ethylene chlorohydrin with chlorine and excess ethylene.

Bis(2-chloroethyl) ether is a particularly good solvent for fats, waxes, or greases, and is extremely resistant to hydrolysis. It acts as a solvent for cellulose esters when mixed with 10-20% of an alcohol. It is used to produce vinyl ether by heating at 200-240°C. With fused potassium hydroxide in a stream of ammonia.

Isopropyl Ether

Isopropyl ether (2-isopropoxypropane, diisopropyl ether) [$(CH_3)_2CHOCH(CH_3)_2$], formula weight 102.11, is a colorless, moderately volatile, flammable liquid with a characteristic ethereal odour.

Physical and Chemical Properties

Constants: F.p., -86.2°C.; b.p., 68.47°C.;

density, 0.7244; refractive index, 1.3681;

Vapour pressure: at 20°C is 12.2 mm Hg Surface tension at 25°C., 19.4 dynes/em.;

viscosity, at 20°C., 0.3220 centipoise;

sp.heat at 30°C., 0.528 gram-cal./(gram) (OC.);

latent heat of vaporization, 68.2 gram-cal./gram;

heat of formation, 841 gram-cal./gram;

heat of combustion, 9,380 gram-cal./gram;

coefficient of cubical expansion, 0.00l44;°C.;

dielectric constant *(E)* at 85.8 kilocycles and 25°C., 4.449;

flash point (closed cup), -9°F.;

explosive limits at 100°C., 1.1-4.5% by vol. in air; auto ignition temp., 443°C.

The solubility of isopropyl ether in water at 20°C. is 0.94% by weight and that of water in isopropyl ether at 20°C. is 0.55% by weight.

Azeotropes of isopropyl ether are given in Table 10.2.

Table 10.2 : Azeotropes of Isopropyl ether

Other component	B.P., °C	Distillate, wt. %	
		Isopropyl ether	Other component
Water	62.2	95.5	4.5
Isopropyl alcohol	66.2	85.9	14.1
Water Isopropyl alcohol	61.0	81.7	6.15 12.15

Reaction

Isopropyl ether undergoes the usual reactions of ethers: oxidation; cleavage with hydriodic acid, sulfuric acid, or phosphorus pentachloride; and chlorination and bromination. It forms explosive peroxides of uncertain composition and constitution. It is believed that isopropyl ether peroxide, (C_3H_7)20.02, does not accumulate in high concentrations but rather decomposes to yield acetone and water. In the presence of water, the peroxide also decomposes to form hydrogen peroxide, acetone, and isopropyl alcohol. Acetone and hydrogen peroxide react to form triacetone peroxide, which is considered to be the explosive material that accumulates during the storage of uninhibited ether. The formation of peroxides can be completely. prevented by the addition of a suitable antioxidant such as I-naphthol to freshly distilled ether in a concentration of about 1 part in 10,000.

Manufacture

Isopropyl ether can be made commercially by the dehydration of isopropyl alcohol with sulfuric acid. However, the quantity made as a by-product in the manufacture of isopropyl alcohol from propylene by the sulfuric acid process has in the past far exceeded the amount required by industry. The process for making iso-propyl alcohol is flexible and by additions and alterations to plant equipment, together with the proper changes in the operating conditions, high percentages of isopropyl ether relative to the isopropyl alcohol could be made, or the entire production might be isopropyl ether. Alternatively, the by-product isopropyl ether can easily be converted into isopropyl alcohol. The manufacture of isopropyl ether from propylene is very similar to the manufacture of ethyl ether from ethylene the first products being the mixed mono- and di-isopropyl sulfates. However, since isopropyl ether forms azeotropic mixtures with isopropyl alcohol and water, it is possible that the alcohol may not be completely removed in the ether tower and the washer. If this is so a portion of the upper phase of the ternary azeotrope from the dehydrating column may be returned to the washer.

Specification, Testing and Standards

A typical inspection of isopropyl ether is as follows:

Colour Max , Hazen	10
Density at 20°C	0.723 – 0.727
Distillation range	66 – 70
Acidity as acetic acid max w%	0.22
Peroxides, max, p.p.m active oxygen	12

The majority of the testing methods employed are the A.S.T.M. D268-49 pro-cedures. A simple qualitative test for the presence of active oxygen is made by treating a sample of 2 volumes of isopropyl ether with 1 volume of 4% potassium iodide solution, slightly acidified. This test may be interpreted semi quantitatively as follows:

Active oxygen, pp.m. of isopropyl ether	Qualitative colour test	Active oxygen, p.p.m. of isopropyl ether	Qualitative colour text
30	Very faint yellow	100	Light brown
50	Faint yellow	145	Brown
70	Yellow to brown	240	Dark iodine color

Uses

Isopropyl ether is of moderate importance as an industrial solvent. Its properties are very similar to those of ethyl ether and, while not as widely used, it does have the advantages of lower volatility and higher ability to dissolve certain oils, fats, and organic acids, which make it valuable for extraction purposes. Acetic and lactic acid are examples of acids that can be efficiently recovered from aqueous solution with isopropyl ether or with isopropyl ether mixed with ethyl or isopropyl acetate Isopropyl ether, alone or in admixture with other solvents, such as acetone aniline or methyl acetate may be used for dewaxing petroleum oils. Isopropyl ether finds application in the preparation of emulsion breakers for oil-field and refinery use and as a solvent in lacquer formulations and in the resin and rubber industry. Also it is used as a source of propylene for use in chemical processes, in a similar manner to the use of ethyl ether as a source of ethylene

Isopropyl ether is a good high-octane blending agent although it has never been widely used. The blending value of isopropyl ether is slightly better than that of technical isooctane, irrespective of the absence or presence of tetraethyl lead in the fuel .

The estimated end-use pattern for the U.S. industry during 1949 was as follows: automotive chemicals, 5%; chemical manufacture, 12%; cosmetics, 2%; drugs and pharmaceuticals, 1%; dyes and textile chemicals, 3%; extraction, 10%; petroleum chemicals, 30%; paint, varnish, and lacquer, 10%; plastics and resins, 17%; rubber industry, 10%.

Bis(2-chloroisopropyl) Ether

Bis(2-chloroisopropyl) ether (β, β-dichlorodiisopropyl ether),

[$ClCH_2CH(CH_3)$-$OCH(CH_3)CH_2Cl$], formula weight 171.07, is a colourless liquid; b.p., 187.1°C.; density, 1.1127 (9.27 lb./U.S. gal. at 20°C.); flash point (open cup), 185°F. It is similar in properties to bis(2-chloroethyl) ether, but is less soluble in water, and has a higher boiling point and lower volatility. It is miscible with practically all oils and organic liquids, and is an excellent solvent and extractant for fats, waxes, and greases. In textile processes bis(2-chloroisopropyl) ether assists the action of soap solutions with-out excessive loss by evaporation from the hot solutions. It is used in paint and varnish removers, spotting agents, and cleaning solutions. It is an intermediate in the manufacture of dyes, resins, and pharmaceuticals.

n-Butyl Ether

n-Butyl ether, [$CH_3(CH_2)_30(CH_2)_3CH_3$], formula weight 130.23, is a colorless stable liquid; m.p., -98°C.; b.p., 142°C.; density, 0.7841 (6.40 lb./U.S. gal. at 20°C.); flash point (open cup), 100°F. Azeotropes of n-butyl ether are given in Table 10.3.

Table 10.3 : Azeotropye of *n*-Butyl Ether

Other component	B.p., °C	Distilate. wt. %	
		n-Butyl ether	Other component
n-Amyl alcohol	135.00	48	52
Ethylene glycol	140.0	90	10
Water } *n*-Butyl alcohol	91	27.7	29.3 42.9

Since *n*-butyl ether tends to form explosive peroxides, they should be destroyed before the ether is distilled.

n-Butyl ether is an excellent extracting agent for use 'with aqueous systems since its solubility in water is less than 1% that of ethyl ether. It is useful as a medium for Grignard and other reactions that require a strictly anhydrous ethereal vehicle. Since it forms many constant-boiling mixtures, it is useful for purifying other solvents. It has a high solvent power for many natural and synthetic gums, resins, oils, organic acids, esters, and alkaloids.

n-Hexyl Ether

n-Hexyl ether, [$CH_3(CH_2)60(CH_2)5CH_3$], formula weight 186.33, is a mild-odoured stable liquid; b.p. 226.2°C.; b9, 95-97°C.; density, 0.7936 (6.61lb./U.S. gal. at 20°C.); flash point (open cup), 170°F. It is much less volatile than lower members of the ether group. It has an extremely low solubility in water. *n*-Hexyl ether is a convenient medium for chemical reactions that are carried out in a strictlyanhydrous ethereal vehicle. It has also been found useful as a component of foam hreakers and as an inert reaction medium.

Benzyl Ether

Benzyl ether, ($C_6H.CH_2OCH_2C_6H_5$), formula weight 198.25, is a colorless oily liquid; m.p. 4-5°C.; b.p., 295-298°C.; density, 1.0428; flash point (closed cup), 275°F. It is insoluble in water, very soluble in hot ethyl alcohol, and soluble in ethyl ether. A process has been developed for the conversion of benzyl ether to benzaldehyde, benzoic acid, and benzyl benzoate in 80-85% yield by non catalytic oxidation by air over a temperature range of 40 to 250°C. It is used as a plasticizer for nitrocellulose.

CHAPTER

11

Carbooxylic Acids

Introduction

The Monocarboxylic acids form a homologous series of the general formula or $C_nH_{2n+1}COOH$. The characteristic properties present in all such R—C(=O)OH acids is the monovalent carbonyl group {—C(=O)OH}. The basicity of the acid is equal to the number of carbonyl group present in the molecule. Acids containing one such group are monobasic those possessing the two are dibasic and so on.

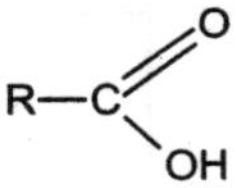

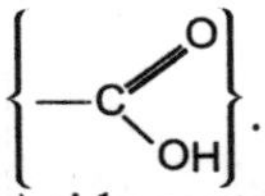

Monobasic acids are commonly known as fatty acids, because some of the higher homologous are obtained by the hydrolysis of fats and oils. Some important members are shown below: -

Formic acid or Methonic Acid	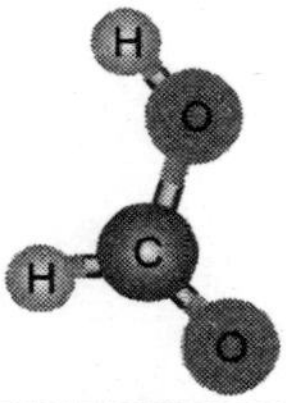 H, O, C, H, O
Acetic Acid or Ethanoic Acid	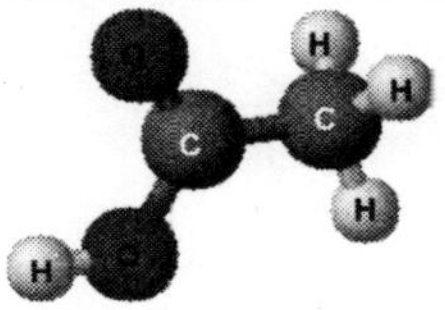 H, C, C, H, H, H

Propionic Acid or Propanoic Acid	
n- butanoic Acid	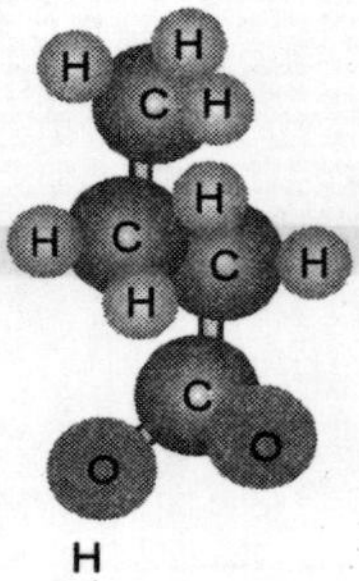
Palmitic acid	

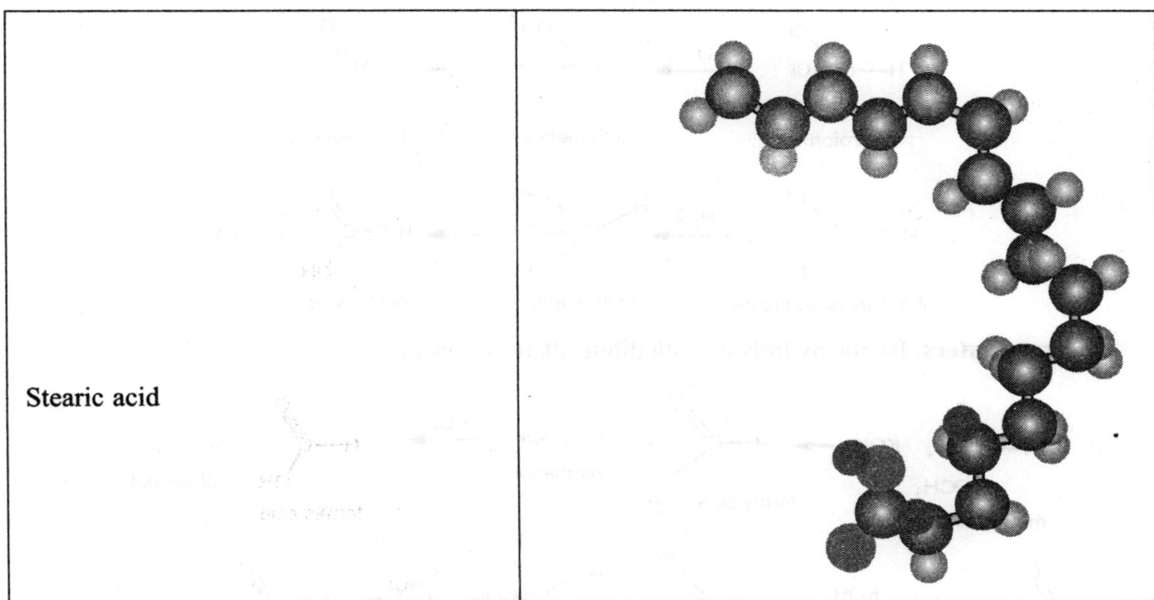
Stearic acid

Are some of the higher fatty acids present in Glyceryl esters in oils and fats

General Methods of Preparation

1. From Primary Alcohols: By oxidation with potassium dichromate and sulphuric acid

$$CH_3OH \xrightarrow{O} HCOH \xrightarrow{O} HCOOH$$

metahnol formaldehyde formic acid

$$C_2H_5OH \xrightarrow{O} CH_3CHO \xrightarrow{O} CH_3COOH$$

ethanol acetaldehyde acetic acid

2. From Alkyl Cyanides: By hydrolysis with hot dilute acids or alkalies

$$H{-}C{\equiv}N + 2H_2O \longrightarrow H{-}C(=O){-}OH$$

nitrilomethane formic acid

$$H_3C{-}C{\equiv}N + 2H_2O \longrightarrow H_3C{-}C(=O){-}OH$$

acetonitrile acetic acid

3. From Trihalogen Paraffin: (with the halogen linked to the same carbon atom)

$$HCCl_3 \xrightarrow{H_2O} HC(OH)_3 \longrightarrow HCOOH + H_2O$$

chloroform — methanetriol — formic acid

$$H_3C{-}CCl_3 \xrightarrow{H_2O} HC(OH)_3 \longrightarrow H_3C{-}COOH + H_2O$$

1,1,1-trichloroethane — methanetriol — acetic acid

4. From Esters: By the hydrolysis with dilute alkalies or acids

$$HCOOCH_3 + KOH \longrightarrow HCOK + H_3C{-}OH \xrightarrow{HCl} HCOOH + KCl$$

methyl formate — formylpotassium — methanol — formic acid — chloropotassium

$$HCOOCH_2CH_3 + KOH \xrightarrow[\text{hydrolysis}]{KOH} H_3C{-}COK + H_3C{-}CH_2OH \xrightarrow{HCl} H_3C{-}COOH + KCl$$

ethyl formate — acetylpotassium — ethanol — acetic acid — chloropotassium

The method is useful for the preparation of higher fatty acids and soaps.

5. From Grignard's Reagents: By the action of the CO_2 and subsequent hydrolysis

$$CO_2 + CH_3MgI + O{=}C(CH_3){-}O{-}Mg{-}I \xrightarrow{H_2O} Mg(OH)I + H_3C{-}COOH$$

Grignard's Reagent — methyl magnesium oxoiodide — Magnesium Methoxy iodide — acetic acid

6. From Sodium Alcoholates: By the action of carbon monoxides under high pressure and temperature.

$$NaOH + CO \longrightarrow HCOONa \longrightarrow HCOOH + NaCl$$

formic acid — chlorosodium

$$H_3C{-}O{-}Na \xrightarrow{CO} H_3C{-}COONa \longrightarrow H_3C{-}COOH + NaCl$$

sodium metoxide — sodium acetate — acetic acid — chlorosodium

7. From Dicarbooxylic Acid: (having both the carbon groups attached to the same carbon atom) by the action of heat.

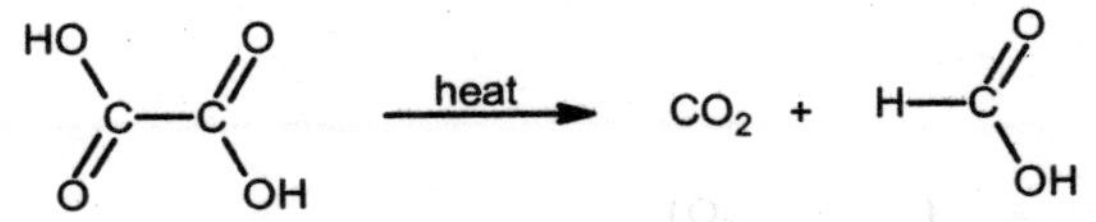

$$HOOC{-}COOH \xrightarrow{heat} CO_2 + H{-}COOH$$

oxalic acid → formic acid

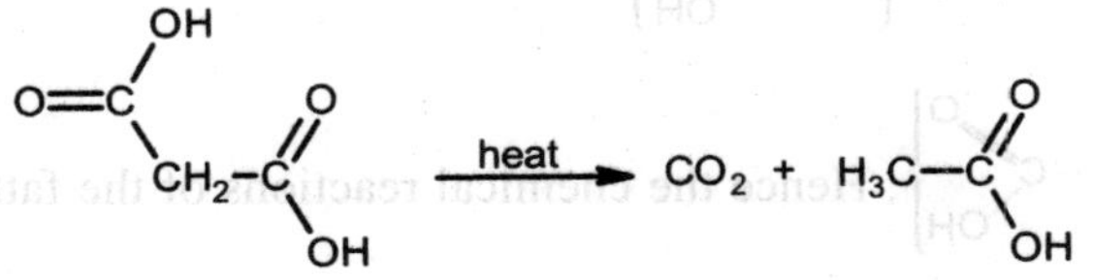

$$HOOC{-}CH_2{-}COOH \xrightarrow{heat} CO_2 + H_3C{-}COOH$$

malonic acid → acetic acid

8. From Acid (hydrides by hydrolysis):

$$H_3C{-}COCl + H_2O \longrightarrow H_3C{-}COOH$$

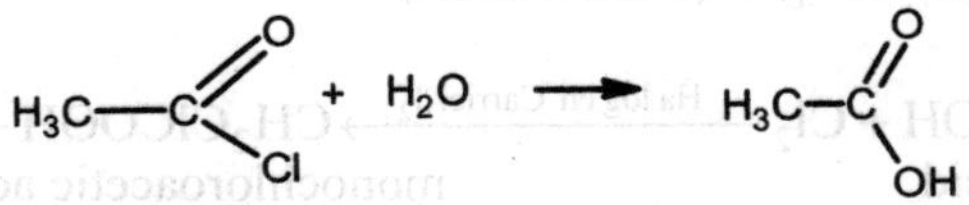

acetyl chloride → acetic acid

9. From Acetic Amides:

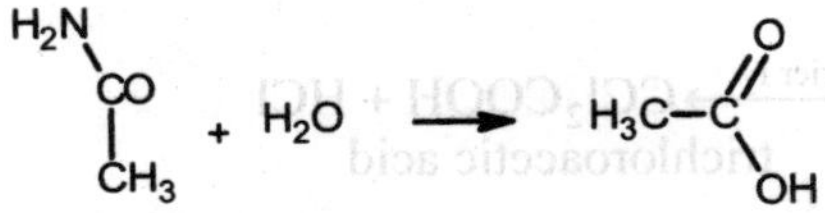

$$H_2N{-}CO{-}CH_3 + H_2O \longrightarrow H_3C{-}COOH$$

acetamide → acetic acid

Formic acid cannot be prepared by this method

10. From Acid Anhydrides: By hydrolysis

$$(H_3C{-}CO)_2O + H_2O \longrightarrow 2CH_3COOH$$

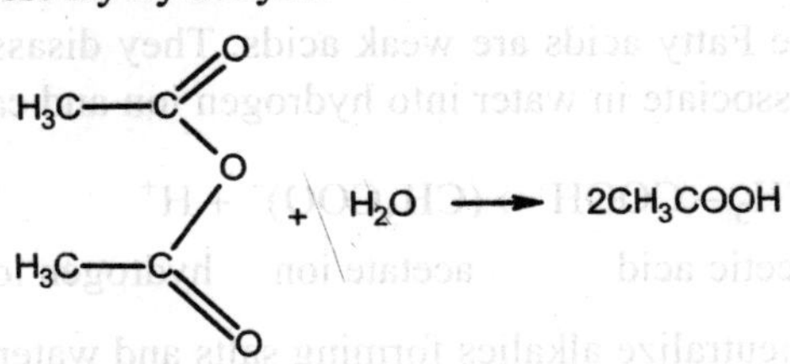

acetic anhydride

Physical Properties

1. The lower members are colourless, corrosive mobile liquids with strong pungent odour. They distill without decomposition and are miscible with water. They slow acidic reactions
2. The middle members (from $C_4 - C_9$) are oily liquids. They are less soluble in water and possess an unpleasant odour. The boiling point rises regularly with increase of carbon.

3. The higher member (from C_{10} onward) are waxy solids, almost insoluble in water, but readily soluble in alcohol and ether. They are volatile in steam.

Chemical Properties

The molecule of carboxylic acid $\left\{R-C\overset{O}{\underset{OH}{\lessgtr}}\right\}$ consists of two distinct groups (a) Alkyl group (R) and (b) carboxyl group $\left\{-C\overset{O}{\underset{OH}{\lessgtr}}\right\}$. Hence the chemical reactions of the fatty acids are mainly due to these two groups.

1. Halogenation: Carboxylic acids (except the formic acid) substitution alkyl group and therefore from substitution products with halogen (chlorination)

$$\underset{\text{acetic acid}}{CH_3COOH} + Cl_2 \xrightarrow{\text{Ha log en Carrier } I_2} \underset{\text{monochloroacetic acid}}{CH_2ClCOOH} + HCl$$

$$\underset{\text{monochloroacetic acid}}{CH_2ClCOOH} + Cl_2 \xrightarrow{\text{Ha log en Carrier } I_2} \underset{\text{dichloroacetic acid}}{CHCl_2COOH} + HCl$$

$$\underset{\text{dichloroacetic acid}}{CHCl_2COOH} + Cl_2 \xrightarrow{\text{Ha log en Carrier } I_2} \underset{\text{trichloroacetic acid}}{CCl_2COOH} + HCl$$

Formic acid, containing no alkyl group cannot be halogenated. Therefore the above reaction (chlorination) may be employed for distinguishing acetic acid from formic acid.

2. Acidic Character: The Fatty acids are weak acids. They disassociate in water into hydrogen ion and carboxyl ion. They dissociate in water into hydrogen ion and carbonyl ion

$$\underset{\text{acetic acid}}{CH_3-COOH} \rightarrow \underset{\text{acetate ion}}{(CH_3COO)^-} + \underset{\text{hydrogen ion}}{H^+}$$

3. With Alkalies: Acids neutralize alkalies forming salts and water.

CH_3COOH	+	NaOH	→	CH_3COONa	+	H_2O
Acetic acid		sodium hydroxide		sodium acetate		water

Formic acid gives formates

$HCOOH$	+	NaOH	→	$HCOONa$	+	H_2O
Formic acid		sodium hydroxide		sodium formate		water
CH_3COOH	+	NH_4OH	→	CH_3COONH_4	+	H_2O
Acetic acid		Ammonium hydroxide		ammonium acetate		water

4. With Carbonates: Acids decompose carbonates to give salts and liberates carbon dioxide

$2CH_3COOH + Na_2CO_3 \rightarrow 2CH_3COONa + H_2O + CO_2$
acetic acid — Sodium acetate

5. With Metals (such as Na, K, Zn etc) salts are formed and hydrogen is liberated.

$2HCOOH + Zn \rightarrow (HCOO)_2Zn + H_2O$
Zinc formate

$2CH_3COOH + 2Na \rightarrow 2CH_3COONa + H_2O$
Acetic acid — Sodium acetate

6. With Alcohols: (esterification) Esters are formed

$H{-}C(=O){-}OH + H_3C{-}CH_2{-}OH \rightleftharpoons H{-}C(=O){-}O{-}CH_2CH_3 + H_2O$
formic acid — ethanol — ethyl formate

$H_3C{-}C(=O){-}OH + H_3C{-}CH_2{-}OH \rightleftharpoons H_3C{-}C(=O){-}O{-}CH_2CH_3 + H_2O$
acetic acid — ethanol — ethyl acetate

The reaction is reversible. It should therefore be carried out in the presence of a suitable dehydrating agent e.g. concentrated sulphuric acid. The reaction is known as esterification

7. With phosphorous pentachloride PCl_5, or phosphorous trichloride PCl_3 or thionyl chloride $SOCl_2$

$H{-}C(=O){-}OH + PCl_3 \longrightarrow 3\,H{-}C(=O){-}Cl + H_3PO_3$
formic acid — formyl chloride

$H_3C{-}C(=O){-}OH + SOCl_2 \longrightarrow H_3C{-}C(=O){-}Cl + SO_2 + HCl$
acetic acid — acetyl chloride

$H_3C{-}C(=O){-}OH + PCl_5 \longrightarrow H_3C{-}C(=O){-}Cl + POCl_3 + HCl$
acetic acid — acetyl chloride

8. With dehydrating Agent:

$2\,H_3C{-}C(=O){-}OH \xrightarrow{P_2O_5} (H_3C{-}C(=O))_2O + 2H_3PO_3$
acetic acid — acetic anhydride — Metaphosphoric Acid

Phosphorous pentoxide is used as a dehydrating agent in this reaction

Formic acid does not form an anhydride

9. Oxidation: Fatty acids (except formic acid) are not easily oxidised

$$HCOOH \xrightarrow{O} CO_2 + H_2O$$

Fatty acids (except HCOOH) are, therefore, not reducing agents

10. Reduction: with hydrogen under pressure and in the presence of a suitable catalyst such as Ni Cu at 400°C or P or HI lithium aluminium hydride, alcohols are formed.

Special Decomposition of Acids

(a) Electrolysis: (Kolbe's reaction) paraffins are produced

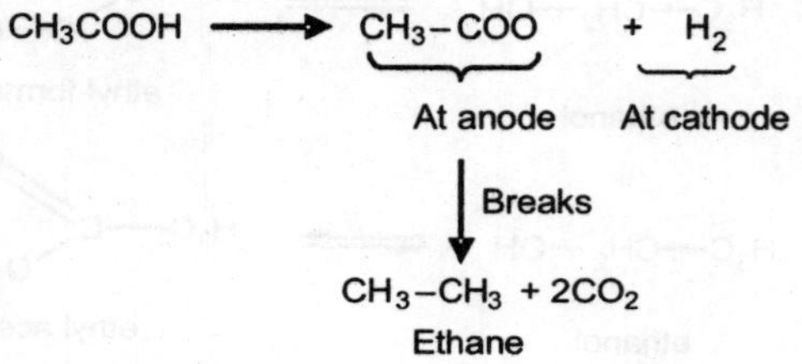

$$CH_3COOH \longrightarrow \underbrace{CH_3-COO}_{\text{At anode}} + \underbrace{H_2}_{\text{At cathode}}$$

Breaks

$$CH_3-CH_3 + 2CO_2$$

Ethane

Formic acid cannot be electrolyzed. Acetic acid or its alkalies salts on electrolysis yield ethane.

$$2CH_3COOH \xrightarrow{\text{electrolysis}} C_2H_6 + 2CO_2 + H_2$$

The reaction can be used to distinguish acetic acid and formic acid.

(b) Heating Sodium Salt of Fatty acids with Sodium Hydroxide

(I) $H_3C-C(=O)-O-Na + NaOH \longrightarrow CH_4 + Na_2CO_3$

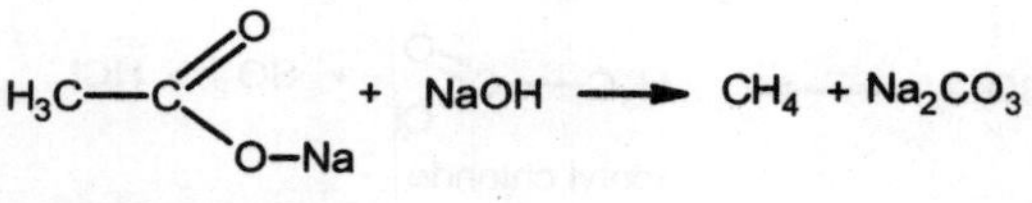

Sodium Acetate

(II) Formic acid does not respond to this reaction (sodium formate, however, on heating alone decomposes to give sodium oxalate and H_2

(c) Heating Calcium Salt Alone: Ketone are formed

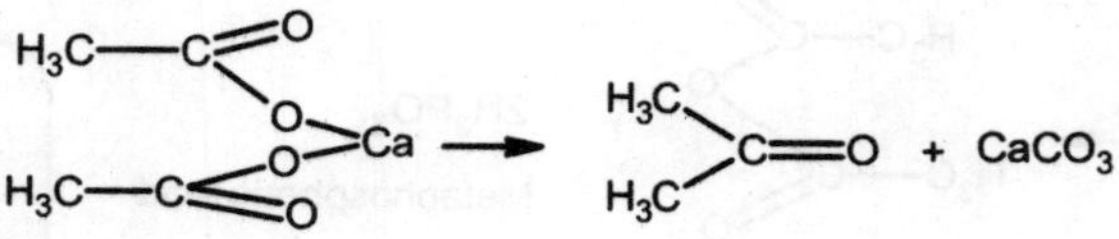

$$(H_3C-COO)_2Ca \longrightarrow (H_3C)_2C=O + CaCO_3$$

(ii) Calcium Formate on distillation gives formaldehyde and not ketone

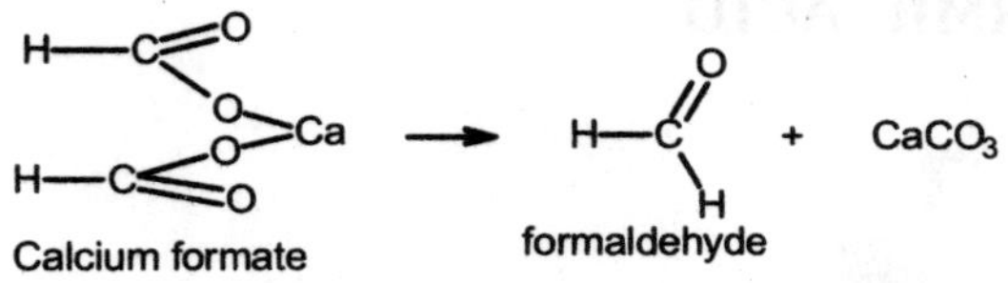

$$2(HCOO)_2Ca \rightarrow HCHO + CaCO_3$$

Calcium formate → formaldehyde

Calcium formate produces formaldehyde and not ketone on heating

The reaction may be used for distinguishing formic acid from acetic acid.

(d) Heating Calcium salt of the fatty acid with Calcium Formate: Aldehydes are produced

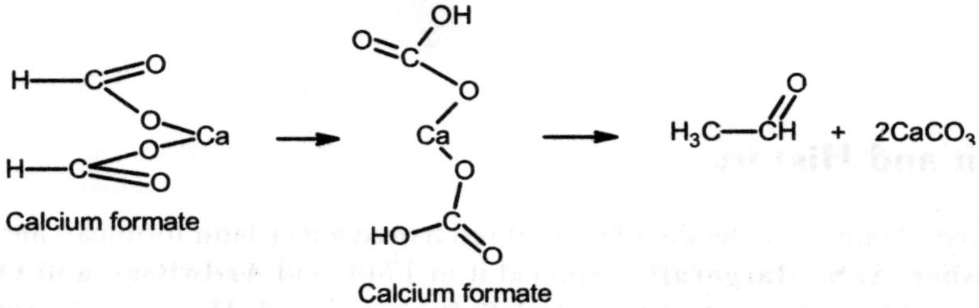

$$(HCOO)_2Ca + (HOOC\text{-}COO)_2Ca \rightarrow H_3C\text{-}CHO + 2CaCO_3$$

Calcium formate → Calcium formate →

(e) Heating Ammonium Salt: Amides are produced

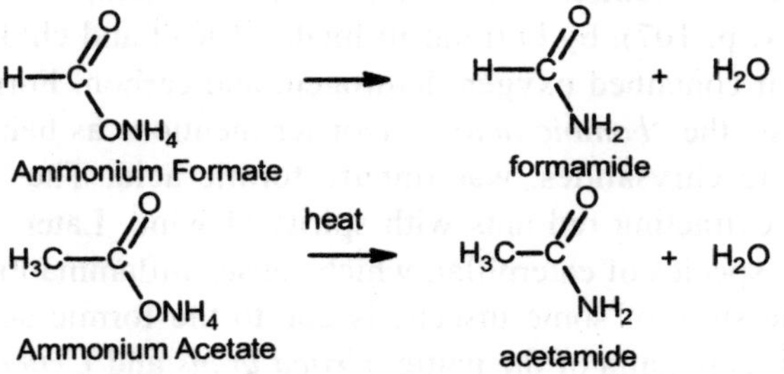

$$HCOONH_4 \rightarrow HCONH_2 + H_2O$$

Ammonium Formate → formamide

$$H_3C\text{-}COONH_4 \xrightarrow{heat} H_3C\text{-}CONH_2 + H_2O$$

Ammonium Acetate → acetamide

Detection of Carboxyl Group

The presence of carbonyl group in a given compound may be detected as given below: -

1. **If the substance is suitable in water or alcohol, its solution will turn blue litmus solution red and evolve carbon dioxide from sodium carbonate.**
2. **If the substance is insoluble in water, it should dissolve in sodium carbonate solution with the liberation of carbon dioxide.**
3. **If the substance is distilled with soda lime an inflammable gas is given off and the residue gives carbon dioxide when treated with dilute acid**
4. **On treatment with phosphorous pentachloride, hydrogen chloride gas is evolved.**
5. **When wormed with ethanol and a little sulphuric acid a fruity odour, the smell of the ester is produced.**

FORMIC ACID

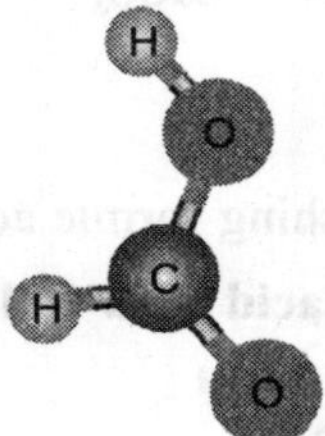

Formula: HCOOH

Introduction and History

This acid was first obtained by the distillation of red haired ants (latin formica= ants rufa=red haired) , by **Samuel Fisher. A. S. Marggraff** examined it in 1749, and **Ardwissan and Oehrn, of Leipsic**, in 1777, fairly establishing its identity as a definite compound. However, in 1802, Fourcroy and Vauquelin endeavoured to prove that it was a mixture of acetic and malic acids; but their opinions were refuted by **Suersen and Gehlen**. Berzelius (1817) first attempted to determine its composition (*Annals of Philosophy*, ix, p. 107), by burning formiate of lead and chlorate of potassium in a glass tube, thus deciding that it contained oxygen, hydrogen, and carbon. Formic acid is found in certain caterpillars, and doubtless, the "*bombic acid*" Lavoisier mentions as being obtained from silk-worm larva that are changing to chrysalides, was impure formic acid. The "*spirit of magnanimity,*" of Hoffman, was made by extracting red ants with spirits of wine. Later, Mr. F. Will has shown that the fluid in the hairs of a species of caterpillar, which causes inflammation of the skin when handled, and the poisoning by the sting of some insects, is due to the formic acid present. It has also been demonstrated that the stinging hairs of the nettle, *Urtica urens* and *Urtica dioica* (more), contain this acid. Formic acid is also found in Pine-tree leaves, and in the blood, bile, urine, perspiration, and muscular tissues of man.

Preparation.—Formic acid derived its name from the fact that it was obtained by distillation of red ants. **Doebereiner** discovered, in the early part of the present century, that it could be prepared artificially by the distillation of tartaric acid, water, sulphuric acid, and black oxide of manganese. At present, it is known that many organic bodies yield this acid, when distilled with black oxide of manganese and sulphuric acid, or with chlorinated lime, or with bichromate of potassium; but decidedly the best process is that based upon the principle discovered by Berthelot (*Comptes Rendus*, March 1856), that oxalic acid and glycerin will produce it, when heated together. 4 parts each of oxalic acid and glycerin are to be mixed with 1 part of water, and a heat of 100° C. (212° F.) applied 12 or 15 hours, or until the acid is decomposed. The residuum is now to be mixed in a still with its bulk of water, and distilled. It is then again mixed with the same quantity of water, and distilled, the operation being repeated, if the formic acid remains in amount to justify. In this process monoformate of glycerin is formed as an intermediary product, which, upon distillation with water, splits into glycerin and formic acid.

Preparation

Formic acid may be obtained by the following methods: -

(1) By the oxidation of methyl alcohol or of formaldehyde using platinum black as a catalyst

CH_3OH Methanol	+	[O]	→	HCHO formaldehyde	+	H_2O
HCHO Formaldehyde	+	[O]	→	HCOOH formic acid		

(2) By the hydrolysis of hydrogen cyanide with hydrochloric acid

H – C.N Hydrogen Cyanide	+	$2H_2O$	+	HCl	→	H – COOH formic acid	$+NH_4Cl$

Reaction is extremely poisonous
Students should never try this

(3) By the reduction of carbon dioxide in aqueous solution with hydrogen in the presence of palladium black as a catalyst.

HO—C(=O)—OH (carbonic acid) + H_2 —palladium black→ H—C(=O)—OH (formic acid) + H_2O

The resulting sodium formate yields free acid when treated with dilute sulphuric acid

(5) By heating crystalline oxalic acid with (catalyst) to 120°C

HO—C(=O)—C(=O)—OH (oxalic acid) —heat→ H—C(=O)—OH (formic acid)

This method is used for the laboratory preparation of formic acid. In actual practice, glycerol is first heated at 120°C. Then a further quantity of hydrated oxalic acid $(COOH)_2 2H_2O$ is added and the temperature is again raised to 120°C . The decomposition of oxalic acid takes place as follows.

H_2C—OH / H_2C—OH (ethylene glycol) + HO—C(=O)—C(=O)—OH (oxalic acid) → glyceryl monoxalate

(HO–C(=O)–O–CH / HC-OH / H_2C–OH, with H–C(=O) bonded to O)

$$\text{glyceryl monoxalate} + H_2O \xrightarrow{120^\circ C} \text{glycerol} + \text{formic acid } (H-COOH)$$

The glycerol set free is used again.

Laboratory Preparation

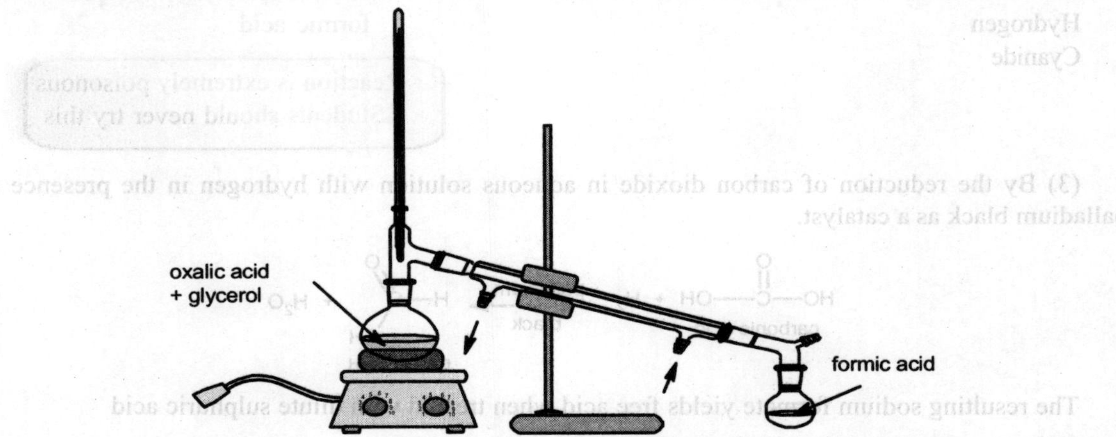

Fig. 11.1 : Showing the Laboratory preparation of Formic Acid

Theory: The formic acid in the lab is prepared by the action of oxalic acid and glycerol and then distillation of the mixture.

Apparatus	Chemicals	Catalyst	Temperature
Distillation apparatus, heating mantle, thermometer, Stands etc.	Oxalic acid		120°C

Reaction Equation

$$\text{ethylene glycol} + \text{oxalic acid} \longrightarrow \text{glyceryl monoxalate}$$

$$\text{glyceryl monoxalate} + H_2O \xrightarrow{120°C} \underset{\text{glycerol}}{H_2C(OH)-HC(OH)-H_2C(OH)} + \underset{\text{formic acid}}{H-C(=O)OH}$$

Precaution: *The temperature should be maintained at the 120°C and the oxalic acid must be pure.*

Anhydrous Formic Acid

Anhydrous (50 ml) and powdered crystalline oxalic acid (40g) are placed in the distillation flask and the apparatus is slowly heated in a heating mantle at a temperature of 120°C, till the evolution of carbon dioxide (marked by effervescence) ceases. The reaction flask is then cooled and a fresh lot of oxalic acid (40g) is added. The mixture is again heated at 120°C, when aqueous solution of formic acid collects in the receiver. This is neutralized with lead carbonate. The solution of lead formate thus produced is filtered and concentrated. The crystals of lead formate are then charged in the inner tube of a water condenser and the apparatus filled up as shown in the fig. 11.2.

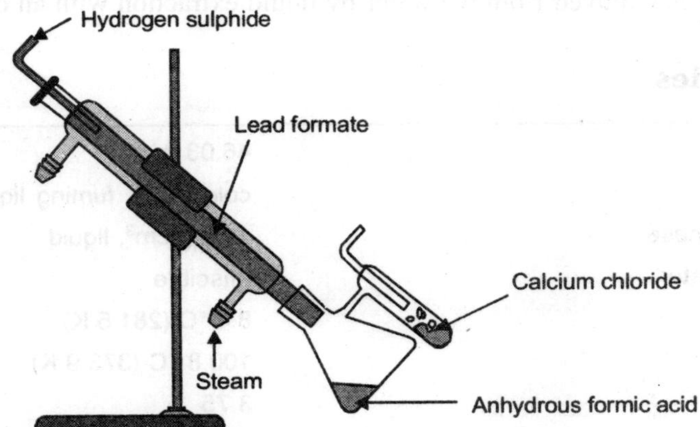

Fig. 11.2. Showing the Lab preparation of Anhydrous Formic Acid

Anhydrous formic acid is also prepared by the reaction of sodium formate and sodium hydrogen sulphate.

$$HCOONa + NaHSO_4 \rightarrow HCOOH + Na_2SO_4$$

Manufacture of Formic Acid

A significant amount of formic acid is produced as a byproduct in the manufacture of other chemicals, especially acetic acid. However, this production is insufficient to meet the present demand for formic acid, and some formic acid must be produced for its own sake.

When methanol and carbon monoxide are combined in the presence of a strong base, the formic acid derivative methyl formate results, according to the chemical equation

$$CH_3OH + CO \rightarrow HCOOCH_3$$

Industrially, this reaction is performed in the liquid phase at elevated pressure. Typical reaction conditions are 80°C and 40 atm. The most widely used base is sodium methoxide. Hydrolysis of the methyl formate produces formic acid:

$$HCOOCH_3 + H_2O \rightarrow HCOOH + CH_3OH$$

However, direct hydrolysis of methanol requires a large excess of water to proceed efficiently, and so some producers perform it by an indirect route, first reacting the methyl formate with ammonia to produce formamide, and then hydrolyzing the formamide with sulfuric acid to produce formic acid:

$$HCOOCH_3 + NH_3 \rightarrow HCONH_2 + H_2O$$

$$HCONH_2 + H_2O + \tfrac{1}{2}H_2SO_4 \rightarrow HCOOH + \tfrac{1}{2}(NH_4)_2SO_4$$

This technique has problems of its own, particularly disposing of the ammonium sulfate byproduct, and so recently, some manufacturers have developed energy-efficient means of separating formic acid from the large excess of water used in direct hydrolysis. In one of these processes, used by BASF, the formic acid is removed from the water by liquid extraction with an organic base.

Physical Properties

Molar mass	46.03 g/mol
Appearance	colourless, fuming liquid
Density and phase	1.22 g/cm³, liquid
Solubility in water	miscible
Melting point	8.4 °C (281.5 K)
Boiling point	100.8 °C (373.9 K)
Acidity (pK_a)	3.75
Viscosity	1.57 cP at 26 °C
Molecular shape	planar
Dipole moment	1.41 D(gas)
MSDS	External MSDS
Main hazards	corrosive
Flash point	69 °C

Chemical Properties

Formic acid is unique amongst simple carboxylic acid because it contains and atom instead of the alkyl group. Thus it has the structure of the both a carboxylic acid and aldehyde.

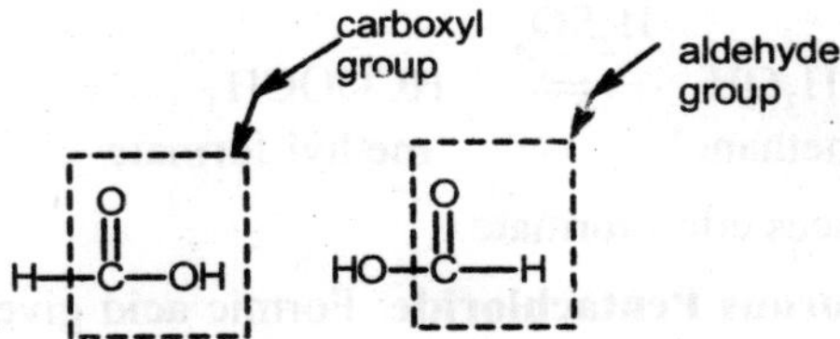

The reactions shown by the formic acid are in keeping with the above aspect of it

1. Reducing Properties: Formic acid can be readily oxidised to form carbon dioxide and water

$$\underset{\text{formic acid}}{H{-}\overset{\overset{O}{\|}}{C}{-}OH} + O \longrightarrow \underset{\text{carbonic acid}}{HO{-}\overset{\overset{O}{\|}}{C}{-}OH} \longrightarrow H_2O + CO_2\uparrow$$

Thus like aldehydes it reduces Fehlling's Solution ammonical silver nitrate (Tollen's reagent) and potassium permanganate solution both Fehlling's Reagent contains sodium hydroxide which converts the formic acid to formate ion which in turn gets oxidised thus,

$$\underset{\text{Formic acid}}{HCOOH} + NaOH \rightarrow \underset{\text{sodium formate}}{HCOONa} + H_2O$$

$$HCOO^- + \underbrace{2Cu^{2+} + 5OH^-}_{\text{Fehlling's Reagent}} \rightarrow Cu_2O\downarrow + CO_3^{2-} + 3H_2O$$

$$HCOO^- + \underbrace{2Ag^+ + 3OH^-}_{\text{Tollen's Reagent}} \rightarrow 2Ag\downarrow + CO_3^{2-} + 3H_2O$$

Similarly acidified potassium permanganate is decolorized:

$$5HCOO^- + 2MnO_4^- + 6H \rightarrow 2Mn^{2+} + 8H_2O + 5CO_2\uparrow$$

Formic acid also reduced mercuric salt to mercurous salt

$$2HgCl_2 + HCOOH \rightarrow Hg_2Cl_2\downarrow + 2HCl + CO_2\uparrow$$

Further it may be pointed out that unlike aldehydes; formic acid does not form oxime, phenylhydrazone and semicarbazone.

2. Formation of Salts: Formic acid is the strongest acid among all the members of the homologous series. It forms salts with alkalies, carbonates and bicarbonates.

$$\underset{\text{Formic acid}}{HCOOH} + \underset{\text{sodium formate}}{NaCO_3} \rightarrow HCOONa + CO_2 + H_2O$$

$$\underset{\text{Formic acid}}{HCOOH} + \underset{\text{potassium formate}}{KOH} \rightarrow HCOOK + H_2O$$

3. Formation of Ester: Formic acid reacts with dilute in presence of mineral acid (as catalyst) to form esters. Thus

$$\underset{\text{Formic acid}}{HCOOH} + \underset{\text{methanol}}{CH_3OH} \overset{H_2SO_4}{\rightleftharpoons} \underset{\text{methyl formate}}{HCOOCH_3} + H_2O$$

With ethyl alcohol it produces ethyl formate.

4. Reaction with Phosphorous Pentachloride: Formic acid gives formyl chloride which being unstable at once decomposes into hydrogen chloride and carbon monoxide.

$$HCOOH + PCl_5 \rightarrow \underset{\substack{\text{Formyl chloride}\\ \text{(unstable)}}}{HCOCl} + POCl_3 + HCl$$

$$HCOCl \rightarrow HCl + CO$$

5. Reaction with Ammonia: Ammonium formate is produced, which decomposes into formamide

$$HCOOH + NH_3 \rightarrow HCOONH_4$$

$$\underset{\text{Ammonium formate}}{HCOONH_4} \rightarrow \underset{\text{formamide}}{HCOONH_2} + H_2O$$

6. Reaction With Sulphuric Acid:

$$HCOOH \overset{H_2SO_4/\text{heat}}{\rightleftharpoons} H_2O + CO$$

This formic acid can be used as means of producing and storing carbon monoxide

7. Reaction of Salts of Formic Acid: Sodium formate when heated above 390°C decomposes to form hydrogen and sodium oxalate.

$$\underset{\text{Sodium Formate}}{H{-}C(=O){-}O{-}Na} \longrightarrow \underset{\text{Sodium Oxalate}}{Na{-}O{-}C(=O){-}C(=O){-}O{-}Na}$$

The free oxalic acid is produced by treatment of the product with a dilute mineral acid

(b) The calcium formate when heated alone gives formaldehyde. However, when heated with calcium salts of other fatty acids it gives higher aldehydes, thus

$$\underset{\text{Calcium formate}}{(HCOO)_2Ca} \overset{\Delta}{\rightarrow} HCHO + CaCO_3$$

$$\underset{\text{Calcium formate}}{(HCOO)_2Ca} + \underset{\text{calcium acetate}}{(CH_3COO)_2Ca} \rightarrow \underset{\text{acetaldehyde}}{2CH_3CHO} + 2CaCO_3$$

8. Decarbooxylation: When heated above 160°C under pressure formic acid decomposes to produce CO_2 and H_2

$$HCOOH \xrightarrow{>160^\circ / \text{under pressure}} CO_2 + H_2$$

Uses

The principal use of formic acid is as a preservative and antibacterial agent in livestock feed. When sprayed on fresh hay or other silage, it arrests certain decay processes and causes the feed to retain its nutritive value longer, and so it is widely use to preserve winter feed for cattle. In the poultry industry, it is sometimes added to feed to kill salmonella bacteria. Some beekeepers use formic acid as a miticide against the Varroa mite.Formic acid is of minor importance in the textile industry and in the tanning of leather. Some formate esters are artificial flavourings or perfumes.

In synthetic organic chemistry, formic acid is often used as a source of hydride ion. The Eschweiler-Clarke reaction and the Leuckart-Wallach reaction are famous examples of this.

Safety

The principal danger from formic acid is from skin or eye contact with liquid formic acid or with the concentrated vapors. Any of these exposure routes can cause severe chemical burns, and eye exposure can result in permanent eye damage. Inhaled vapors may similarly cause irritation or burns in the respiratory tract. Since carbon monoxide may be also be present in formic acid vapors, care should be taken wherever large quantities of formic acid fumes are present. The US OSHA Permissible Exposure Level (PEL) of formic acid vapor in the work environment is 5 parts per million parts of air (ppm).Formic acid is readily metabolized and eliminated by the body. Nonetheless, some chronic effects have been documented. Some animal experiments have demonstrated it to be a mutagen, and chronic exposure may cause liver or kidney damage. Another possibility with chronic exposure is development of a skin allergy that manifests upon re-exposure to the chemical. In addition, The methanol in aspartame (Equal, NutraSweet) breaks down into formaldehyde and formic acid

ACETIC ACID

Formula: CH_3COOH

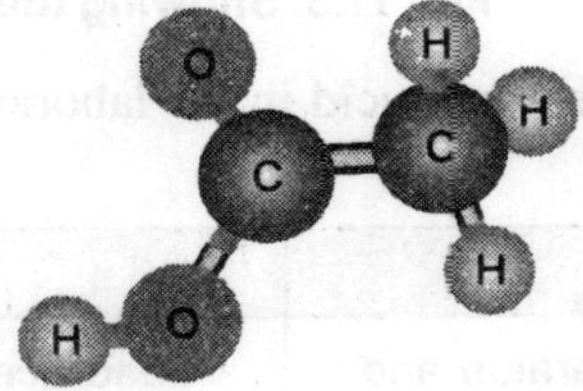

Introduction and History

The chemical compound **acetic acid** (from the Latin word *acetum*, meaning "vinegar"), Vinegar is as old as civilization itself, if not older. Acetic acid-producing bacteria are universally present, and any culture practicing the brewing of beer or wine inevitably discovered vinegar as the natural result of these alcoholic beverages being exposed to air. The use of acetic acid in chemistry extends into antiquity. The Greek philosopher **Theophrastos** described in the third century BC how vinegar acted on metals to produce pigments useful in art, including *white lead* (lead carbonate) and *verdigris* (a green mixture of copper salts including cupric acetate). Ancient Romans boiled soured wine in lead pots to produce a highly sweet syrup called *sapa*. Sapa was rich in lead acetate, a sweet substance also called *sugar of lead* or *sugar of Saturn*, which contributed to lead poisoning among the Roman aristocracy. Renaissance-era alchemists prepared glacial acetic acid through the dry distillation of metal acetates. The 16th century German alchemist **Andreas Libavius** described such a procedure, and he compared the glacial acetic acid produced by this means to vinegar. The presence of water in vinegar has such a profound effect on acetic acid's properties that for centuries glacial acetic acid and the acid found in vinegar were believed to be two different substances. The French chemist **Pierre Adet** proved them to be identical, and in 1847, the German chemist **Hermann Kolbe** synthesized acetic acid from inorganic materials for the first time.

Laboratory Perparation

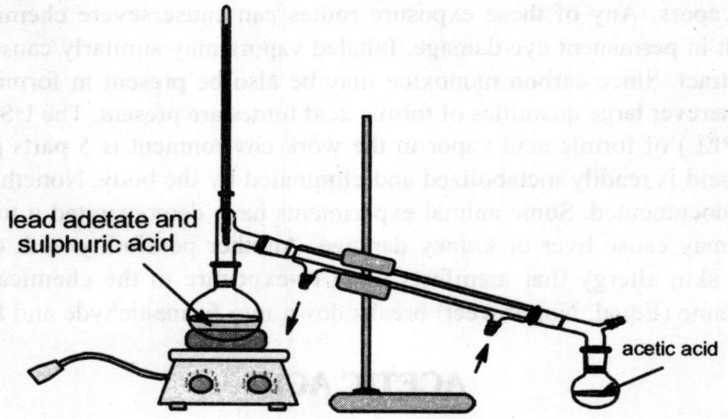

Fig. 11.3. Showing the Laboratory preparation of Acetic Acid

Theory: The acetic acid in the laboratory is prepared from the reaction of lead acetate and dilute sulphuric acid.

Apparatus	Chemicals	Catalyst	Temperature
Distillation apparatus and heating mantle reservoir flask stands and clamps	Lead acetate and dilute sulphuric acid		100°C

Precaution: The lead acetate must be pure and

Manufacture of Acetic Acid

Raw pyroligneous acid is settled to remove insoluble tars and then distilled in "primary coppers" to remove soluble tars. The vapour from the primaries is fed into the 8th plate of a 28 plate column.

The column reflux consists of 4% methanol from dephlegmator plus a mixture of woods oil obtained by vacuum distillation of the settled tar.

The descending wood oil reflux (45.35 - 68.025Kg per kg of acid) scrubs the primary vapours essentially free of acetic acid (0.3 – 0.4%) and carries them into re-boiler, while it allows a weak methanol to distil at the top. The wood oil – acetic acid mixture is then fed to a 7 plate exhausting column, operating under vacuum which recovers the oil at the base for reuse. The vapours from the exhauster are fed into the base of a 29 plate rectifying column fitted with its own reboiler and equipt with large

Condensers to handle a heavy reflux. The acid reaches its maximum concentration on 12th plate (90 – 95%) and is withdrawn as a liquid and discharged to vacuum receivers. The small amount of condensate from the top of the rectifier contains 15% acetic acid this is returned to primary still for rewarding. The final acid is clear but has a yellow tinge and is about 90 – 95% in strength.

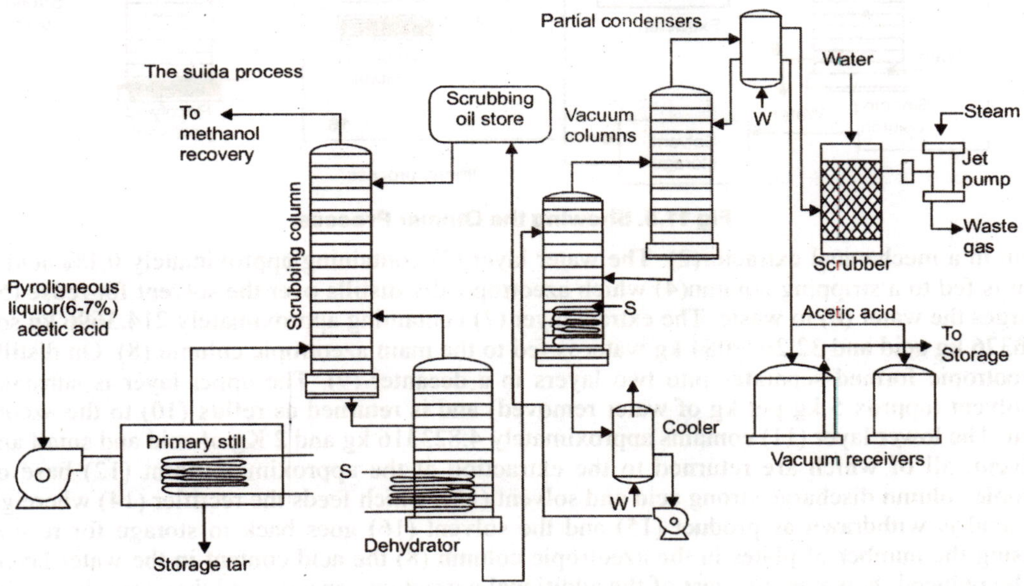

Fig 11.4. Showing the Suida Process

Acid Concentration and Recovery

The concentrated acid is required for the laboratory used under the name of the GR(granted reagent), AR (analytical reagent) and EXLAR(exclusive analytical reagent) trade names. For this a special process called Othmer Process by Prof D.F Othmer is used. This is a combination of extraction and azeotropic* distillation process that materially reduces the steam consumption. In this process (see fig. 11.5) a solvent (1) Boiling 102 - 150°C (usually amyl acetate) is used to extract the acid from its water

*Zeotropic distillation is separation of two compound formed by chemical union of two or more elements or ingredients in definite proportion by weight with the addition of a third compound that causes the formation of a heterogeneous (two liquid phase) zetrope with one of the two compounds to be separated. An example of this is the addition of benzene to separate ethanol and water.

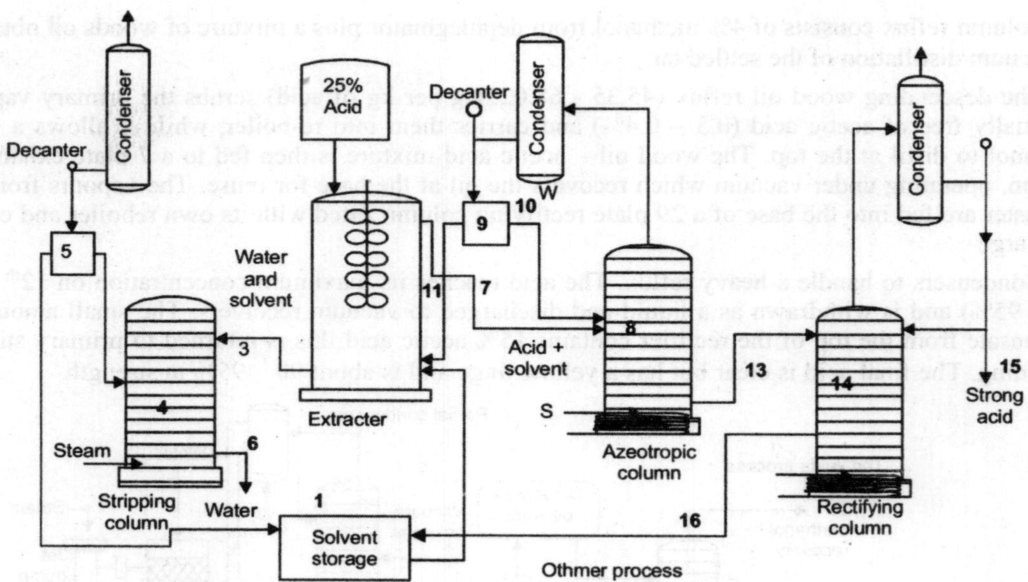

Fig 11.5. Showing the Othmer Process

solution in a mechanical extractor(2). The water layer (3) containing approximately 0.1% acid 0.2% solvent is fed to a stripping column(4) which azeotropically distills over the solvent for reuse (5) and discharges the water (6) to waste. The extract layer (7) containing approximately 214.3696 kg solvent 9.9286376 kg acid and 32.2679084 kg water is fed to the main azeotropic column (8). On distillation the azeotropic formed separates into two layers in a decanter (9). The upper layer is substantially pure solvent (approx 5 kg per kg of water removed) and is returned as reflux (10) to the azeotropic column. The lower layer (11) contains approximately 4.823316 kg and 2 Kg of acid and small amount of solvent, all of which are returned to the extraction at the approximate point (12) base of the azeotropic column discharge strong acid and solvent (13), which feeds the rectifier (14) where glacial acetic acid is withdrawn as product (15) and the solvent (16) goes back to storage for re use. By increasing the number of plates in the azeotropic column (8) the acid content in the water layer (11) could be reduced; however, the cost of the additional extraction capacity and therefore the azeotropic column is used essentially as an extractor. Theoretically 4.80885664 kg is required per pound of water removed from 28% acid this would amount to $\frac{72}{28} \times 1.36 = 3.5$ lb or 12.375734 kg steam per kg of acid concentrated. In practice when steam required for heating the feed, radiation losses etc are considered, the consumption amounts to about 17.67971 kg of steam per Kg of acid, depending somewhat on the particular plant. The solvent less in practice varies form 0.007071884 kg per kg of acid produced to 0.01767971 kg per kg of acid depending more on the intelligence of the operation than on any other factor.

Vinegar

FRINGS, the world market leader in vinegar technology, designs and supplies new vinegar plants or adopts existing ones to the latest standard of technology, including the heart of a plant, the Acetator and sophisticated variety of peripheral equipment, as well as computerized process control system. With these plants our customers can produce biological vinegar with maximum acidities of more than 20%. And up to now FRINGS has offered services for 70% vinegar plan over the world. Frings offers technology and equipment for four main processes:

(*a*) Continuous fermentation Raw material: Maximum Acidity:	pure alcohol, wine, rice wine, special fruit wine 9%
(*b*) Normal batch process Raw material: Maximum Acidity:	pure alcohol, wine, rice wine, special fruit wine 14%
(*c*) Single stage high strength process Raw material: Maximum acidity:	pure alcoho 117-18%
(*d*) Dual stage high strength process Raw material: Maximum Acidity:	pure alcohol Over 20%

Fring's Process

Due to the constant development of the vinegar production process by Frings so far a productivity and yield level is reached, which no other company can offer. 95% yield and 28 Wh energy consumption per litre 10% vinegar may lead to production cost below 0,4 RMB per litre 10% vinegar. Combined with the advantages as a natural, healthy and biological made product, white vinegar production with Frings technology is the only alternative instead of using chemical made acetic acid with all it's disadvantages.

Acetic Acid (Physical Properties)

Melting Point	16.6°C
Boiling Point	118.1°C
Density	1.0492
Refractive Index	1.37182
Surface tension	23.5 10^{-6} Newton meter
Viscosity	0.013 centipoises at 18°C
Critical temperature	321.6°C
Critical Pressure	57.2 atmos
Critical Density	0.351 gm/cm^3
Heat of Combustion forming Carbon dioxide	209 Kgcal/gm mol at 20°C
Flash point	41.6°C
Co-efficient of Expansion	0.001433 units per °C
Heat of Fusion	44.7 cal/gm at 16.58
Heat of Vaporization	96.8 cal/gm at 118.3°C

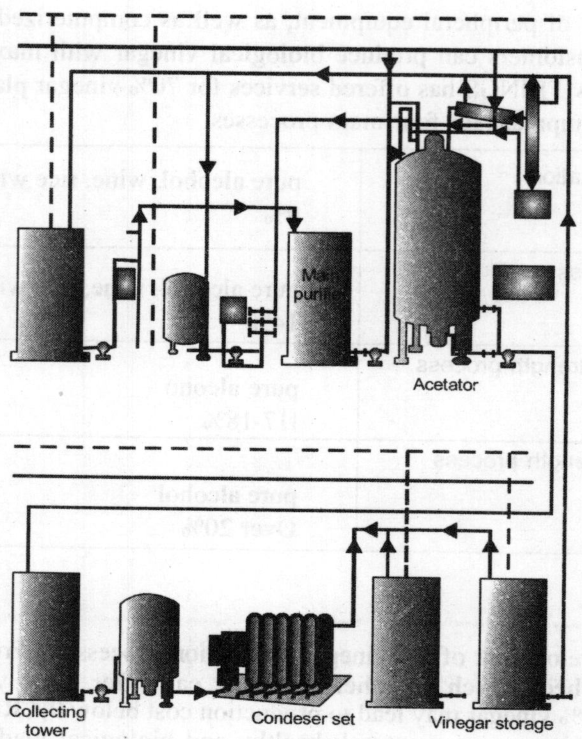

Fig. 11.6. Showing the Fring's Process

Chemical Properties

1. Action of Chlorine: Chlorine reacts with acetic acid in the presence of a catalyst (iodine, sulphur or phosphorous) giving mono, di, and trichloroacetic acid.

$$CH_3COOH \xrightarrow[P]{Cl_2} CH_2ClCOOH \xrightarrow[P]{Cl_2} CHCl_2COOH \xrightarrow[P]{Cl_2} CCl_3COOH$$

2. Formation of Salts: It's reacts with oxides hydroxides and carbonates to form acetates

$$CH_3COOH + NaOH \rightarrow CH_3COONa + H_2O$$

$$2CH_3COOH + (NH_4)_2CO_3 \rightarrow 2CH_3COONH_4 + H_2O + CO_2$$

The reaction is reversible and is catalyst by strong mineral acids

3. Formation of Anhydride: When the vapours of acetic acid are passed over a heated phosphate catalyst, it yields acetic anhydride.

$$2CH_3COOH \rightarrow \underset{\text{acetic anhydride}}{(CH_3CO)_2O} + H_2O$$

4. Action of Phosphorous Pentachloride:

CH_3COOH + PCl_5 → CH_3COCl + $POCl_3$ + HCl
Acetyl chloride

5. Formation of Amide: Ammonium acetate decomposes on heating to form Acetamide.

CH_3COONH_4 → CH_3CONH_2 + H_2O
Ammonium acetate acetamide

6. Decomposition of Calcium salt: - Calcium acetate, decomposes upon heating acetone

$(CH_3COO)_2Ca$ → $(CH_3)_2CO$ + $CaCO_3$
calcium acetate acetone

7. Decomposition with Salts of sodium: Sodium acetate when

CH_3COONa + NaOH → CH_4 + $NaCO_3$
heated with sodium hydroxide methane is produced

8. Electrolysis: The electrolysis of a solution of potassium acetate forms ethane

$2CH_3COONa$ → C_2H_6 + $2CO_2$ + 2Na
ethane

Uses

In the form of vinegar, acetic acid is used directly as a condiment, and also in the pickling of vegetables and other foodstuffs. Acetic acid is also sprayed onto silage as a preservative to discourage bacterial and fungal growth.The glacial acetic acid produced by the chemical industry is used in the manufacture of photographic films and stop bath and sometimes in the production of the plastic polyethylene terephthalate (PET). It is also used as an intermediate for the production of vinyl acetate, an important chemical in the paint and adhesives industry, and for cellulose acetate, a synthetic textile.Dilute solutions (4% - 6%) of acetic acid are extremely useful in treating the sting of the box jellyfish; the acid disables the stinging cells of the jellyfish, and can prevent serious injury or death if immediately applied. It is also used in treating outer ear infections in people in preperations such as Vosol.Some of the esters of acetic acid are commonly used solvents and artificial flavorings.

Safety

Concentrated acetic acid (also known as *glacial acetic acid*) is corrosive and has to be handled with extra care since it can cause skin burns, permanent eye damage, and irritation to the mucous membranes. It can also penetrate the skin, and cause acidic burns or blisters to appear several hours after exposure.Dilute acetic acid (in the form of vinegar) is harmless and has been consumed for millennia. However, ingestion of stronger solutions is dangerous. It can cause severe damage to the digestive system, and a potentially lethal change in the acidity of the blood. Acetic acid poses no known cancer risk.

Formic Acid and Acetic Acid (Comparison)

Since formic acid and acetic acid contain the same functional group, COOH, they resemble each other in respect of the reactions of these groups. Formic acid however, differs from acetic acid because it has only one (H) atom attached to the carbonyl group, while in acetic acid we have a methyl group, (– CH_3) attached to the carbonyl group. The methyl group in acetic acid shows the usual substitution reaction but the atom of formic acid does not. Nevertheless, the atom of formic acid can be easily oxidised to – OH and hence it shows reducing properties. The chief points of resemblance and difference in behaviour of the two acids are listed blow in the table: -

Formic Acid	Acetic Acid
Point of Resemblance	
1. Form the salts with alkalies and certain metals decompose carbonates and bicarbonates with evolution of carbon dioxide.	1. Forms salts with metals and alkalies decompose carbonates and bicarbonates with evolution of carbon dioxide.
2. Yields esters readily when treated with alcohols even in the absence of conc. Sulphuric acid of dry hydrogen chloride gas.	2. Yields esters when treated with alcohols and conc. sulphuric acid and hydrogen chloride gas.
3. Reacts with PCl_5 to from formyl chloride which being unstable decomposes to give CO + HCl	3. Reacts with PCl_5 to form acetyl chloride
4. Its ammonium salts (ammonium formate) when heated produces formamide	4. Ammonium acetate on heating gives acetamide.
Point of difference	
1. When heated at 160°C under pressure it decomposes to produce.$CO_2 + H_2$. On heating with sulphuric acid it gives $CO + H_2O$	1. It is stable to heat and conc. sulphuric acid
2. It is stronger than acetic acid	2. It is weaker than the formic acid
3. Does not undergo substitution reactions when treated with chlorine in presence of red.	3. It forms mono, di and tri chloroacetic acid, when heated with chlorine and red phosphorous.
4. Gives formaldehyde when heated its calcium salt	4. Gives acetone when its calcium salt is heated
5. Sodium formate when heated at 390°C gives sodium oxalate and hydrogen	5. No action
6. Electrolysis of aqueous solution of sodium or potassium salt yields hydrogen	6. Ethane is produced when aqueous solution of sodium or potassium salt is electrolyzed.
7. 'Reduces Tollen's Reagent or Fehlling's Solution or mercuric salt to mercurous salt	7. No action

PROPIONIC ACID

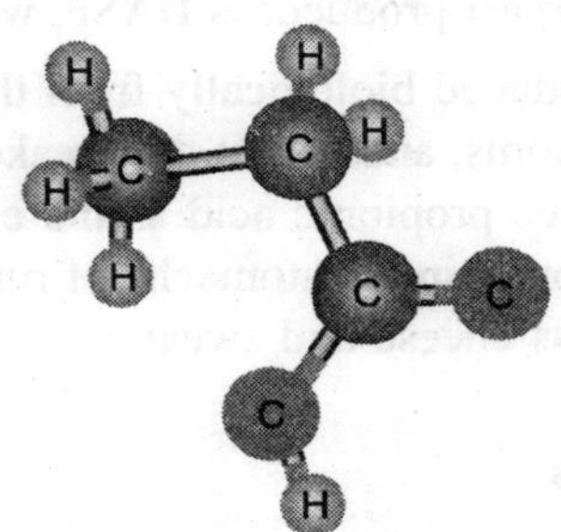

Formula: CH_3CH_2COOH

Introduction History

Propionic acid was first described in 1844 by **Johann Gottlieb**, who found it among the degradation products of sugar. Over the next few years, other chemists produced propionic acid in various other ways, none of them realizing they were producing the same substance. In 1847, the French chemist Jean-Baptiste Dumas established that all the acids were the same compound, which he called propionic acid, from the Greek words *protos* "first", because it was the smallest acid that exhibited the properties of the other fatty acids, such as producing an oily layer when salted out of water and having a soapy potassium salt

Preparation

It is prepared (i) by the oxidation of n- propyl alcohol with sodium dichromate and sulphuric acid

$$\underset{\text{n-propyl alcohol}}{CH_3CH_2CH_2OH} \xrightarrow{Na_2Cr_2O_7} \underset{\text{propionic acid}}{CH_3-CH_2-COOH}$$

(ii) By the reaction of ethylene carbon monoxide and water at high temperature and pressure in presence of phosphoric acid as catalyst.

$$CH_2=CH_2+CO+H_2O \xrightarrow[H_3PO_4]{\text{High temp/high preseure}} CH_3CH_2COOH$$

Industrial Preparation

Industrially, propionic acid is usually made from the air oxidation of propionaldehyde. In the presence of cobalt or manganese ions, this reaction proceeds rapidly even at mild termperatures. Usually, the industrial process is carried out at 40-50°C, and is represented by the chemical equation

$$CH_3CH_2CHO + \tfrac{1}{2}O_2 \rightarrow CH_3CH_2COOH$$

Large amounts of propionic acid were once produced as a byproduct of acetic acid manufacture, but changes in the way acetic acid is made have made this a very minor source of propionic acid today. Current world's largest producer is BASF, with approx. 80 ktpa production capacity.

Propionic acid is produced biologically from the metabolic breakdown of fatty acids containing odd numbers of carbon atoms, and also it the breakdown of some amino acids. Bacteria of the genus *Propionibacterium* produce propionic acid as the end product of their anaerobic metabolism. These bacteria are commonly found in the stomachs of ruminants, and their activity is partially responsible for the odor of both Swiss cheese and sweat.

Physical Properties

Molar mass	74.08 g/mol
Appearance	colourless liquid
Density and phase	0.99 g/cm^3, liquid
Solubility in water	miscible
Melting point	-21 °C (252 K)
Boiling point	141 °C (414 K)
Acidity (pK_a)	4.88
Viscosity	10 mPa·s
Flash point	54°C

Uses

Propionic acid inhibits the growth of mold and some bacteria. Accordingly, most propionic acid produced is used as a preservative for both animal feed and food for human consumption. For animal feed, it is used either directly or as its ammonium salt. In human foods, especially bread and other baked goods, it is used as its sodium or calcium salt. Similar usage occurs in some of the older anti-fungal foot powders.Propionic acid is also useful as a chemical intermediate. It can be used to modify synthetic cellulose fibers. It is also used to make pesticides and pharmaceuticals. The esters of propionic acid are sometimes used as solvents or artificial flavourings.

Safety

The chief danger from propionic acid is chemical burns that can result from contact with the concentrated liquid. In studies on laboratory animals, the only adverse health effect associated with long-term exposure to small amounts of propionic acid has been ulceration of the esophagus and stomach from consuming a corrosive substance. No toxic, mutagenic, carcinogenic, or reproductive effects have ever been observed. In the body, propionic acid is readily oxidized, metabolized, and eliminated from the body as carbon dioxide in the Krebs cycle, and so it does not bioaccumulate

BUTYRIC ACIDS

There are two isomeric butyric acids known

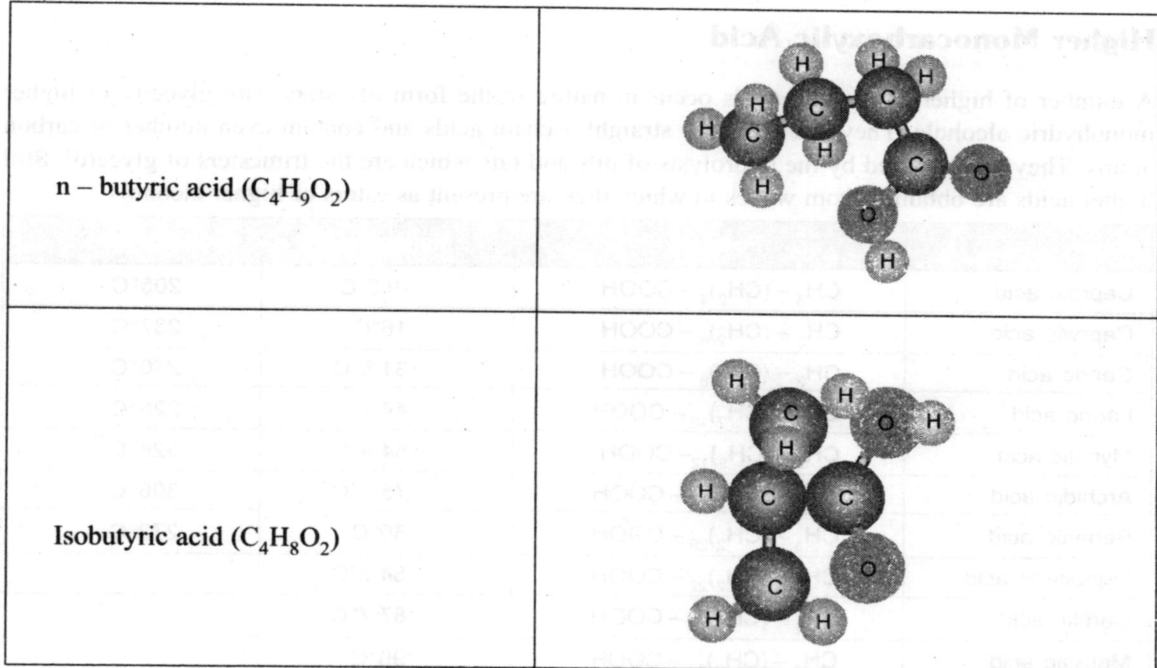

n-**Butyric acid**, IUPAC name ***n*-Butanoic acid**, or normal butyric acid, is a carboxylic acid with structural formula $CH_3CH_2CH_2$-COOH. It is notably found in rancid butter, parmesan cheese, or vomit and has an unpleasant odor and acrid taste, with a sweetish aftertaste (similar to ether).

Butyric acid can be detected by mammals with good scent detection abilities (e.g. dogs, at 10 ppb, while humans can detect it in concentrations above 10 ppm.

Normal butyric acid or fermentation butyric acid is found in butter, as an hexyl ester in the oil of *Heracleum giganteum* (cow parsnip) and as an octyl ester in parsnip (*Pastinaca sativa*); it has also been noticed in the fluids of the flesh and in perspiration.

It may be prepared by the hydrolysis of ethyl acetoacetate, or by passing carbon monoxide over a mixture of sodium acetate and sodium ethylate at 205°C. It is ordinarily prepared by the fermentation of sugar or starch, brought about by the addition of putrefying cheese, calcium carbonate being added to neutralize the acids formed in the process. The butyric fermentation of starch is aided by the direct addition of Bacillus subtilis.

The acid is an oily colorless liquid that solidifies at -5°C; it boils at 164°C. It is easily soluble in water, ethanol and ether, and is thrown out of its aqueous solution by the addition of calcium chloride. Potassium dichromate and sulfuric acid oxidize it to carbon dioxide and acetic acid, while alkaline potassium permanganate oxidizes it to carbon dioxide. The calcium salt, $Ca(C_4H_7O_2)_2 \cdot H_2O$, is less soluble in hot water than in cold.

There is an isomer, isobutyric acid, which has the same chemical formula C_4H_8 O_2 but a different structure. It has similar chemical properties but different physical properties.

Higher Monocarboxylic Acid

A number of higher carboxylic acids occur in nature in the form of esters with glycerol or higher monohydric alcohols. They are generally straight – chain acids and contain even number of carbon atoms. They are prepared by the hydrolysis of oils and fats which are the trimesters of glycerol. Still higher acids are obtained from waxes in which they are present as esters of higher alcohols.

Name	Formula	Melting point	Boiling Point
Caproic acid	$CH_3-(CH_2)_4-COOH$	-9.5°C	205°C
Caprylic acid	$CH_3-(CH_2)_6-COOH$	16°C	237°C
Capric acid	$CH_3-(CH_2)_8-COOH$	31.5°C	270°C
Lauric acid	$CH_3-(CH_2)_{10}-COOH$	44°C	225°C
Myristic acid	$CH_3-(CH_2)_{12}-COOH$	54.4°C	326°C
Archidic acid	$CH_3-(CH_2)_{18}-COOH$	75.4°C	306°C
Behenic acid	$CH_3-(CH_2)_{20}-COOH$	80°C	272°C
Lignoceric acid	$CH_3-(CH_2)_{22}-COOH$	84.2°C	
Carolic acid	$CH_2-(CH_2)_{24}-COOH$	87.7°C	
Melissic acid	$CH_3-(CH_2)_{28}-COOH$	90°C	

Unsaturated Monocarboxylic Acid

The unsaturated Monocarboxylic acids having a single bond in their molecule are represented by the general formula $C_nH_{2n+1}COOH$. Most of these acids are known by their common names their IUPAC names are derived by considering the longest carbon chain bond with respect to the carbonyl group is indicated by a number. The common and IUPAC names of a few important Monocarboxylic acids are in the table below:

Formula	Common Name	IUPAC Name
$CH_2=CH-COOH$	Acrylic acid	Prop – 2 – enoic – acid
$CH_3=CHOOH$	Crotonic acid	But – 2 – enoic acid
$H_2C=C(CH_3)-C(=O)HO$	Methyl acrylic acid	2 methyl prop – 2 – enoic acid

Formula	Common Name	IUPAC Name
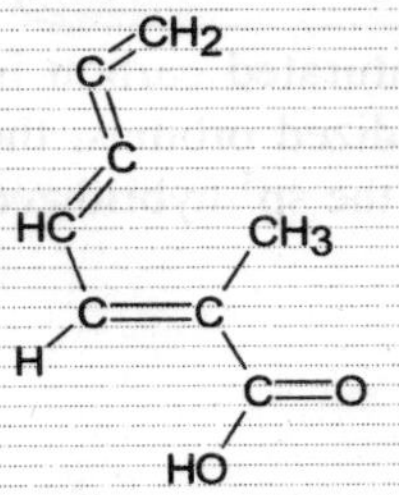 $H_2C=C=CH-CH=C(CH_3)-C(=O)OH$ (H and C=O on same side)	Tiglic acid	cis – 2 methyl but – 2 – enoic acid
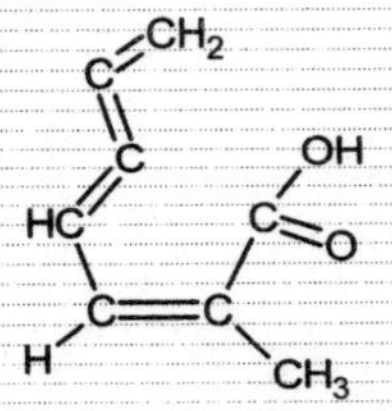 $H_2C=C=CH-CH=C(CH_3)-C(=O)OH$ (H and CH_3 on same side)	Angetic acid	Trans – 2 methyl but – 2 enoic acid

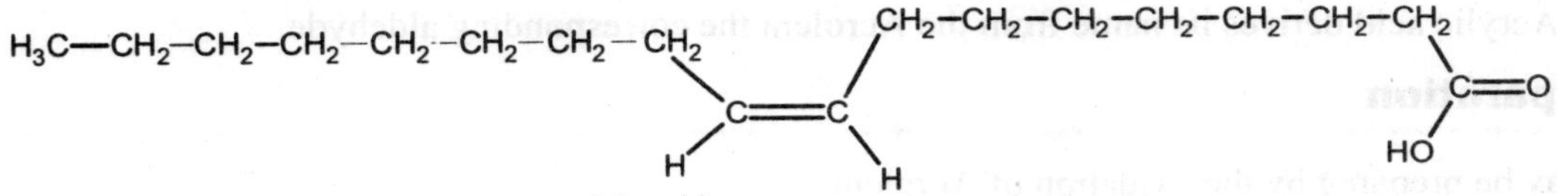

$H_3C-CH_2-CH_2-CH_2-CH_2-CH_2-CH_2-CH_2-CH=CH-CH_2-CH_2-CH_2-CH_2-CH_2-CH_2-CH_2-COOH$ (H, H cis)

oleic acid

$H_3C-CH_2-CH_2-CH_2-CH_2-CH_2-CH_2-CH_2-CH=CH-CH_2-CH_2-CH_2-CH_2-CH_2-CH_2-CH_2-COOH$ (H, H trans)

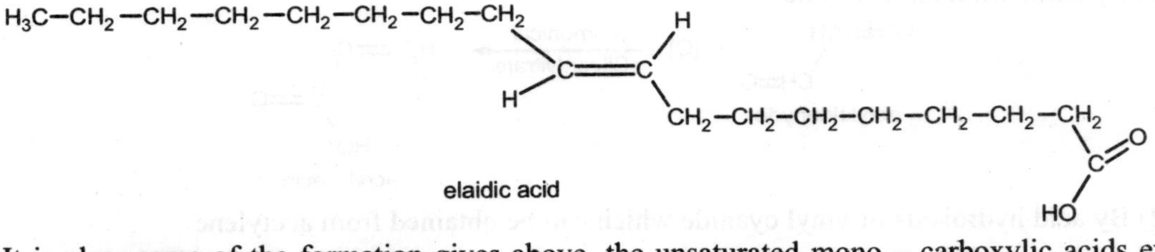

elaidic acid

It is clear same of the formation gives above, the unsaturated mono – carboxylic acids exhibit geometrical isomerism.

Unsaturated carboxylic acid of the type $RCH = CH(CH_2)_n$ COOH in which the double bond and the carboxyl group are far apart (that is n is large) usually exhibit the proper cis characteristic of isolated double bonds and isolated carboxyl group are sufficiently close together to insert strongly as in α, β - unsaturated acids.

$$\underset{}{R-\overset{\beta}{C}H=\overset{\beta}{C}H-COOH}$$

e.g. acrylic acid $CH_2 = CHCOOH$ and exceptional behaviour is seen

Acidity: The acid strength of the unsaturated Monocarboxylic acids in which the double bond and the carbonyl group are far apart, is approximately the same as that of the correspond saturated

Acrylic acid is stronger than propionic acid

analogues. But if the double bond is adjacent to the unsaturated acid is greater that the corresponding saturated acid. For example acrylic acid $CH_2 = CHCOOH$ has a K_a value of 5.6×10^{-5} compared with the 1.4×10^{-5} for the saturated propionic acid. This is due to the fact that a - unsaturated carbon in acrylic acid has sp^3 hybridized orbitals, which in propionic acid it has sp^3 hybridized orbitals, the electrons are closer to nucleus because of greater S characters than in sp^3 orbitals, the sp^3 hybridized carbon atom consequently acrylic acid is a stronger acid than propionic acid.

ACRYLIC ACID

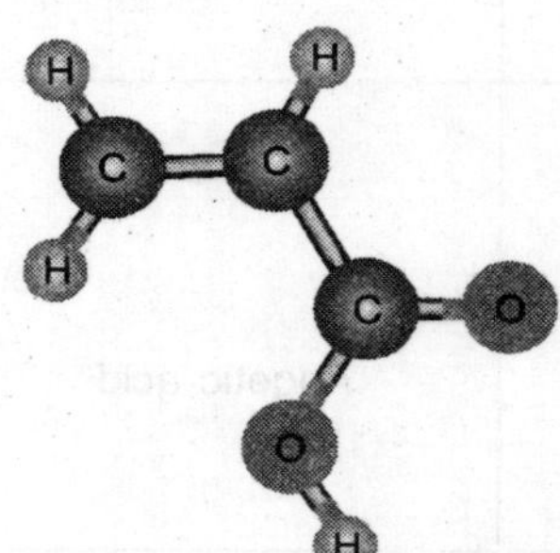

Formula: $C_3H_4O_2$

Acrylic acid derives its name from the Acrolein the corresponding aldehyde.

Preparation

It may be prepared by the oxidation of Acrolein

(1) By ammonical silver nitrate

$$H_2C{=}CH{-}CH{=}O \text{ (acrylaldehyde)} + [O] \xrightarrow[\text{Silver Nitrate}]{\text{ammonical}} H_2C{=}CH{-}C({=}O)OH \text{ (acrylic acid)}$$

(2) By acid hydrolysis of vinyl cyanide which can be obtained from acetylene

$$HC{\equiv}CH \text{ (acetylene)} + HCN \xrightarrow[90^\circ]{Cu,\ NH_4Cl} H_2C{=}CH{-}C{\equiv}N \text{ (acrylonitrile)} \xrightarrow{H^+/H_2O} H_2C{=}CH{-}C({=}O)OH \text{ (acrylic acid)}$$

(3) By refluxing α, β dibromopropionic acid with zinc and ethanol

$$H_2C(Br){-}CH(Br){-}C({=}O)OH \text{ (2,3-dibromopropanoic acid)} \xrightarrow{Zn/C_2H_5OH} H_2C{=}CH{-}C({=}O)OH \text{ (acrylic acid)} + ZnBr_2$$

(4) By distillation of β - hydroxypropionic acid with $ZnCl_2$

$$HO-CH_2-CH_2-COOH \xrightarrow{\Delta\ ZnCl_2} CH_2-CH-COOH + 2H_2O$$

β - hydroxypropionic acid → Acrylic acid

β - hydroxypropionic acid is obtained from ethylene in the following manner

$$CH_2=CH_2 \xrightarrow{HOCl} HOCH_2-CH_2Cl \xrightarrow{KCN} HOCH_2-CH_2CN \rightarrow HOCH_2-CH_2COOH$$

Ethylene → Ethylene chloride → ethylene cyanohydrin → β-hydroxypropionic acid

This is probably the best method of preparation available for acrylic preparation.

Manufacture

(i) From ethylene cyanohydrin:

oxirane $\xrightarrow{HCN}$ 3-hydroxypropanenitrile ($H_2C(OH)-H_2C-C\equiv N$) $\xrightarrow[-H_2O]{Conc.\ H_2SO_4}$ $H_2C=CH-C\equiv N$ (acrylonitrile) $\xrightarrow[H_2O]{H_2SO_4}$ $H_2C=CH-C(=O)HO$ (acrylic acid)

(ii) By hydrocarboxylation of acetylene in the presence of nickel carboxyl as catalyst

$$CH\equiv CH + CO + H_2O \xrightarrow[HCl]{Ni(CO)_4} H_2C=CH-C(=O)HO$$

acetylene → acrylic acid

Physical Properties

Boiling Point (°C)	141
Freezing Point (°C)	13
Density (20/20 °C)	1.049
Flash Point (Tag Closed Cup °C)	50
Relative density (water = 1)	1.05
Solubility in water	Miscible
Vapour pressure, Pa at 20°C	413
Relative vapour density (air = 1)	2.5
Auto-ignition temperature	360°C
Explosive limits, vol% in air:	2.4-8
Octanol/water partition coefficient as log P_{ow}	0.36 (estimated)

Chemical Properties and Reactions

Acrylic acid behaves both as an alkene and a carboxylic acid

A. Reactions of the Double Bond

(i) On catalytic hydrogenation it gives propionic acid

$$H_2C{=}CH{-}C(=O){-}OH + H_2 \xrightarrow{Ni} H_3C{-}CH_2{-}C(=O){-}OH$$

acrylic acid propionic acid

(ii) With chlorine and bromine: It forms addition compounds

$$H_2C{=}CH{-}C(=O){-}OH + Br_2 \xrightarrow{CCl_4} H_2C(Br){-}CH(Br){-}C(=O){-}HO$$

acrylic acid 2,3-dibromopropanoic acid

(iii) With halogen acid, it gives β- halopropionic acid

$$CH_2 = CH - COOH + HBr \rightarrow Br - CH_2 - CH_2 - COOH$$

Acrylic acid β- halopropionic acid

This addition takes place in rebelliousness of Markovnikov rule.

(iv) On hydration in presence of mercuric sulphate and sulphuric acid acrylic acid forms β- Hydroxypropionic acid and not the Markovnikov product i.e. α - Hydroxypropionic acid

$$CH_3 = CH - COOH + H_2O \rightarrow HO - CH_2 - CH_2 - COOH$$

β-Hydroxypropionic acid

(v) On hydroxylation with dilute potassium permanganate solution, it gives glyceric acid which on vigorous oxidation decomposes to oxalic acid.

$$H_2C{=}CH{-}C(=O){-}OH + H_2O + [O] \xrightarrow{\text{dil } KMnO_4} H_2C(OH){-}CH(OH){-}C(=O){-}HO \longrightarrow HO{-}C(=O){-}C(=O){-}OH$$

acrylic acid 2,3-dihydroxypropanoic acid oxalic acid

(vi) On standing it decomposes to glassy solid

B. Reactions of the Carbonyl Group

(i) It forms salts with metals carbonates and an alkalies

$$2CH_2 = CHCOOH + Na_2CO_3 \rightarrow 2CH_3 = CHCOONa + CO_2 + H_2O$$

acrylic acid — sodium acrylate

$$CH_2 = CHCOOH + KOH \rightarrow CH_2 = CHCOOK + H_2O$$

(ii) On esterification with absolute ethanol in presence of acid it forms with ammonia gives acrylamide.

$$H_2C{=}CH{-}C({=}O)OH \xrightarrow{PCl_5} H_2C{=}CH{-}C({=}O)Cl \xrightarrow{NH_3} H_2C{=}CH{-}C({=}O)NH_2 + NH_4Cl$$

acrylic acid — acryloyl chloride — acrylamide

Uses

The primary use of acrylic acid is in the production of acrylic esters and resins, which are used primarily in coatings and adhesives. It is also used in oil treatment chemicals, detergent intermediates, water treatment chemicals, and water absorbent polyacrylic acid polymers. Acrylic acid is used widely for polymerization, including production of polyacrylates. It is a monomer for polyacrylic and polymethacrylic acids and other acrylic polymers. It is used in the manufacture of plastics, as a tackifier, as a flocculant, in the production of water-soluble resins and salts, as a comonomer in acrylic emulsion and solution polymers and in molding powder for signs, construction units, decorative emblems and insignias. It is used in polymer solutions for coatings applications, in paint formulations, in leather finishings, in paper coatings, in polishes and adhesives and in general finishes and binders.

Polymerization of poly(acrylic acid) and poly(methyl methacrylic acid)

$$\left[\!-CH_2{-}CH(C({=}O)OH)-\!\right]_n$$

Both polymers are water soluble.

$$H_2C{=}CH{-}C({=}O)OH \longrightarrow -\left[H_2C{-}CH{-}C({=}O)OH\right]-$$

acrylic acid — poly acrylic acid

$$\text{2-methylacrylic acid } (CH_2{=}C(CH_3){-}COOH) \longrightarrow \text{poly metaacrylic acid } [-CH_2{-}C(CH_3)(COOH)-]_n$$

A free radical process using solution polymerization is possible using water as the solvent. Note that for solution polymerization there will be a correlation between the growing polymer molecular weight and viscosity.

This differs from the emulsion and suspension polymerizations in water discussed previously Initiator- potassium persulphate is added to the silver solution of the acid, with an activator (sodium thiosulphate) and a transfer agent (mercaptosuccinic acid), at a temperature from 50 to 100°C.)If solid polymer is required, peroxide can be used for the initiator, and benzene may be used as the solvent. Salts of acrylic and methacrylic acid may also be polymerized in aqueous solution by treatment with FT initiators.

Both acrylic acid and methacrylic acid undergo the carboxylic acid reactions discussed in organic chemistry.

Alternative routes to acrylic acid ethylene and propylene feedstocks

$$\underset{\text{ethylene}}{H_2C{=}CH_2} + H_2O + CO \longrightarrow \underset{\text{acrylic acid}}{H_2C{=}CH{-}C({=}O){-}OH}$$

$$\underset{\text{ethylene}}{H_2C{=}CH_2} \xrightarrow{\text{oxidation}} \underset{\text{acrylaldehyde}}{H_2C{=}CH{-}C({=}O){-}H} \longrightarrow \underset{\text{acrylic acid}}{H_2C{=}CH{-}C({=}O){-}OH}$$

The above is carried out in one step by passing a mixture of 10% propylene, 50% air, and 40% steam over a cobalt molybdenum catalyst at 400-500 deg C. **The above is carried out in one step by passing a mixture of 10% propylene, 50% air, and 40% steam over a cobalt molybdenum catalyst at 400-500°C.**

Synthesis of ethacrylate **The synthesis of ethacrylate from ethylene is shown**

$$\underset{\text{ethylene}}{H_2C{=}CH_2} + NC(CO)_4 + \underset{\text{ethanol}}{H_3C{-}CH_2{-}OH} \longrightarrow \underset{\text{ethyl acrylate}}{H_2C{=}CH{-}C({=}O){-}O{-}CH_2{-}CH_3}$$

The above reaction is carried out in one step by passing a mixture of 10% propylene, 50% air, and 40% steam over a cobalt molybdenum catalyst at 400-500°C.

CROTONIC ACID

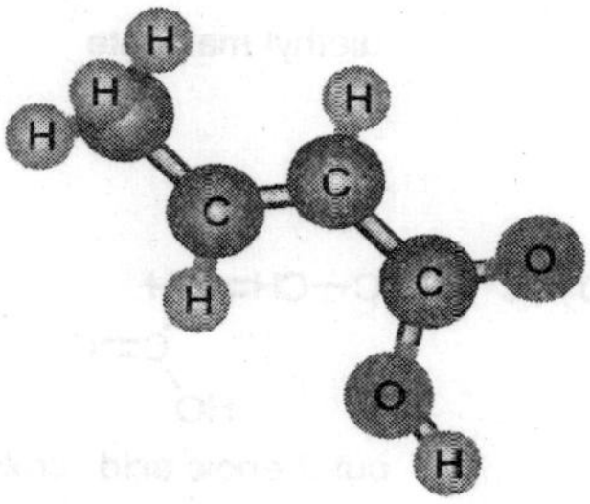

Formula: $C_4H_6O_2$

It is also known as 2 butenoic acid.

It is the Trans form of the structure $CH_3CH = CHCOOH$, while the cis form exists as isocrotonic acid.

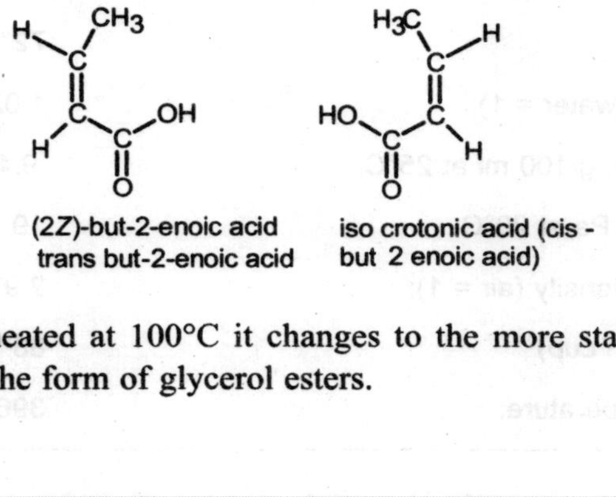

(2Z)-but-2-enoic acid trans but-2-enoic acid

iso crotonic acid (cis - but 2 enoic acid)

Isocrotonic acid when heated at 100°C it changes to the more stable Crotonic acid. Crotonic acid occurs in Croton oil in the form of glycerol esters.

Preparation

(i) By the oxidation of crotonaldehyde with ammonical silver nitrate

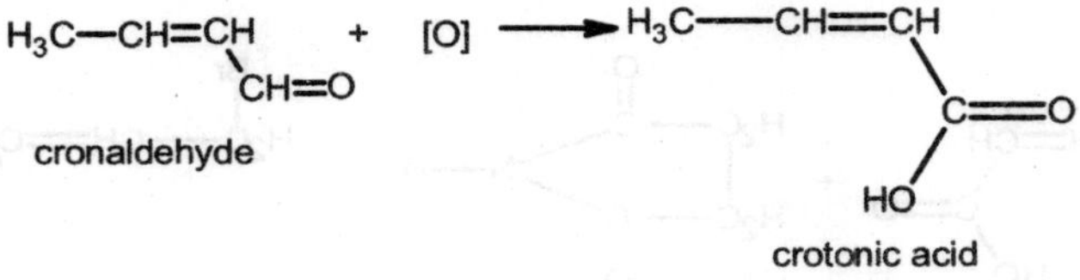

$$H_3C-CH=CH-CH=O + [O] \longrightarrow H_3C-CH=CH-C(=O)OH$$

cronaldehyde → crotonic acid

(2) By hating acetaldehyde and diethyl malonate in the presence diethyl amine or piperidine (Knovengal reaction) followed by hydrolysis and Decarbooxylation.

acetaldehyde + diethyl malonate —(piperidine / dimethyl amine)→ diethyl ethylidenemalonate —(aq KOH)→ —(acidify)→ CO_2 + but-2-enoic acid crotonic acid

Properties (Physical)

Boiling point:	189°C
Melting point:	72°C
Relative density (water = 1):	1.02
Solubility in water, g/100 ml at 25°C:	9.4
Vapour pressure, Pa at 20°C:	19
Relative vapour density (air = 1):	2.97
Flash point: (open cup)	88°C o.c.
Auto-ignition temperature:	396°C

Chemical Properties

When reacts with N- bromosuccinamide (NBS) it forms γ-bromocrotonic acid.

but-2-enoic acid + 1-bromopyrrolidine-2,5-dione → 4-bromobut-2-enoic acid bromocrotonic acid

OLEIC ACID

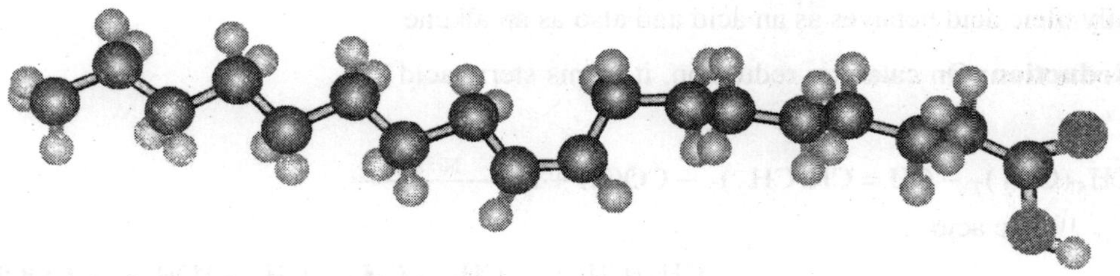

Formula: $C_6H_{31}O_2$

It is also known as cis – heptadec – 8 – ene – 1 carboxylic acid

As the name suggests the world oleic comes from the Latin word oleum = oil. It occurs mainly in fats and oils, olive oil cotton seed oil

$$
\begin{array}{l}
H_2C-O-\overset{O}{\overset{\|}{C}}-C_{17}H_{35} \\
HC-\overset{O}{\overset{\|}{C}}-C_{17}H_{35} \\
H_2C-\overset{O}{\overset{\|}{C}}-C_{15}H_{31}
\end{array}
+ 3NaOH \longrightarrow
\begin{array}{l}
H_2C-OH \\
HC-OH \\
H_2C-OH
\end{array}
\longrightarrow
\begin{array}{l}
H_{35}C_{17}-COOH \\
H_{35}C_{17}-\underset{O}{\underset{\|}{C}}-\underset{O}{\underset{\|}{C}}-O-Na \\
H_{31}C_{15}-\underset{O}{\underset{\|}{C}}-\underset{O}{\underset{\|}{C}}-O-Na
\end{array}
$$

olepalmitostearin; glycerol; sodium searate; sodium oleate; sodium palmitate

The water soluble minutes of sodium salts obtained above is treated with lead acetate when the corresponding lead salts are precipitated out. The mixture of lead salts so formed is then treated with either with dissolves lead oleate leaving behind lead palmate and lead stearate. The ethereal solution of lead oleate is evaporated and resulting lead oleate is treated with mineral acid to get free oleic acid. The solution is then extracted with either and dried over fused calcium chloride. Ether is distilled off and the residue upon further distillation under reduced pressure yields pure oleic acid

Purification of oleic acid

Physical Properties

Melting point °C	13.5(alpha),16.3(beta)
Boiling point °C (mm Hg)	334.7 (400)
Density D_4^t (°C)	0.8634 (60)
Viscosity mPa·s (°C)	27.64 (25), 4.85 (90)
Refractive Index n_D^t (°C)	1.4449 (60)
Specific Heat J/g (°C)	2.046 (50)

Chemical Properties

Chemically oleic acid behaves as an acid and also as an alkene

(*i*) **Reduction**: On catalytic reduction, it forms steric acid

$$CH_5(CH_2)_7 - CH = CH(CH_2)_7 - COOH + H_2 \xrightarrow{Ni/heat}$$

oleic acid

$$CH_3(CH_2)_7 - CH_2 - CH_2 - CH_2 - (CH_2)_7 - COOH$$

stearic acid

(*ii*) **Bromination**: On bromination, it gives 9,10 di bromo stearic acid

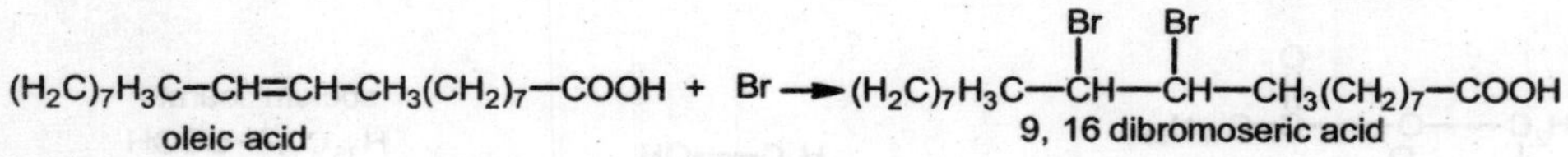

$$(H_2C)_7H_3C-CH=CH-CH_3(CH_2)_7-COOH + Br \rightarrow (H_2C)_7H_3C-\underset{}{\overset{Br}{CH}}-\overset{Br}{CH}-CH_3(CH_2)_7-COOH$$

oleic acid — 9, 16 dibromoseric acid

(*iii*) **Oxidation**: Oleic acid when treated with cold dilute potassium permanganate gives 9, 10 dihydroxide stearic acid which upon further oxidation yields pelargonic acid and azelaric acid.

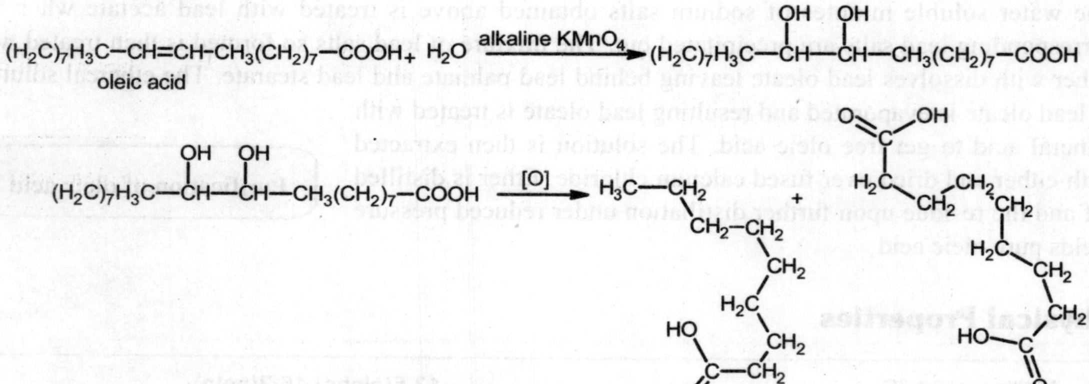

$$(H_2C)_7H_3C-CH=CH-CH_3(CH_2)_7-COOH + H_2O \xrightarrow{\text{alkaline } KMnO_4} (H_2C)_7H_3C-\overset{OH}{CH}-\overset{OH}{CH}-CH_3(CH_2)_7-COOH$$

oleic acid

$$(H_2C)_7H_3C-\overset{OH}{CH}-\overset{OH}{CH}-CH_3(CH_2)_7-COOH \xrightarrow{[O]} H_3C-CH_2-CH_2-CH_2-CH_2-H_2C-CH_2-CH_2-C(=O)-HO \; + \; HO-C(=O)-CH_2-H_2C-CH_2-CH_2-H_2C-CH_2-CH_2-C(=O)OH$$

pelargonic acid — azelaic acid

(*iv*) **Ozonolysis**: When reacted with ozone oleic acid forms the corresponding ozonides which when decomposed with zinc and water gives pelargonic aldehyde azelaric acid half aldehyde.

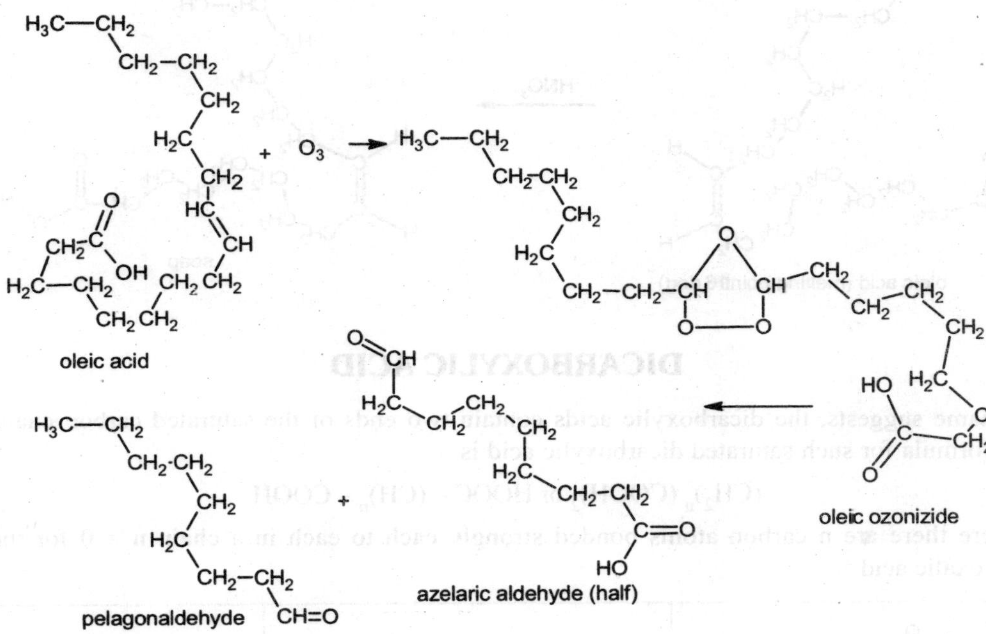

oleic acid

oleic ozonizide

pelagonaldehyde

azelaric aldehyde (half)

(v) Conversion To *trans* form: Oleic acid exhibits geometric isomerism. Ordinary acid is the *cis*- form and when this is treated with nitric acid it is converted into its trans- form elaidic acid.

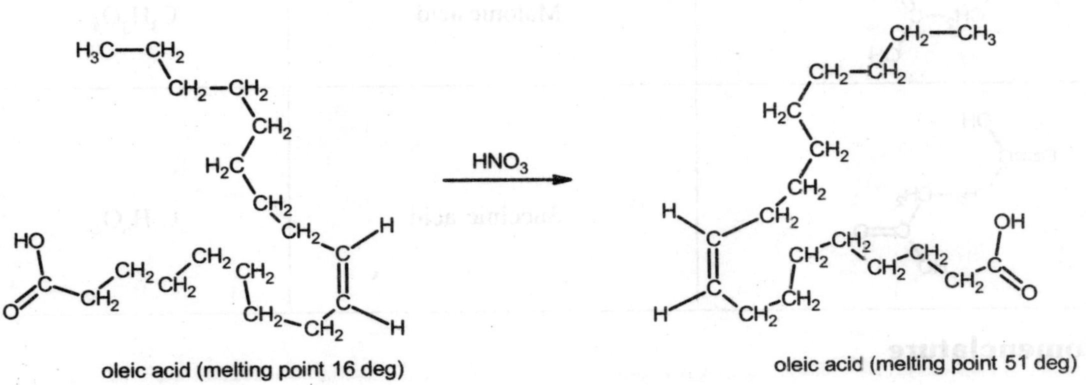

oleic acid (melting point 16 deg)

oleic acid (melting point 51 deg)

(vi) Addition with alkalies: Oleic acid reacts with NaOH or KOH to form salts known as soaps

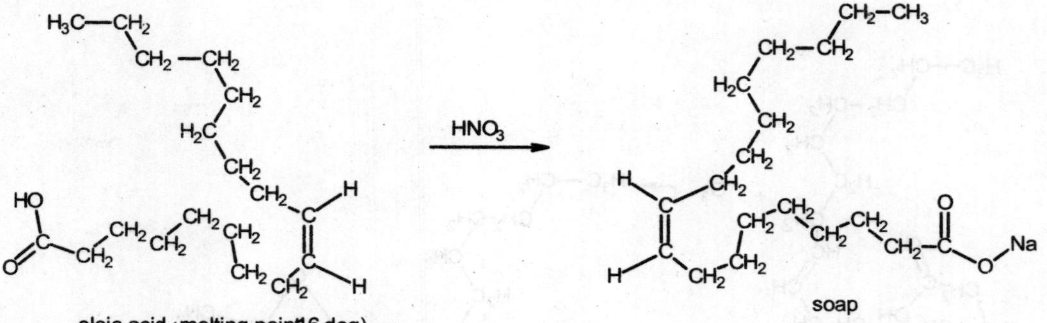

$\xrightarrow{HNO_3}$

oleic acid (melting point16 deg)

soap

DICARBOXYLIC ACID

As the name suggests, the dicarboxylic acids contain two ends of the saturated carbon chain. The general formula for such saturated dicarboxylic acid is

$$(CH_2)_n\ (COOH)_2 \text{ or } HOOC - (CH)_n - COOH$$

Where there are n carbon atoms bonded strongly each to each in a chain n = 0 for the first member oxalic acid

Structure	Name	Formula
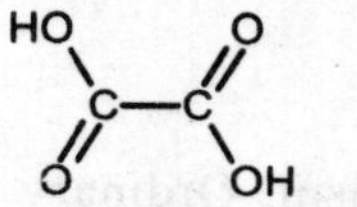 HO–C(=O)–C(=O)–OH	Oxalic acid	$C_2H_2O_4$
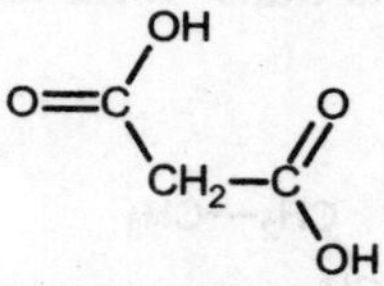 HO–C(=O)–CH$_2$–C(=O)–OH	Malonic acid	$C_3H_3O_4$
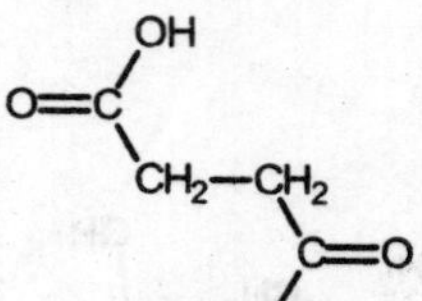 HO–C(=O)–CH$_2$–CH$_2$–C(=O)–OH	Succinic acid	$C_4H_6O_4$

Nomenclature

Most of the dicarboxylic acid is known by the common names which indicate the source from which they have been isolated. In the IUPAC system they are named as **Alkenedioic acids**. The suffix – dioic is appended to the name of the alkane containing the same number carbon atoms as the dicarboxylic acid. The position of the two carboxylic groups is indicated by Arabic numerical, if necessary while naming the substituted acids for example:

$$\overset{4}{HOOC}-\overset{3}{C}H_2-\overset{2}{C}H_2-\overset{1}{C}OOH$$

1-4-butanedionic acid
butane 1 4 dionic acid
butanedioic acid

$$HOO\overset{6}{C}-\overset{5}{C}H_2-\overset{4}{C}H_2-\overset{3}{C}H_2-\overset{2}{C}H_2-\overset{1}{C}OOH$$

1-6 hexane dioic acid
1- hexane 6 -dioic acid
hexane dioic acid

An alternative scheme is to regard the carboxyl group as substituent when the name of the acid is obtained by adding the suffix carboxylic acid.

$$HOOC-\overset{3}{C}H_2-\overset{2}{C}H_2-\overset{1}{C}OOH$$

1-3-propanedicarboxylic acid

butane 1 3 dicarboxylic acid

In these common system the names of the substituent are indicated by the Greek alphabets α and β and γ etc. The meaning of substituted acids according to the two systems is illustrated below:

$$HOO\overset{5}{C}-\underset{\alpha'}{\overset{4}{C}}H_2-\underset{\beta}{\overset{3}{C}}H_2-\underset{\alpha}{\overset{\overset{Br}{2}}{C}}H-\overset{1}{C}OOH$$

common name:- α- bromoglutaric acid
IUPAC name: - 2 - bromopentane -1 -5dioic acid

$$HOO\overset{5}{C}-\underset{\alpha'}{\overset{4}{C}}H_2-\underset{\underset{CH_3}{\beta}}{\overset{\overset{CH_3}{3}}{C}H}-\underset{\alpha}{\overset{2}{C}}H_2-\overset{1}{C}OOH$$

Common name: - β,β- dimethyl glutaric acid
IUPAC name: - 3,3 - dimethylpentane 1, 5 dioic acid

$$HOOC-\underset{\alpha}{\overset{\overset{Cl}{5}}{C}}H-\underset{\beta}{\overset{4}{C}}H_2-\underset{\beta}{\overset{3}{C}}H-\underset{\alpha}{\overset{\overset{Cl}{2}}{C}}H-\overset{1}{C}OOH$$

common name: - α,α' dichloroadiphic acid

Physical Properties

1. All the dicarboxylic acids are crystalline solids their boiling points being much higher than the corresponding saturated Monocarboxylic acid

2. The lower members are sufficiently soluble in water but are only soluble in organic solvents.

The solubility in water goes on decreasing with increase in molecular weight. This is attributed to the fact that hydrogen bonding with water molecules becomes less effective as the insoluble by hydrocarbon part of the molecule predominates with rising molecular weight this reducing the solubility. None of the dicarboxylic acid is steam volatile.

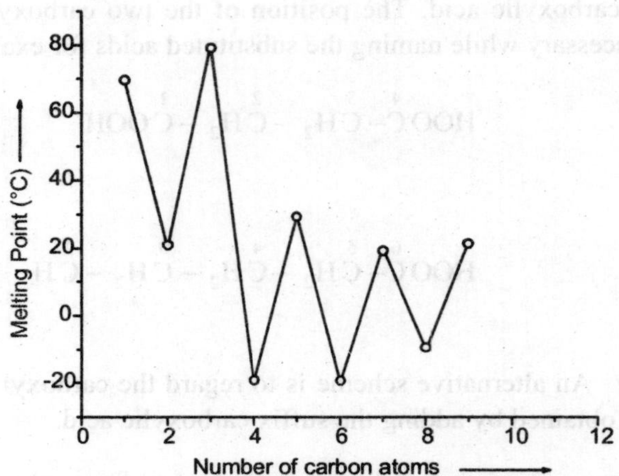

Fig. 11.7. Showing the alternation of melting point in the series of the dicarboxylic acids

3. The melting points of the dicarboxylic acids follow the **Alternation** or **oscillation rule** from one member to the other. The melting point of an 'even number of carbon atoms) is always higher than that of the 'odd acid' lying immediately below and above it in series: -

The phenomenon of alternation id probably due to the fact that the carbon chains of dicarboxylic acids are arranged in the zigzag fashion see fig. 11.8. This view has been confirmed by X – Ray examination of dicarboxylic acids. As a result of this zigzag orientation, the carboxyl groups of an acid 'odd acid' lie on the same side of the carbon chain while an 'even acid' lie on the opposite side of the chain.

This latter arrangement permits either packing of the molecules in the crystals lattice and evidently stronger intermolecular force exists between the chains. This makes the melting point of 'even acid' higher than there of 'odd acids'. It is further interesting to note from fig. 11.7 that a maximum melting point is found with three carbon dicarboxylic acids. This minimum may be consequence of two opposing types of intermolecular attraction stabilizing the crystal lattice. As the carbon chain increases its length, the molecules become more alkane like; the hydrogen bonding becomes less effective in holding the large molecules in position in the lattice, but the forces of attraction between the hydrocarbon chains (Van der Waal's forces) become more and more predominant as the chain lengthens. At the dicarboxylic acid with five carbon, there opposing forces counter and melting points is maximum.

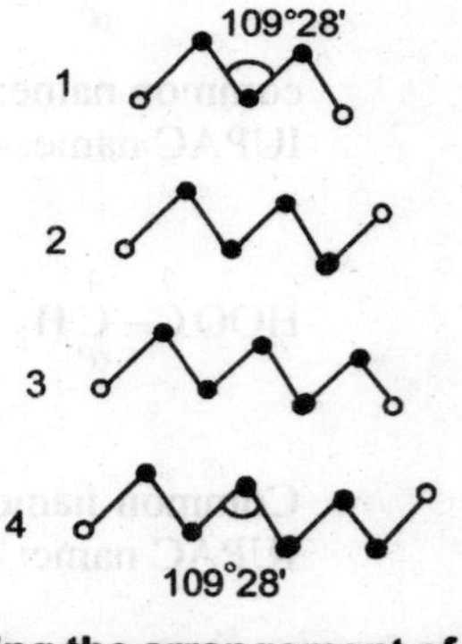

Fig. 11.8. Showing the arrangement of carbon chains odd and even acids 1-3 odd aid 2-4 even acids

Acidity: The dicarboxylic acid are much stronger then the Monocarboxylic acid. This is expected since a carbonyl group

> Dicarboxylic acids are stronger than Monocarboxylic acids

being electron –attracting its presence close to another carbonyl group facility the release of the first proton. The inductive effect falls off shapely as soon as the carbonyl group is separated by more than one saturated carbon atom (-CH_2 -). Thus oxalic acid (K_a = 54000 × 10^{-5}) is stronger than malonic acid in which the carbonyl group have little effect on each other. For example succinic acid (K_a = 6.4 × 10^{-5}). Hence the acid strength of dicarboxylic acids decreases with the increase in molecular weight.

Dicarboxylic acid ionize in two steps

$$HOOC - COOH + H_2O \rightleftharpoons HOOC - COO^- + H_3O^+ \quad (1)$$

$$^-OOC - COOH + H_2O \rightleftharpoons {}^-OOC - COO^+ + H_3O^+ \quad (2)$$

Since in the second ionization a proton has to be removed from a negatively charged species containing an electron donating substituent i.e. –COO^- the equilibrium lies to the left. As a result of this equilibrium, oxalic acid malonic acid succinic acid is weaker in the second ionization than the formic acid and propionic acid respectively.

Methods of Preparation

The methods of preparation of di-carboxylic acids are similar to those used for mono-carboxylic acid. Here we have to develop two COOH groups instead of one for the Monocarboxylic acids. Consequently di functional substrates would be the most suitable starting materials.

1. By the oxidation of glycols or hydroxy acids with potassium permanganate of potassium dichromate in the presence of sulphuric acid.

$$H_2C(OH)-CH_2(OH) + 4[O] \xrightarrow[H_2SO_4]{K_2Cr_2O_7} HOOC-COOH$$

ethylene glycol → oxalic acid

$$HOCH_2-CH_2-COOH + 2[O] \xrightarrow[H_2SO_4]{K_2Cr_2O_7} HOOC-CH_2-COOH + H_2O$$

3-hydroxypropanoic acid → malonic acid

2. By the hydrolysis of dinitrites or cyano-monocarboxylic acids having the cyano group at the other end of the chain, with boiling dilute hydrochloric acid

$$N{\equiv}C-CH_2-CH_2-C{\equiv}N + 4H_2O + 2HCl \xrightarrow{boil} HOOC-CH_2-CH_2-COOH$$

succinonitrile → succinic acid

This provides a method for obtaining dicarboxylic acids of alkanes as follows:

$$CH_2{=}CH_2 \xrightarrow{Br_2 \text{ in } CCl_4} BrH_2C{-}CH_2Br \xrightarrow{KCN} N{\equiv}C{-}CH_2{-}CH_2{-}C{\equiv}N \xrightarrow{H_2O/HCl} HO{-}CO{-}CH_2{-}CH_2{-}CO{-}OH$$

ethylene → 1,2-dibromoethane → succinonitrile → succinic acid (poison)

$$O{=}C(OH){-}CH_2{-}C{\equiv}N + H_2O + HCl \xrightarrow{reflux} O{=}C(OH){-}CH_2{-}C(OH){=}O + NH_4Cl$$

cyanoacetic acid → malonic acid

3. By the action of silver or zinc and halogenated monocarboxylic esters. A Wurtz type reaction takes place for example:

$$2Ag + I{-}CH_2CH_2{-}C(=O){-}CH_2{-}CH_3 \longrightarrow CH_3CH_2{-}C(=O){-}CH_2CH_2{-}CH_2CH_2{-}C(=O){-}CH_2CH_3 + 2AgI$$

1-iodopentan-3-one, ethyl B - iodopropionate → diethyl adipate

$$\text{diethyl adipate} + 2H_2O \xrightarrow[hydrolysis]{H^+} HOOC{-}CH_2{-}CH_2{-}CH_2{-}CH_2{-}COOH + 2C_2H_5OH$$

hexanedioic acid, adiphic acid

4. By Electrolysis of salts of acids esters of lower dicarboxylic acids (Crum brown and Walter electrolysis method)

$$2KOOC(CH_2)_nCOOC_2H_5 \rightarrow 2K^+ + 2OOC(CH_2)_nCOOC_2H_5$$

At cathode: $2K^+ \; 2e \rightarrow 2K; \; 2K + 2H_2O \rightarrow 2KOH + H_2$

At anode:

$$2OOC\,(CH_2)_n\,COOC_2H_5 \xrightarrow{-2e} 2OOC(CH_2)_n\,COOC_2H_5\,COOC_2H_5 \rightarrow \begin{array}{c}(CH_2)_nCOOC_2H_5\\ |\\ (CH_2)_nCOOC_2H_5\end{array} + 2CO_2$$

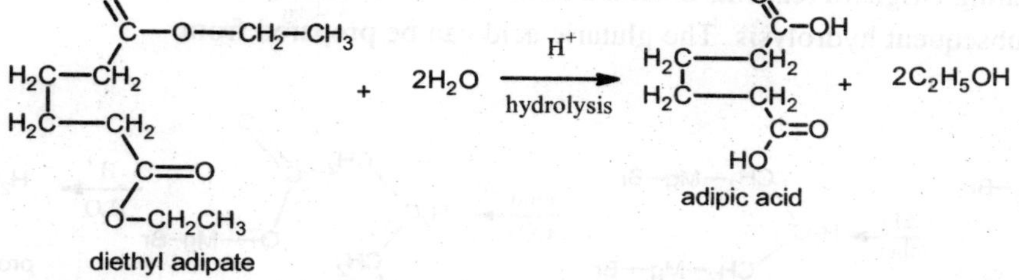

diethyl adipate + $2H_2O$ $\xrightarrow[\text{hydrolysis}]{H^+}$ adipic acid + $2C_2H_5OH$

(4) By electrolysis of salts of acid esters of lower dicarboxylic acids (Crum- Brown and Walter electrolysis method)

$$2KOOC(CH_2)_nCOOC_2H_5 \rightarrow 2K^+ + 2OOC(CH_2)_nCOOC_2H_5$$

At cathode: -

$$2k^+ + 2e \rightarrow 2k;\ 2k + 2H_2O \rightarrow 2KOH + H_2$$

At anode

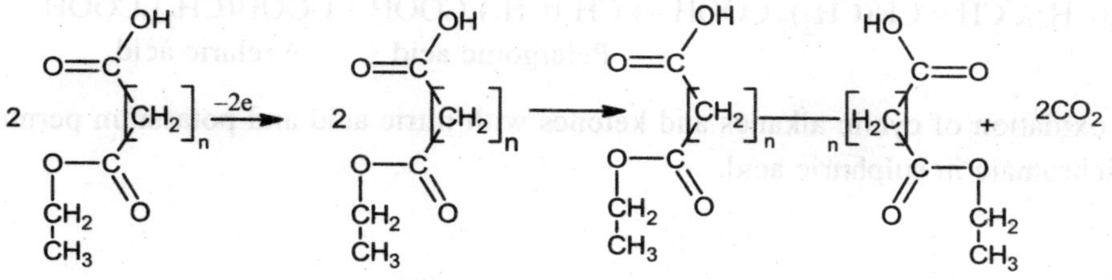

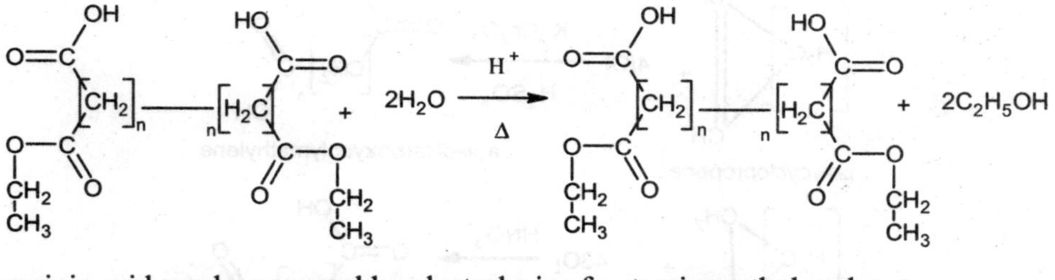

$$2\,HOOC(CH_2)_nCOOCH_2CH_3 \xrightarrow{-2e} 2\,HOOC(CH_2)_nCOOCH_2CH_3 \longrightarrow HOOC(CH_2)_n(H_2C)_nCOOCH_2CH_3\ (\text{diester}) + 2CO_2$$

$$\text{diester} + 2H_2O \xrightarrow[\Delta]{H^+} \text{diacid} + 2C_2H_5OH$$

This succinic acid can be prepared by electrolysis of potassium ethyl malonate

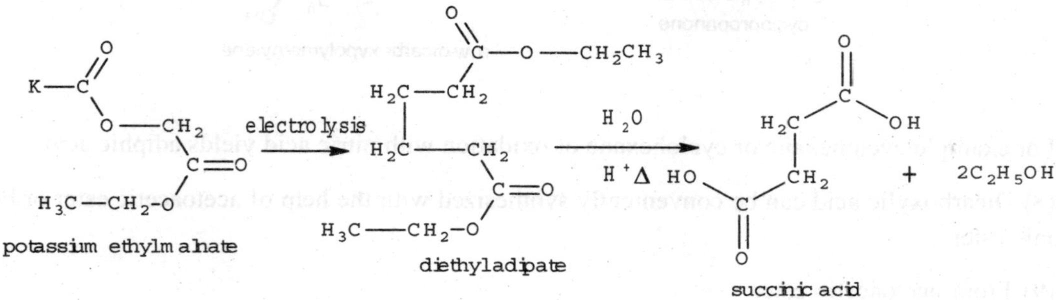

potassium ethylmalnate $\xrightarrow{\text{electrolysis}}$ diethyladipate $\xrightarrow[H^+\Delta]{H_2O}$ succinic acid + $2C_2H_5OH$

(8) By treating Grignard reagents obtained from the corresponding dihaloalkanes, with solid carbon dioxide and subsequent hydrolysis. The glutaric acid can be prepared from

$$\text{H}_2\text{C}(\text{CH}_2\text{-Br})_2 \xrightarrow[\text{ether}]{\text{Mg}} \text{H}_2\text{C}(\text{CH}_2\text{-Mg-Br})_2 \xrightarrow{\text{solid } CO_2} \text{H}_2\text{C}(\text{CH}_2\text{-C(=O)-O-Mg-Br})_2 \xrightarrow[H_2O]{H^+} \text{H}_2\text{C}(\text{CH}_2\text{-OH})_2$$

1,3-dibromopropane; propane-1,3-diol; Glutaric Acid

(6) By oxidation of unsaturated fatty acids: - For example oleic acid when treated with Nitric acid or potassium permanganate solution or ozone gives pelargonic acid and azelaric acid.

$$CH_3(CH_2)_7\,CH = CH(CH_2)_7\,COOH \xrightarrow{\text{Pot permanganate}} \underset{\text{Pelargonic acid}}{CH_3(CH_2)_7COOH} + \underset{\text{Azelaric acid}}{COOH(CH_2)_7COOH}$$

(7) By the oxidation of cyclic alkanes and ketones with nitric acid and potassium permanganate or potassium dichromate in sulphuric acid.

$$\underset{\text{polycyclopropene}}{[\text{H}_2\text{C}(\text{CH=CH})]_n} + 4[O] \xrightarrow[H_2SO_4]{K_2Cr_2O_7} \underset{\text{a,w-dicarboxypolymethylene}}{\text{HO-C(=O)-[CH}_2]_n\text{-C(=O)-OH}}$$

$$\underset{\text{cyclopropanone}}{[\text{H}_2\text{C}(\text{CH}_2)\text{C=O}]_n} + 43O] \xrightarrow{HNO_3} \underset{\text{a,w-dicarboxypolymethylene}}{\text{HO-C(=O)-[CH}_2]_n\text{-C(=O)-OH}}$$

For example cyclohexane or cyclohexane or oxidation with nitric acid yields adiphic acid

(8) Dicarboxylic acid can be conveniently synthesized with the help of acetoacetic ester or better malonic ester.

(9) From acetoacetic ester:-

acetoacetic ester $\xrightarrow{C_2H_5ONa}$ $CH_3COCHNaCOOC_2H_5$ (sodium acetoacetic ethylate) $\xrightarrow[(NaCl)]{ClCH_2COOC_2H_5}$ $CH_3COCHCOOC_2H_5$ $\xrightarrow{2H_2O}$ succinic acid + acetic acid

(10) From malonic ester:

2 ethyl methyl malonate $\xrightarrow{C_2H_5Na}$ bis(ethoxycarbonyl)(methylene)sodate(3-) sodium diethyl malonate

Sodium diethyl malonate is then reached with 1, 2 dibromoethane, the resulting ester is hydrolyzed and the free tetracetic acid heated to yield the desired dicarboxylic acid.

1,2-dibromoethane + sodium diethyl malonate $\rightarrow$ diethyl butylmalonate $\xrightarrow{2H_2O}$ butane-1,1,4,4-tetracarboxylic acid $\rightarrow$ adipic acid

Chemical Properties

When the two COOH groups are separated by a chain of more than five carbon atoms, they react more or less independently. Therefore the dicarboxylic acid $(CH_2)_n(COOH)_2$ in which n>5 gives normal reactions of monocarboxylic acids twice over. However, in the lower members where the COOH groups are closed together the possibilities of natural interactions increase. Thus many of their apart from the COOH groups i.e. on the value of n

(1) Reactions of the COOH groups: One or both the carbonyl group may react with appropriate reagents to form mono or di-derivative. Thus oxalic acid gives the following salts esters and acid chloride.

oxalic acid + NaOH → sodium hydrogen oxalate → disodium oxalate

oxalic acid + $2C_2H_5OH/H^+$ → ethoxy(oxo)acetic acid → diethyl oxalate

oxalic acid + NH_4OH → Ammonium hydrogen oxalate + amino(oxo)acetic acid

oxalic acid + $SOCl_2$ → ethanedioyl dichloride $\xrightarrow{NH_4OH}$ oxalamide

The two COOH groups can also be made to react successively with different reagents to yield mixed derivatives. For example:

oxalic acid $\xrightarrow[H^+]{C_2H_5OH}$ ethoxy(oxo)acetic acid $\xrightarrow{SOCl_2}$ ethyl chloro(oxo)acetate

oxalic acid $\xrightarrow{NH_4OH}$ ammonium hydrogen oxalate $\xrightarrow[-H_2O]{\Delta}$ amino(oxo)acetic acid $\xrightarrow{SOCl_2}$ amino(oxo)acetyl chloride

The mixed derivatives are useful in organic synthesis.

(2) Action of Heat on Dicarboxylic Acids: The nature of product formed when a dicarboxylic acid is heated depends on the member of carbon atoms intervening the two carboxyl groups. Thus, these acids can be divided into three groups on the basis of the effect of heat on then

(i) Those with one intervening carbon: The dicarboxylic acids with the two COOH groups are linked directly to each other or to the same carbon atom, when heated spit out a molecule of carbon dioxide from one of the COOH groups to form monocarboxylic acids. Thus oxalic and acetic acid respectively.

$$HOOC{-}COOH \longrightarrow HCOOH + CO_2$$

oxalic acid → formic acid

$$HOOC{-}CH_2{-}COOH \longrightarrow H_3C{-}COOH + CO_2$$

acetic acid

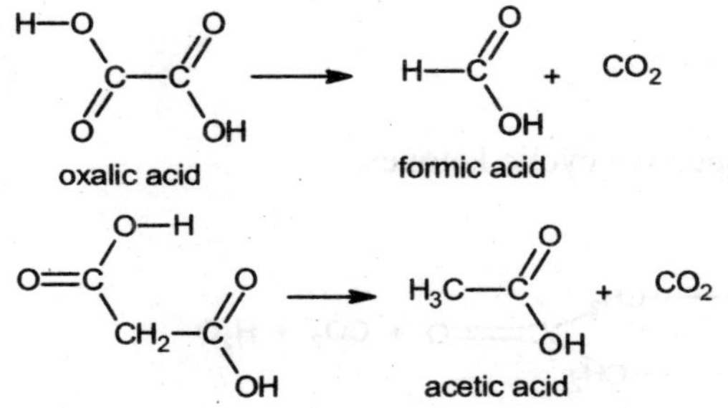

Oxalic acid and malonic acid decomposes to give formic acid and acetic acid respectively

(ii) Those with two or three intervening carbons: The dicarboxylic acid in which the two COOH groups are separated by two or three carbon atoms eliminate a molecule of water to form cyclic anhydrides. Thus succinic acid and glutaric acid when heated alone or preferably with acetic anhydride yield succinic anhydride respectively.

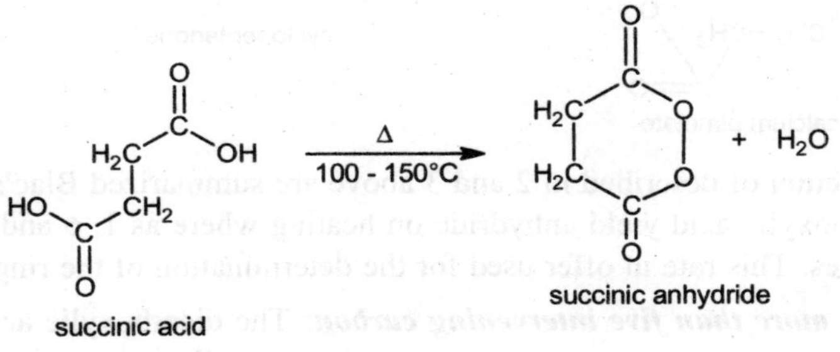

$$HOOC{-}CH_2{-}CH_2{-}COOH \xrightarrow[100-150°C]{\Delta} \text{succinic anhydride} + H_2O$$

succinic acid → succinic anhydride

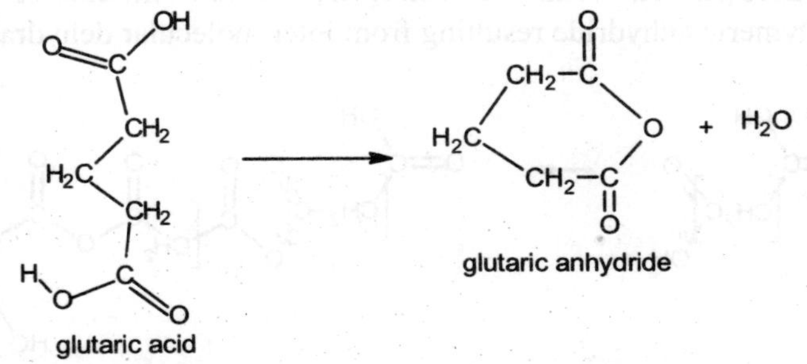

$$HOOC{-}CH_2{-}CH_2{-}CH_2{-}COOH \longrightarrow \text{glutaric anhydride} + H_2O$$

glutaric acid → glutaric anhydride

Such reactions occur where five or six memberd ring anhydrides can be formed.

(iii) Those with four or five intervening carbon: **The dicarboxylic acid in which the two COOH groups are separated by four or five carbon atoms upon heating alone or distilling with acetic acid pimetic acid. When heated at 300°C yields cyclopentane and cyclohexane respectively.**

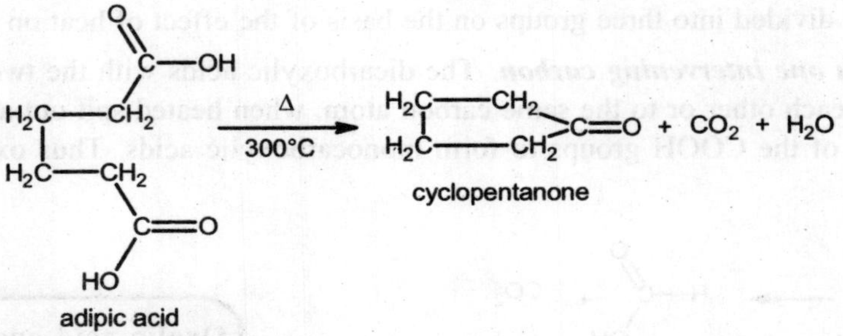

$$\text{adipic acid} \xrightarrow[300°C]{\Delta} \text{cyclopentanone} + CO_2 + H_2O$$

adipic acid

cyclopentanone

The calcium salt of the above acids give the respective cyclic ketones.

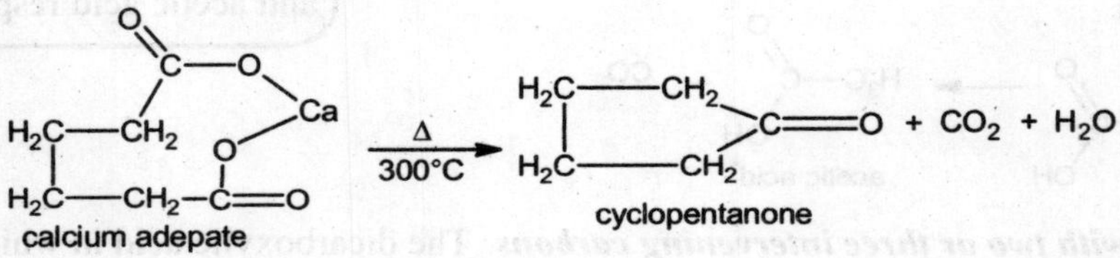

$$\text{calcium adepate} \xrightarrow[300°C]{\Delta} \text{cyclopentanone} + CO_2 + H_2O$$

calcium adepate

cyclopentanone

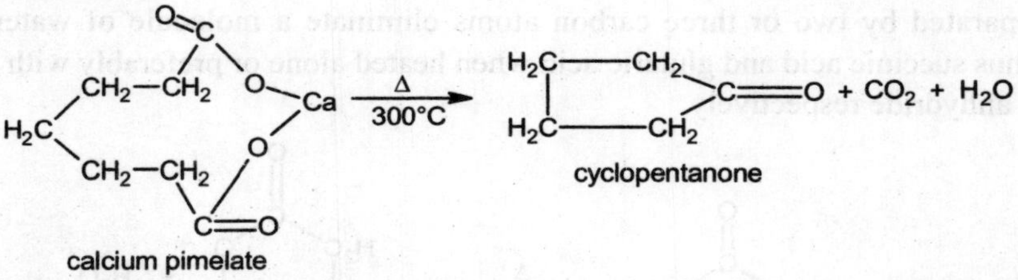

$$\text{calcium pimelate} \xrightarrow[300°C]{\Delta} \text{cyclopentanone} + CO_2 + H_2O$$

calcium pimelate

cyclopentanone

The results of action of described in 2 and 3 above are summarized Blac's rule which states that 1, 4, and 1, 5 dicarboxylic acid yield anhydride on heating where as 1, 6 and 1, 7 dicarboxylic acid offered cyclic ketones. This rate in offer used for the determination of the ring size.

(iv) Those with more than five intervening carbon: **The dicarboxylic acids having two COOH separated by six or more carbon atoms on heating from small amounts of ketones but the major product in a linear polymeric anhydride resulting from inter molecular dehydration.**

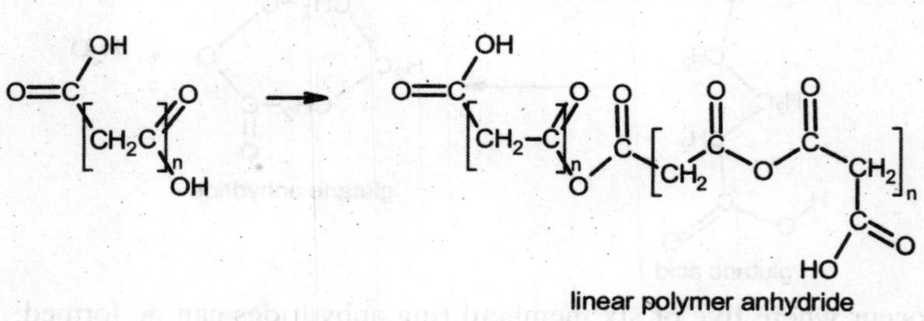

linear polymer anhydride

(3) Oxidation: The dicarboxylic are in general stable to oxidizing agents. However oxalic acid is oxidised with acidified potassium permanganate solution to give carbon dioxide and water.

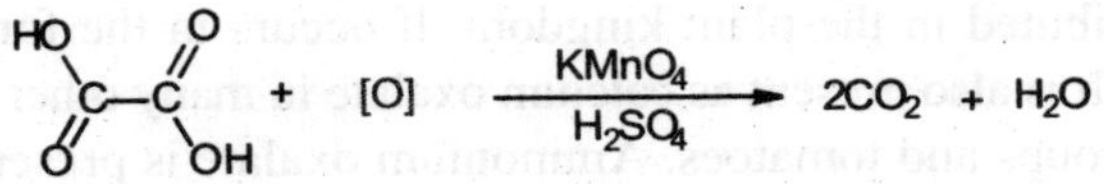

$$HOOC{-}COOH + [O] \xrightarrow[H_2SO_4]{KMnO_4} 2CO_2 + H_2O$$

Exp: - Make two individual solutions (A) oxalic acid + manganese dioxide and (B) sulphuric acid + potassium permanganate. Mix them together and allow to stand and see the colour change from violet to colourless, the oxidation of oxalic and reduction of potassium permanganate (it is also known as chemical clock reactions.

(4) Acyloin Reaction: Higher dicarboxylic acid form acyloins with sodium and xylene followed by acid hydrolysis.

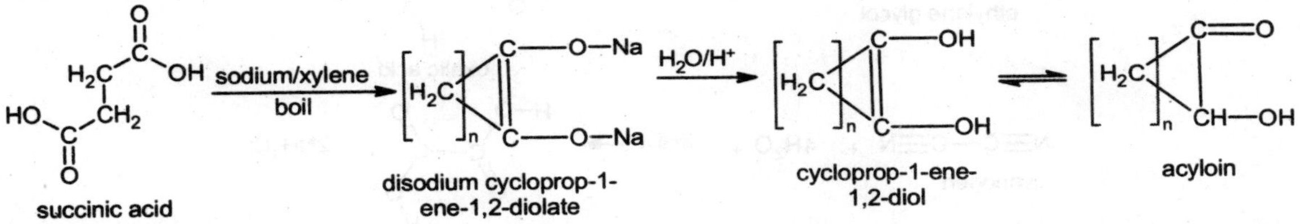

succinic acid $\xrightarrow[\text{boil}]{\text{sodium/xylene}}$ disodium cycloprop-1-ene-1,2-diolate $\xrightarrow{H_2O/H^+}$ cycloprop-1-ene-1,2-diol ⇌ acyloin

(6) Halogenation: When a hot higher acid is treated with Cl_2 or Br_2 the corresponding halogenated acids are obtained the equations:

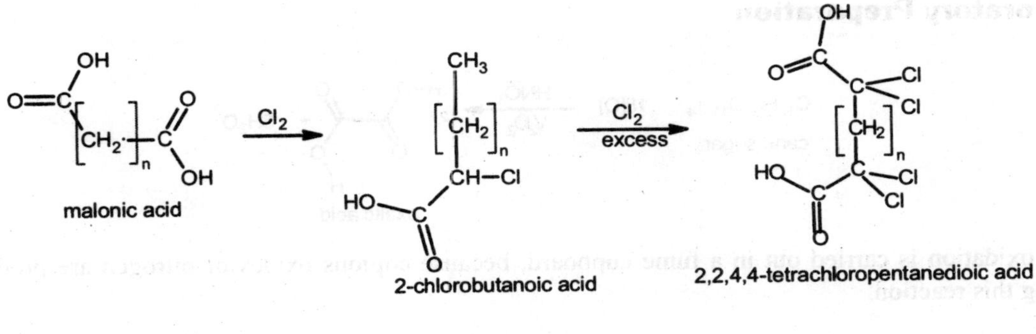

malonic acid $\xrightarrow{Cl_2}$ 2-chlorobutanoic acid $\xrightarrow[\text{excess}]{Cl_2}$ 2,2,4,4-tetrachloropentanedioic acid

OXALIC ACID

Formula: $C_2H_2O_4$

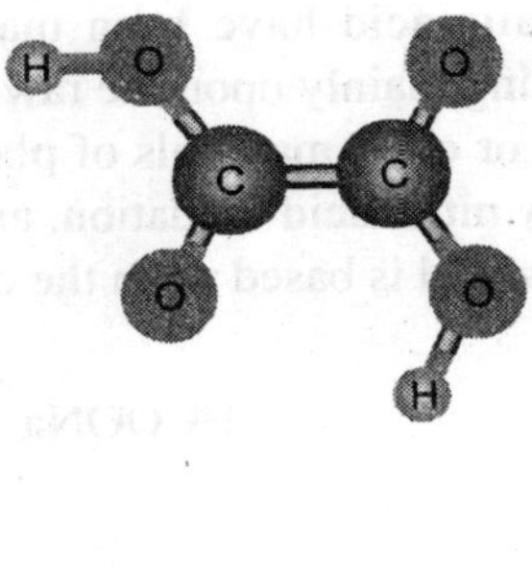

Introduction and History

The Greek word Oxn meaning acid, from which the oxalic has been derived. The acid us long been known and is widely distributed in the plant kingdom. It occurs in the form of potassium hydrogen oxalate (oxalis acetocella). It is also present as calcium oxalate in many other plants as spinach, rhubarb, sweet potatoes, cabbage groups and tomatoes. Ammonium oxalate is present in Urine.

Preparation

Oxalic acid can be obtained by the oxidation of a glycol and also by the hydrolysis of cyanogen

$$\underset{\text{ethylene glycol}}{\begin{matrix} H_2C\text{—}OH \\ H_2C\text{—}OH \end{matrix}} + 4[O] \xrightarrow[H_2SO_4]{K_2Cr_2O_7} \underset{\text{oxalic acid}}{HO\text{—}CO\text{—}CO\text{—}OH} + 2H_2O$$

$$\underset{\text{cyanogen}}{N{\equiv}C\text{—}C{\equiv}N} + 4H_2O + 2HCl \longrightarrow \underset{\text{oxalic acid}}{HO\text{—}CO\text{—}CO\text{—}OH} + 2NH_4Cl$$

Laboratory Preparation

$$\underset{\text{cane sugar}}{C_{12}H_{22}O_{11}} + 18[O] \xrightarrow[V_2O_5]{HNO_3} 2\ \underset{\text{oxalic acid}}{HO\text{—}CO\text{—}CO\text{—}OH} + 5H_2O$$

This oxidation is carried out in a fume cupboard, because copious oxides of nitrogen are produced during this reaction.

Manufacture of Oxalic Acid

Oxalates and oxalic acid have been manufactured by four general methods, of which each has variations depending mainly upon the raw materials selected. Three processes involve the degradation of carbohydrates, or other materials of plant origin, in different ways: (1) by fusion with alkali metal hydroxides, (2) by nitric acid oxidation, and (3) by fermentation with certain molds. A fourth process involves synthesis, and is based upon the conversion of sodium formate, by heating, to sodium oxalate and hydrogen:

$$2HCOONa \rightarrow (COONa)_2 + H_2$$

Alkali Fusion of Cellulose. This method was discovered by Gay-Lussac; the process, now obsolete, was first used commercially near the middle of the nineteenth century. Cellulosic materials, usually waste products such as sawdust, grain hulls, and the like, were "fused," by heating to 240-285°C with concentrated aqueous sodium or potassium hydroxide, or a mixture of the two. A substantial part, perhaps 50%, of the salts so formed was the oxalate, and the remainder mainly the carbonate. Leaching the reaction product with hot water and cooling the concentrated solution gave sodium oxalate crystals, effecting a separation of the oxalate and the carbonate. Sodium hydroxide was regenerated by causticizing separately, the sodium carbonate and the sodium oxalate by reaction with lime the soda was thus used cyclically. Causticization of alkali carbonates by lime is a familiar process the process is equally adaptable to the alkali oxalates:

$$Na_2C_2O_4 \quad + \quad Ca(OH)_2 \quad + \quad CaC_2O_4 \quad + \quad 2NaOH$$

From calcium oxalate, oxalic acid was prepared by reaction with sulphuric acid, essentially the same procedure as that currently used in some of the modern processes to be described:

$$CaC_2O_4 \quad + \quad H_2SO_4 \rightarrow H_2C_2O_4 \quad + \quad CaSO_4$$

Othmer and co-workers (9) give the following optimum conditions for the alkali fusion of sawdust in a batch operation: ratio of caustic (NaOH) to sawdust, 3: 1 ; concentration of caustic, 50%; final fusion temperature, 200°C.; depth of fusion mass, 0.25-0.5 in. Under these optimum conditions the following results were obtained:

Raw material, 1b.	
Sawdust (dry)	100
NaOH (not recovered)	9
CaO (lime)	34.7
H_2SO_4 (100%)	61.1

Product formed, 1b	
Oxalic acid	45.5
Acetic acid	11.7
Formic acid	2.48
Methanol	5.5
$CaSO_4$ (waste)	85
Wood oils	3

Their investigations of a continuous-fusion process indicated that the amount of caustic could be decreased from that found necessary in the batch operation, and that higher yields of oxalic and acetic acids could be obtained.

The alkali fusion process for oxalates cannot compete with modern methods of manufacture, particularly the synthetic (via formate), to which it bears some degree of resemblance, because of the relatively low yields per cycle and the inevitable loss of alkali; although alkali losses occur in both processes, they are undoubtedly greater in the fusion process.

Oxidation, by Nitric Acid, of Sugars. This, the oldest method for artificially preparing oxalic acid, was discovered by Scheele in 1776 when he oxidized glucose with nitric acid. The potentialities of the method as a means of manufacturing this product have long been recognized, but it was not successfully developed as a commercial process until the second quarter of the twentieth century.

Technical advances in the manufacture of nitric acid, particularly in the recovery of nitrogen oxides in a form suitable for use cyclically, were mainly responsible for the attainment of an economic situation favourable to the successful development of this process.

When sugars, such as glucose, or related carbohydrates are oxidized by nitric acid under favourable conditions oxalic acid is produced in high yields, and it is then practically the only organic acid among the end products of the oxidation. It is worth noting, however, that under different conditions organic acids other than oxalic acid may comprise substantial or major proportions of the products. Conditions favourable to the production of oxalic acid in high yields include conducting the oxidation in an aqueous sulphuric acid solution, the addition of one or more catalytic agents (compounds of vanadium, especially V_2O_5, molybdenum, iron, and manganese are among those claimed), careful temperature control, and maintenance of an excess of the carbohydrate during most of the oxidation. Large quantities of nitrogen oxides are evolved and, for efficient operation, must be reclaimed for re-use. Oxalic acid is itself oxidized and destroyed by the mixture which produces it; this loss is minimized by keeping the temperature within a narrow optimum range, by avoiding substantial excesses of nitric acid, and by maintaining an optimum sulphuric acid concentration.

Many products of plant origin have been suggested as raw materials, such as sugars, starch, dextrin, molasses, cellulose wastes from the lumber industries, agricultural products such as corn-cobs and other grain wastes, peat, lignin and the like. There is little doubt that the process is adaptable to the utilization of such materials, although the pretreatment of some might be an important cost item. In any case, choice of the raw material is influenced by somewhat complex economic factors. If the raw material chosen is not essentially a monosaccharide solution it must, as a practical matter, be degraded by hydrolysis to this stage before use.

Simpson (18) and Brooks (20) describe in detail the oxidation process for oxalic acid starting with 60% glucose solution. The process is adaptable to either batch or continuous operation. Mother liquor from previous operations, containing about 50% sulphuric acid, approximately 5% oxalic acid, and soluble catalytic substances, is used to dilute the glucose solution; for convenience cold mother liquor may also be used to dilute the nitric acid. In batch operation the glucose solution so prepared is heated to the optimum temperature range of 150 to 160°F. (65.6-71.1°C.); if vanadium pentoxide and ferric salts are the catalysts used, the concentrations are adjusted to fall within the ranges of 0.001 to 0.005% vanadium pentoxide, and 0.39 to 0.8% iron. The sulphuric acid concentration is so adjusted that the mother liquor obtained after completion of the operation will contain about 50%. The nitric acid (60-90%), diluted with mother liquor or undiluted, is added slowly with stirring to the prepared, heated glucose mixture while maintaining the temperature within the specified range. Large quantities of nitrogen oxides, mixed with other gases and water vapor, are evolved. The nitrogen oxides are eventually recovered, principally as nitric acid, in a form suitable for use in subsequent batches. After completion of the oxidation, which requires several hours, the solution is cooled, whereupon the oxalic acid crystallizes and is separated by conventional methods.

The continuous process is essentially similar to the batch process. The two solutions are metered concurrently into a quantity of reaction product, maintained within optimum temperature limits, at such a rate that oxidation is always practically complete, and the reaction product is withdrawn continuously. Under favourable conditions there is produced from 1 ton of glucose about 1 tons of oxalic acid, indicating a yield of more than 2 molecules oxalic acid per molecule of glucose (theoretical yield would be 3 molecules). Flow Diagram is shown in the fig.

Flow Diagram 1. Oxalic Acid by Nitric Acid Oxidation of Glucose

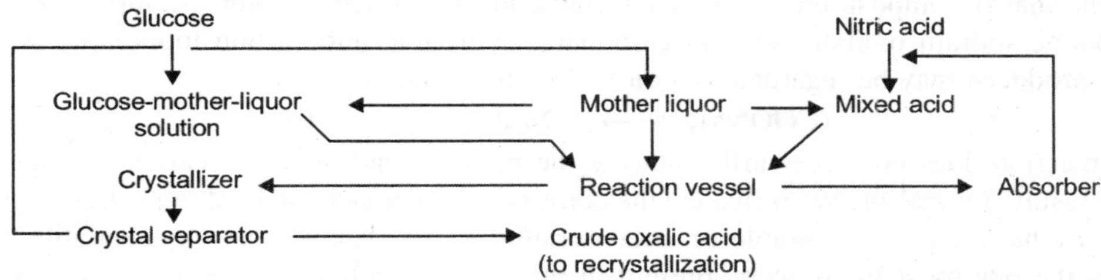

Showing the Flow diagram of Oxalic Acid Manufacture

Fermentation Process. The occurrence of oxalates in the foliage of some plants has been noted; rather recently it has been discovered that certain lower forms of plants which feed upon carbohydrates can produce the acid. The commercial exploitation of this phenomenon is an outgrowth of an earlier process used in the manufacture of citric acid by the mold fermentation of sugars (see *Citric acid,* mycological production). Among the products from this fermentation, oxalic acid may be absent, or it may be produced concurrently with citric acid, depending upon certain controlling factors, one of which is the form in which nutrient nitrogen is supplied. The ratio of oxalic acid to citric acid may be changed within certain limits by controlling these factors. The fermentation process is probably not economically sound for the production of oxalic acid primarily; this is always an incidental by-product. To the extent that it is controllable, the ratio of oxalic acid to citric acid is governed by economic considerations; its production, at the expense of lower yields of the higher priced citric acid would not, in ordinary commercial practice, be encouraged. Despite its secondary position, oxalic acid from this source is a significant, although minor, proportion of the total production.

Synthetic Process, from Formates. The quantities of oxalic acid manufactured synthetically from formates greatly exceed those from other sources. Sodium formate is available commercially from several sources. It is a by-product in the manufacture of pentaerythritol and other synthetic organic compounds, and it is produced directly from sodium hydroxide and carbon monoxide. However, although sodium formate from such sources may be used in oxalate manufacture, the salt is usually prepared specifically for the process, and the soda is cycled. In this process, the primary raw materials are carbon monoxide, lime, and sulphuric acid, and the end products are oxalic acid, hydrogen, and calcium sulphate. Soda is not used up, but there are minor losses during the various operations, and the deficit must be made up by periodically adding alkali or formate from an outside source at the appropriate stage. The cyclic process is based upon four reactions:

Preparation of Sodium Formate by reaction Because of the relatively slow reaction between carbon monoxide and alkali hydroxide there is an economic advantage in using a gas rich in the desired component. Oxalic acid manufacture is therefore often associated with a process which produces high-grade by-product carbon monoxide, such as that obtained in the electro thermal production of phosphorus.

Conversion of Formate to Oxalate by reaction (2). This is the most important of the reactions not only because it is the fundamental one, but because it requires the most critical control; the efficiency of the entire process depends largely upon the success of this operation.

The thermal decomposition of sodium formate always proceeds with the formation of at least four products, sodium oxalate, sodium carbonate, hydrogen, and carbon monoxide. Any sodium carbonate produced may be regarded as a direct loss of sodium oxalate:

$$(COONa)_2 \rightarrow Na_2CO_2 + CO$$

This reaction does not necessarily indicate the mode of carbonate formation but represents the economic result. Several factors influence the conversion efficiency attained impurities in the sodium formate may have either favourable or unfavourable effects. Among the objectionable impurities, calcium is the one most likely to be present in practice; contamination, even in trace amounts, by this element may significantly increase the carbonate-oxalate ratio. Adverse effects due to other common impurities are less serious. Free sodium hydroxide, on the other hand, improves conversion to oxalate, while some other impurities commonly present in technical sodium formate seem to be completely inert.

Aside from the adverse or favourable effects of impurities, the rate of heating the formate, and the maximum temperature attained in the product are important factors in the yield of oxalate. When pure sodium formate is gradually heated, it melts without decomposition at about 253°C.; technical grades melt at somewhat lower temperatures because of impurities, including alkali and water. If the temperature is gradually increased to approximately 300°C., slow decomposition begins but, if this temperature is maintained, the products are carbonate, hydrogen, and carbon monoxide, with but minor amounts of oxalate. If, however, the temperature is increased promptly to about 380°C. or higher, there ensues a vigorous exothermic reaction of a different character, whereby oxalate and hydrogen are produced in high yields. At still higher temperatures, above about 440°C., sodium oxalate begins to decompose. From this discussion it is evident that, for maximum yields, the sodium formate must be heated rapidly through the unfavorable temperature range to the optimum conversion temperature, yet the heat supplied must not be excessive; otherwise the mass may become destructively overheated during the rapid, uncontrollable, exothermic conversion.

The desired objectives may clearly be accomplished most effectively by introducing as much heat as possible prior to the beginning of any reaction. The sodium formate is therefore melted slowly and carefully to obviate supplying heat of fusion in the subsequent rapid heating, while any approach to the critical temperature in any part of the charge is avoided since otherwise a premature and dangerous conversion might begin. The liquid is then heated quickly by intimate contact with a heat reservouir (for example, a mass of hot metal) adequate to raise the temperature above the critical value. Provision for the safe and rapid disposal of hydrogen is essential, for the conversion is, indeed, energetic and uncontrollable; the apparatus should be designed, preferably, to conserve and utilize the heat of conversion. In suitable equipment, each increment of sodium formate is converted within a very few minutes from the molten salt to a voluminous, usually light-brown, sodium oxalate powder. In batch operation the time cycle may be as short as a quarter of an hour. The process is, obviously, adaptable to either batch or continuous operation.

Sodium oxalate has a relatively low solubility in water as compared to sodium carbonate. By simple extraction of the crude product with water good separation from soluble impurities is effected, and nearly pure sodium oxalate is produced.

Causticization of Sodium Oxalate by reaction (3). The partially purified sodium oxalate is added to lime slurry, whereupon very insoluble calcium oxalate and a solution of sodium hydroxide are formed. This operation is not necessarily conducted separately but may be combined with the first. Lime slurry and sodium oxalate may comprise the charge in the carbon monoxide absorbers, and the effluent from these will then be a suspension of calcium oxalate in sodium formate solution. Filtration effects a complete separation of these salts. The calcium oxalate is obtained as a "mud," and is, ordinarily, used in the next step without drying. The sodium formate solution is carefully purified and evaporated.

Decomposition of Calcium Oxalate with Sulphuric Acid by reaction (4). Sulphuric acid is diluted with mother liquor and wash water from previous operations, heated, and mixed with calcium oxalate, usually in lead-lined equipment. The reaction proceeds readily, producing a solution of oxalic acid and, of course, insoluble calcium sulphate. The hot solution is separated by filtration and cooled, whereupon oxalic acid crystallizes and is recovered by conventional methods.

Reversion of Calcium Sulphate. Oxalic acid is relatively strong, and calcium oxalate is a very insoluble salt; for these reasons the reaction of calcium oxalate with sulphuric acid is reversible:

$$(COO)_2Ca \quad + \quad H_2SO_4 \rightleftharpoons (COOH)_2 \quad + \quad CaSO_4$$

When, therefore, equivalent quantities of these reactants are mixed, the reaction does not proceed to completion. In practice, it is necessary to use a substantial excess of sulphuric acid, and to maintain an excess whenever oxalic acid and calcium sulphate might undergo reversion with detrimental results. The concentration of sulphuric acid in these solutions is not critical and may be varied within wide limits; if it is high the stable solid phase is anhydrous calcium sulphate, if it is in a certain intermediate range, $CaSO_4.\ 7H_2O$, and if it is still lower, $CaSO_4.\ 2H_2O$. The preferred concentration may depend upon economic and technical factors, but the lower concentrations are usually chosen. In a typical operating plant the mother liquor contains about 12% sulphuric acid.

Flow Diagram 2. Oxalic Acid from Carbon Monoxide, via Formate

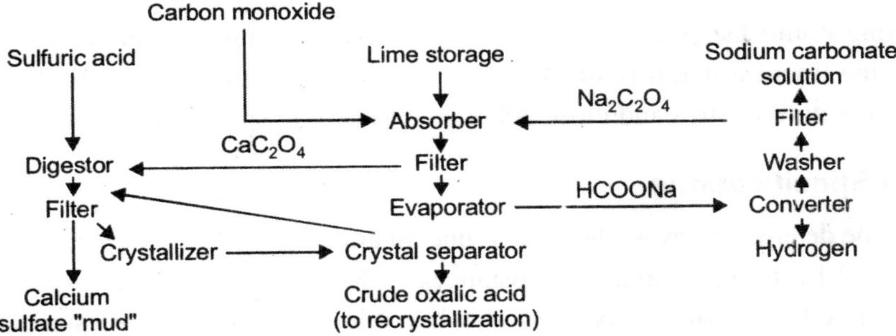

The crystalline oxalic acid, from liquor containing such a high concentration of sulphuric acid, is too impure to meet commercial specifications and must be recrystallized. In this step, the product is not only purified, but the crystal size is controlled to conform with current requirements.

The synthetic oxalic acid process is shown diagrammatically in Flow Diagram 2.

Proposed Modifications. Interesting variations in the formate process are possible. If, for instance, it is desirable to use by-product sodium formate as the raw material, the first step is eliminated and the soda is withdrawn as sodium hydroxide after reaction (3). In effect, sodium formate is converted to caustic soda and an equivalent quantity of oxalic acid. soda:

Similarly, by using sodium sulphate as the source of

$$Na_2SO_4 + Ca(OH)_2 + 2\,CO \xrightarrow{\text{heat}} 2\,HCOONa + CaSO_4$$

$$2\,HCOONa \rightarrow (COON\,a). + H2$$

$$(COONa)_2 + Ca(OH)_2 \quad (COO)_2Ca + 2NaOH$$

sodium sulphate may be, indirectly, causticized by lime, the soda being withdrawn as sodium hydroxide and the lime as calcium sulphate.

A cycle based upon electrolytic caustic and chlorine has been proposed. From sodium chloride there is produced sodium hydroxide, chlorine, and hydrogen, by conventional methods (see *Alkali and chlorine industries)*. Sodium formate is made by reaction of the alkali with carbon monoxide and converted to oxalate. Chlorine and hydrogen from the electrolytic process are combined by burning (see *Hydrochloric acid),* and the sodium oxalate is decomposed by hot hydrochloric acid solution. The sodium chloride, which is relatively insoluble in the presence of excess hydrochloric acid, is removed by filtration and cycled. The filtrate is cooled, whereupon oxalic acid crystallizes and is recovered by conventional methods. In this process the only raw material is carbon monoxide, although a substantial amount of electric power is required.

Anhydrous Oxalic Acid.

Commercially important quantities of anhydrous oxalic acid are not often required. For laboratory use two methods have been recommended: (1) distillation of the water with the aid of a low-boiling, inert, organic liquid, such as carbon tetrachloride, and (2) heating in air to a temperature about 3°C. below the melting point. Details are given in reference (2). In the first method, water is separated from the entraining liquid which is returned to the flask. The second method is preferable, but requires rather exact temperature control and close adherence to certain specified conditions.

Analysis and Specifications

Oxalic acid may be determined by acidimetry (if uncontaminated by acidic impurities) or by oxidimetry (if uncontaminated by foreign reducing substances). The two methods are approximately equal in accuracy and simplicity. Titration of oxalic acid in hot, dilute sulfuric acid with standard permanganate is a universally used analytical procedure, and is usually preferred as an assay method for the product. Technical oxalic acid, as supplied by a typical manufacturer, conforms to the minimum standards listed in Table 1.

Table 11.1. Specifications for Oxalic Acid

	Coarse crystal	Fine crystal
$(COOH)_2 2H_2O$,%	99.4	99.8
Free moisture, %	0.50	0.10
Sulfates, %	Trace	Trace
Impuritiesa	Trace	Trace
Ash (calcined at 900°C), %	0.10	0.10

*Impurities usually present in trace amounts include Pb and Fe.

Uses

A substantial part of the oxalic acid of commerce is consumed in the processing of textiles. It has long been established as a laundry sour to neutralize detrimental alkalinity in the rinsing of washed fabrics, in dyeing and, especially, printing of cotton, and in the bleaching of plant fibers, particularly straw for hats; it is also used in bleaching and tanning of leather. ,Its many miscellaneous uses include removal of rust and ink stains, the formulation of metal polishes, inks, wood cleaners, and the like, refining of tall oil, and the purification of crude natural gums, It is also used as an intermediate or component in the manufacture of dyes, and numerous other chemical products, salts, and esters.

A comparatively new, large-scale use for oxalic acid is in acidic compositions for removing the rust and scale which often accumulate in cooling systems, especially those of water-cooled internal-combustion engines; this acid comprises all, or a major part, of the active ingredients in several compositions of this type. The peculiar behaviour of oxalic acid toward iron and its corrosion product account for its successful use in these radiator cleaners; it is a strong acid, strong enough to react with and dissolve iron rust, since ferric oxalate is very soluble; it is also strong enough to attack metallic iron, and thus tends to loosen or dislodge scale which might adhere to it, but this attack proceeds slowly and superficially because ferrous oxalate is insoluble in oxalic acid solution, and forms an adherent, protective layer upon the metal. If contact with the acidic solution is not unduly prolonged, the rust is largely dissolved and the scale loosened or partially dissolved, while the metal itself is not significantly attacked.

Derivatives

Amides

There are two ami des of oxalic acid, the monoamide, oxamic acid, $HOOCCONH_2$ and the diamide, oxamide $(CONH_2)_2$ In contrast to the amide of the simplest monocarboxylic acid, formamide, both of the oxalic acid amides are insoluble, high-melting, and relatively inert. Cyanogen, $(CN)_2$, a gas, is oxalonitrile. No important industrial uses for oxalic amides have been developed.

Salts

Oxalic acid forms not only the usual neutral and acid salts expected of a dicarboxylic acid, but also more complex salts. The simple oxalates of practically all inorganic and many organic bases are known. Several of the compounds in this class both simple and complex, have important uses although the total production in tons is small.

Sodium oxalate, $Na_2C_2O_4$, formula weight 134.01, is obtainable in such high purity, and is so stable that it is supplied for use as a primary standard, especially in oxidimetry (National Bureau of Standards). This salt is not very soluble in water; the saturated solution, at 0°C. contains 2.6% and at 100°C. 6.1% $Na_2C_2O_4$; within this range the solubility may be calculated from a simple formula, $S = 2.6 + 0.0331$. in which S = % $Na_2C_2O_4$, and t = °C.

Sodium acid oxalate, $NaHC_2O_4.H_2O$, formula weight 130.04, is often called sodium binoxalate. Its physical and chemical properties are intermediate between those of sodium oxalate and oxalic acid; its solubility in cold water is lower than that of the neutral salt but increases greatly at higher temperatures; the following formula, in which S = % NaHC2O., and t = cC., gives results corresponding closely to published values: $S = 1.1 + 0.0016t2$.

No sodium salt corresponding to potassium tetroxalate is known.

Potassium oxalate, $K_2C_2O_4\,H_2O$, formula weight 184.23, is preferred as a reagent in analytical chemistry and in miscellaneous uses principally because of its high solubility as compared with other simple neutral oxalates; its saturated solution, at 0C. contains about 20%, and at 20°C. about 25% $K_2C_2O_4$. The solubility at any temperature between freezing and boiling may be calculated, approximately only, by the formula $S = 0.386t - 6.75$, in which S = % $K_2C_2O_4$, t = °C. Potassium. oxalate was once extensively used in photography, in the formulation of ferrous oxalate developers, but the process is now obsolete.

Potassium acid oxalate, or binoxalate, $KHC_2O_4H_2O$, formula weight 146.15, is of historical interest. It is the salt of sorrel found in vegetation, and it was the first oxalate isolated. A study of the system potassium oxalate-oxalic acid-water indicates clearly that the acid oxalate is not stable in contact with a saturated solution of potassium acid oxalate at temperatures below 50cC. Potassium acid oxalate is, therefore, decomposed to some extent if brought into contact with cold water. In the preparation of the pure salt, it is necessary to maintain a very substantial excess of neutral potassium oxalate in the mother liquor or, alternatively, to operate within limits well above ordinary room temperatures. These facts explain, no doubt, the observation that commercial potassium binoxalate is, occasionally, not pure, but a mixture of the binoxalate and the tetroxalate. The salt is stable in contact with its saturated solution at 50°C., at which the solution contains 14% KHC_2O_4 At any temperature between 50 and 100°C., the approximate solubility may be calculated by the formula $S = 0.386t - 6.75$, in which S = % KHC_2O_4, and t = °C.

Potassium tetroxalate, $KHC_2O_4(COOH)_2.2H_2O$, formula weight 254.20, is strongly acidic; it is the least soluble of the potassium oxalates at the lower temperatures. The formula $S = 1.1 + 0.046t + 0.0019t2$, in which S = % $KHC_2O_4(COOH)_2$ and t = °C., is applicable with accuracy only at temperatures below about 70cC. The solubility is 18% at 80°C., and rises sharply above this temperature to 42% at 100 CC. It is used for the removal of rust marks.

**Ammonium oxalate, (NH4HC2O$_4$2H$_2$O,** formula weight 142.12, dissolves in water, forming, at OCC., a saturated solution containing 2.22% ($NH_4HC_2O_4$) The formula $S = 2.3 + 0.064t. + 0.0017t^2$, in which S = % ($NH_4HC_2O_4$ and t = °C., gives the approximate composition of saturated solutions between the freezing and the normal boiling points.

Ammonium acid oxalate, or binoxalate, $NH_4HC_2O_4H_2O$, formula weight 125.09, is obtainable in pure form, like the potassium salt, by crystallization from solutions containing some of the neutral oxalate, or by operating in a range above a critical temperature, around 40cC. The solution, saturated at 45cC., contains about 15% acid oxalate.

Ammonium tetroxalate, $NH_4HC_2O_4(COOH)_2 2H_2O$, formula weight 233.14, resembles the corresponding potassium salt j it is not particularly important.

Calcium oxalate, $CaC_2O_4H_2O$, formula weight 146.12, is of importance principally as an intermediate in oxalic acid manufacture and in analytical chemistry j it is the form in which calcium is frequently quantitatively isolated. Its solubility in water is very low, lower than that of the other alkaline earth oxalates. The approximate solubilities of this and several related salts are indicated in Table II.

Table 11.2. Solubilities of Alkaline Earth Oxalates

Formula of solid phase	Anhydrous salt in saturated solution, %	
	0°C	100°C
$CaC_2O_4.H_2O$	0.0005	0.0015
$SrC_2O_4.H_2O$	0.006	0.015
$BaC_2O4.1/2H_2O$	0.009	0.021
$MgC_2O_4.2H_2O$	0.026	0.041

Ferrous oxalate, $FeC_2O..2H_2O$, formula weight 179.90, is a yellow crystalline powder. Its solubility in water is of the same order as that of the alkaline earth oxalates. The salt is very soluble in solutions of alkali oxalates, especially potassium oxalate, and such solutions were once used extensively in photography in the formulation of developers. The salt may be used, conveniently, as an intermediate in the preparation of the industrially useful ferric oxalate and the oxalatoferrates.

Ferric oxalate, Fe2(C_2O.h.6H_2O, formula weight 483.85, is very soluble in water, in striking contrast to the ferrous salto When the solution is evaporated, the ferric oxalate dries, forming greenish-yellow granules or scales. Ferric oxalate is sensitive to light which reduces it to ferrous oxalate:

$$Fe_2(C_2O_4)_3 \rightarrow 2FeC_2O_4 + 2CO_2$$

It must therefore be suitably protected. Ferric oxalate may be used in the manufacture of blueprint paper, but for this purpose the oxalatoferrates are more commonly employed.

Ferric sodium oxalate (sodium oxalatoferrate, $Na_3[Fe^{III}(C_2O_4)3].41H_2O$, formula weight 470.0, ferric potassium oxalate (potassium oxalatoferrate(III», $K_3[Fe^{III}(C_2O_4)_3].3H_2O$, formula weight 491.25, and ferric ammonium oxalate (ammonium oxalatoferrate) $(NH_4)_3[Fe^{III}(C_2O_4]_3.3H_2O$, formula weight 428.1, are all crystalline, emerald-green compounds, which, like ferric oxalate, are easily soluble in water and sensitive to light:

$$2\,M2Fe^{III}(C_2O_4)_3] \rightarrow 2FeC_2O_4 + 3\,M1C_2O_4 + 2CO_2$$

All are used in photographic processes, especially in the manufacture of blueprint paper.

Many hundreds of other complex compounds of oxalic acid have been described, and they include virtually all possible combinations of metallic elements and radicals as well as a great variety of anionic groups and nonmetallic elements. Trivalent chromium, manganese, cobalt, and some other metals form complex oxalato compounds analogous to the alkali oxalatoferrates

Aluminum oxalate, $Al_2(C_2O_4)_3.H_2O$, and two oxalato complexes of titanium, $(NH_4)_2[OTiC_2O_4].H_2O$ and $K_2[OTiC_2O_4].2H_2O$, have been used as mordants in the fabrication of cotton and leather goods.

Cerium oxalate (cerii oxalas, N.F IX) is a mixture of the oxalates of cerium, neodymium, praseodymium, lanthanum, and other associated elements. It is a fine, white or slightly pink, odorless, tasteless powder; it is insoluble in water and in cold alkaline or acidic solutions, but is dissolved by hot aqueous sulfuric or hydrochloric acid.

Typical prices of various oxalates are listed in Table III.

Table 11.3. Typical Prices of Various Oxalation in 1951

Name of salt	Price, c/1b
Sodium oxalate, neutral	10.5
Potassium oxalate, neutral	30-32
Ammonium oxalate, neutral	25-31
Cerium oxalate, N.F.	70-72
Iron (ferric) oxalate	83-85
Ferric ammonium oxalate	25.5-28.5
Ferric sodium oxalate	25.5-28.5
Ferric potassium oxalate	30.5-33.5
Titanium potassium oxalate	72-85

Esters

Oxalic acid is readily esterified by reaction with alcohols using conventional pro-cedures (see *Esterification);* it is sufficiently acidic to serve as its own esterification catalyst. If the commercial acid, the dihydrate, is used without preliminary dehydra-tion it is necessary to remove twice the usual proportion of water; it is, however, usu-ally more economical to dehydrate and esterify in a single operation than to add the relatively costly step of preparing anhydrous oxalic acid.

Dehydration and esterification may be regarded as proceeding in three stages:

$$(COOH)_2\,H_2O \rightarrow (COOH)_2 + 2H_2O$$

$$(COOH)_2 + ROH \rightarrow HOOCCOOR + H_2O$$

$$HOOCCOOR + ROH \rightarrow (COOR). + H_2O$$

From oxalic acid two series of esters are obtainable, the monoalkyl oxalates, or alkyl acid oxalates, HOOCCOOR, and the dialkyl oxalates, or neutral esters, $(COOR)_2$. The alkyl acid oxalates are strongly acidic, nonvolatile, and thermally unstable, although they form stable salts. The neutral esters are usually oily liquids resembling in a general way others in the large class of organic esters. They are generally thermally stable, and the lower members of the series may be distilled at atmospheric pressure. Two of the neutral esters, diethyl oxalate and di-n-butyl oxalate, have important uses and are manufactured in fairly large quantities; the ethyl ester is the more important. According to the United States Tariff Commission the production of diethyl oxalate in 1949 was 461,000 lb. In 1951 the tank-car price of diethyl oxalate was 40c/lb

Physical Properties of Esters. Dimethyl oxalate is a crystalline solid at ordinary temperatures; all of the other lower esters are colorless, oily liquids. The physical properties of three common dialkyl oxalates are shown in Table IV.

Table 11.4. Physical Properties of Esters

	Dimethyl oxalate	Diethyl oxalate	Di-n-butyl oxalate
Formula	$(COOCH_3)_2$	$(COOC_2H_5)_7$	$(COOC_4H_9)_7$
Formulate weight	118.08	146.13	202.24
M.p., °C	54	-40.6	-29.6
b_{760}, °C	163.5	185.4	245.5
d_4	1.148^{54}	1.0785^{20}	0.9873^{20}
n_D^{20}	—	1.4104	1.424
Viscosity at 20°C, centipoises	—	2.01	—
Flash point, °F	—	168	265

Analysis and Specification: - Technical diethyl oxalate, as made by a typical manufacture, conform to the minimum standards listed in the table V:

Table 11.5. Typical Minimum Standards of Technical Diethyl and Dibutyl Oxalates

	Diethyl oxalate	Dibutyl oxalate
Acidity [as $(COOH)_2$]	0.05 max	0.05
Color	Water-white	Water-white
Odor	Mild, nonresidual	Mild
Nonvolatile	0.005 gram/00 ml. max.	0.005 gram/100 ml.
d_{20}^{20}	1.078-1.082	0.989-0.993
Purity (easter content)	99% min	99%
Water[a]	None	None
Distillation range	Below 180°C, none	Below 240°C, 5% mx.
	Below 182°C, 10% max.	Below 248°C, 90% min.
	Below 188°C, 90% min.	Above 255°C, none
	Above 190°C, none	

Uses

The neutral oxalic esters are good solvents for nitrocellulose and have limited uses in special lacquers when a slowly evaporating solvent is desired. General use in such applications is precluded by their low evaporation rates and also by their relatively unfavourable costs as compared with suitable competitive solvents. Instances in which technological advantages outweigh other considerations include the formulation of special lacquers for electronic tube manufacture, and of certain types of printing inks in which the function of the ester is to prevent drying of the ink on the hot press rolls; diethyl oxalate is used in the first-named special application while di-n-butyl oxalate is especially," valuable in t second. See *Solvents.*

Oxalic esters are particularly useful in synthetic organic chemistry because of condensations which they undergo under appropriate conditions; this is the basis for an important segment of industrial chemistry. Although these condensation reactions may be carried out with oxalic esters generally, diethyl oxalate is usually preferred, and this accounts for the relatively greater importance of this ester. The condensation of diethyl oxalate with ethyl acetate is typical:

$$(COOC2H_5)_2 + CH_3COOC_2H_5 \rightarrow \underset{\text{oxalacetic ester}}{C_2H_5OOCCOCH_2COOC_2H_5} + C_2H_5OH$$

Sodium ethylate is the condensing agent, and anhydrous conditions are required. The product actually obtained is the sodium salt of the enol form of oxalacetic ester. $C_2H_5OOCC(ONa)$: $CHCOOC_2H_5$. This salt is stable, and is a commercial commodity; it is used in the production of pyrazolone dyes. The ester, obtained by treatment of the sodium salt with an acid, is some, that unstable and may be converted to diethyl malonate:

$$\underset{\text{oxalacetic ester}}{C_2H_5OOCCOCH_2COOC_2H_5} \rightarrow \underset{\text{malonic ester}}{C_2H_5OOCCH_2COOC_2H_5} + CO$$

This condensation reaction is applicable, generally, to substituted acetic esters, $RR'CHCOOC_2H_5$, as is the subsequent conversion to malonic ester derivatives. These substituted malonic esters may in turn become intermediates in further condensations; with urea, for example, the final products are the very important, medically useful, barbiturates.

Oxalic esters and condensation derivatives can participate in a considerable variety of condensation reactions other than those mentioned, and in reactions other than condensations, too; they can thus serve as raw materials for the synthesis of a diverse group of organic compounds.

Physical Properties

Physical State	white crystals
Melting Point	101 - 102°C
Boiling Point	149 - 160 C (sublimes)
Specific Gravity	1.6 - 1.7
Solubility In Water	1 G/7 ml
Ph	1.00
Vapor Density	4.4
Polarizability	5.72 Cm^3
Refractive Index	1.480
Molar Volume	50.8 Cm^3
Flash Point	163°C
Surface Tension	87.3 dynes/cm

Chemical Properties

1. Acidity: Oxalic acid is the strongest of all the dicarboxylic acids being even stroger that phosphoric acid.

2. Formation of Salts: It gives the usual replacement reactions of the two carboxylic group forming two series of salts esters, amides and acid halides (see general reactions and derivatives of oxalic acid)

3. Action of Heat: when heated alone, oxalic acid decomposes above melting point giving formic acid and carbon dioxide.

$$HCOOH - COOH \xrightarrow{190 - 200°C} HCOOH + CO_2$$

Formic acid

4. Action of Sulphuric Acid: - Oxalic acid when heated with sulphuric acid decomposes to yield mixture of carbon monoxide and carbon dioxide.

$$HCOOO - COOH \underset{D}{\xrightarrow{H_2SO_4}} CO_2 + CO + H_2O$$

5. Action of Glycol: It reacts with glycerol to form formic acid or alkyl alcohol depending on experimental condition (see glycerol)

6. Action of Ethylene Glycol: When heated with ethylene glycol, it forms ethylene oxalate, acylic compounds.

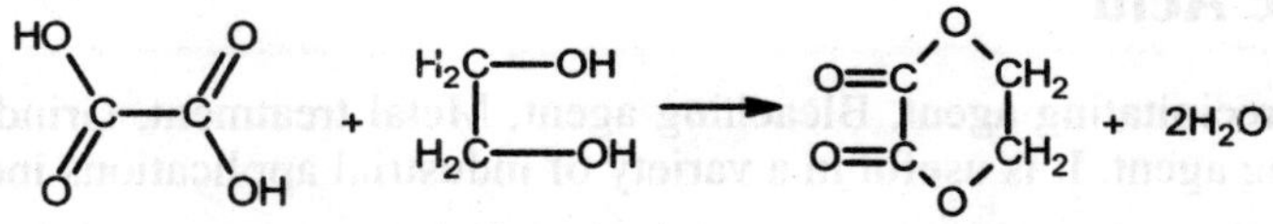

oxalic acid + ethylene glycol → 1,4-dioxane-2,3-dione (ethylene oxalate) + $2H_2O$

7. Action with Phosphorous Pentachloride: It forms oxalyl chloride, but if an insufficient quantity of phosphorous pentachloride is used with oxalic acid it decomposes via the intermediate half acid chloride.

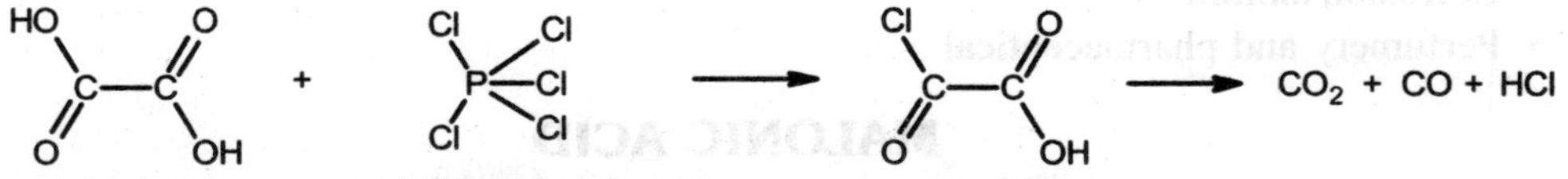

oxalic acid + pentachlorophosphorane → chloro(oxo)acetic acid → CO_2 + CO + HCl

oxalic acid $\xrightarrow[\text{excess}]{PCl_5}$ ethanedioyl dichloride

oxylyl chlroide

8. Oxidation: Oxalic acid when treated with potassium permanganate and dilute sulphuric acid at 60°C is oxidised to carbon dioxide and water.

$$2KMnO_4 + 5H_2C_2O_4 + 3H_2SO_4 \rightarrow K_2SO_4 + 2MnSO_4 + 10CO_2 + 8H_2O$$

9. With Silver Nitrate: Silver oxalate as a white precipitate is produced when a neutral solution of an oxalate is treated with silver nitrate solution. The solution is soluble in excess of ammonia solution on warming gives no silver mirror.

Oxalic acid does not reduce Tollen's reagent

K—O(C=O)—C(=O)—O—K + 2AgNO$_3$ ⟶ Ag—O(C=O)—C(=O)—O—Ag + 2KNO$_3$

dipotassium oxalate; silver oxalate white ppt

10. With Calcium Oxalate: A neutral solution of an oxalate gives a precipitate of calcium oxalate, with calcium chloride solution. This solution is insoluble in acetic acid and but soluble in hydrochloric acid.

K—O(C=O)—C(=O)—O—K + Cl—Ca—Cl ⟶ calcium oxalate (Ca) + 2 K—Cl

dipotassium oxalate; Calcium Chloride; calcium oxalate; potassium chloride

Uses of Oxalic Acid

Purifying agent, Precipitating agent, Bleaching agent, Metal treatment, Grinding agent, Waste water treatment, Reducing agent. It is useful in a variety of industrial applications include;

- Plasticizer for polymers
- Biodegradable solvents and lubricants
- Engineering plastics
- Epoxy curing agent
- Adhesive and powder coating
- Corrosion inhibitor
- Perfumery and pharmaceutical

MALONIC ACID

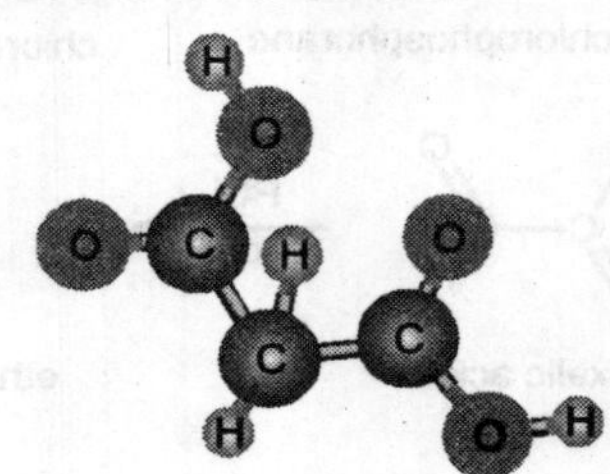

Formula: $C_3H_4O_4$

Occurrence

Malonic acid (from Latin malum = apple) (1,3-propanedioic acid) is a dicarboxylic acid with structure $CH_2(COOH)_2$. The ion form of malonic acid, as well as its esters and salts, are known as **malonates.** For example, diethyl malonate is malonic acid's ethyl ester

Preparation

(I) From Monochloroacetic acid:- By treatment with KCN and subsequent hydrolysis.

$$H_3C-COOH \xrightarrow{Cl_2} HO-CO-CH_2-Cl + HCl$$

acetic acid → chloroacetic acid + hydrogen chloride

$$HO-CO-CH_2-Cl \xrightarrow{KCN} HOOC-CH_2-C\equiv N + NH_3$$

chloroacetic acid → cyanoacetic acid + ammonia

Physical Properties

Melting point (decomposes):	135°C
Density:	1.6 g/cm3
Solubility in water, g/100 ml at 20°C:	7.3
Octanol/water partition coefficient as log Pow:	0.91/-0.18 (calculated)

Chemical Properties

(i) Action of Heat: When heated at 140 - 150°C or when refluxed in sulphuric acid solution it looses a molecule of CO_2 to produce acetic acid.

$$HOOC-CH_2-COOH \xrightarrow{140 - 150°C} H_3C-COOH + CO_2$$

malonic acid → acetic acid

The decarboxylation probably occurs by the following mechanism

$$HO{-}C(=O){-}CH_2{-}C(=O){-}OH \xrightarrow{\Delta} CO_2 + [CH_2{=}C(OH)_2] \xrightarrow{\text{Tautomerises}} H_3C{-}C(=O){-}OH$$

ethylene-1,1-diol; acetic acid

(ii) Action of Phosphorous Pentoxide: When heated with phosphorous pentoxide malonic acid gives a small amount of carbon suboxide.

$$\underset{\text{malonic acid}}{HO{-}C(=O){-}CH_2{-}C(=O){-}OH} \xrightarrow{P_2O_5} \underset{\text{carbon suboxide}}{O{=}C{=}C{=}C{=}O} + 2H_2O$$

Carbon suboxide is a diketone and combines with water to give back the malonic acid and also with ethanol to yield malonate.

(iii) Action with Aldehydes: When a pyridine solution of malonic acid and an aldehyde is heated $\alpha - \beta$ unsaturated acid is obtained.

(iv) Addition with Nitrous Acid: Malonic acid on treatment with succinic acid, followed by hydrolysis gives mesoxalic acid.

$$\underset{\text{nitrous acid}}{HO{-}N{=}O} + \underset{\text{malonic acid}}{HOOC{-}CH_2{-}COOH} \xrightarrow{-H_2O} HO{-}N{=}C(COOH)_2 \xrightarrow{H_2O/H^+} \underset{\text{oxomalonic acid / mesoxalic acid}}{O{=}C(COOH)_2}$$

(v) Action of Bromine: A suspension of malonic acid in ether when treated with bromine affects monobromomalonic acid.

$$\underset{\text{malonic acid}}{HOOC{-}CH_2{-}COOH} + \underset{\text{bromine}}{Br{-}Br} \longrightarrow \underset{\text{bromomalonic acid}}{HOOC{-}CHBr{-}COOH} + \underset{\text{hydrogen bromide}}{HBr}$$

Uses of Malonic Acid

In a well known reaction malonic acid condenses with urea to barbituric acid. Malonic acid is frequently used as an enolate in Knoevenagel condensations or condensed with acetone to form Meldrum's acid.

This chemical is relatively unstable and has few uses, but its ester derivative, diethyl malonate, is used to synthesize useful compounds such as barbiturates, flavors, fragrances, and vitamins (B1 and B6).

SUCCINIC ACID

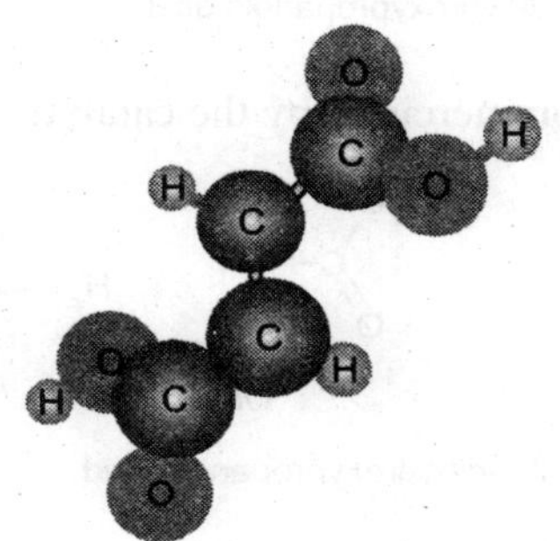

Formula: $C_4H_6O_4$

Introduction

Succinic acid was first obtained as a small of distillation of the fossil resin amber and hence its name (sucinum = amber Latin) It occurs widely in nature in numerous plants and unripe fruits. It is produced in small amount during the alcohol fermentation of sugar.

Preparation

(i) Succinic acid is prepared from ethylene bromide by treating with potassium cyanide and subsequent hydrolysis of ethylene cyanide.

$$\underset{\text{1,2-dibromoethane}}{BrH_2C{-}CH_2Br} \xrightarrow{2KCN} \underset{\text{succinonitrile}}{N{\equiv}C{-}CH_2{-}CH_2{-}C{\equiv}N} \xrightarrow{H_2O/HCl} \underset{\text{succinic acid}}{HOOC{-}CH_2{-}CH_2{-}COOH}$$

(ii) By heating malic or tartaric acid in a sealed tube with hydrochloric acid phosphorous

$$\underset{\text{2,3-dihydroxypropanoic acid}}{HOOC{-}CH(OH){-}CH_2{-}OH} + 2HI \xrightarrow[\Delta]{\text{red P}} \underset{\text{succinic acid}}{HOOC{-}CH_2{-}CH_2{-}COOH} + I_2 + H_2O$$

2,3-dihydroxypropanoic acid + 4HI —(red P, Δ)→ succinic acid

(iii) It is prepared commercially by the catalytic reduction of maleic acid.

2,3-dihydroxypropanoic acid + H_2 —(Ni)→ succinic acid

Physical Properties

Physical State	Odorless White Crystals
Melting Point	187 - 189 C
Boiling Point	235 C
Specific Gravity	1.552
Solubility in Water	Moderately Soluble
pH	
Vapor Density	3.04
Autoignition	630°C
NFPA Ratings	Health: 2; Flammability: 1; Reactivity: 0
Refractive Index	1.405
Flash Point	206°C
Stability	Stable Under Ordinary Conditions

Chemical Properties

(i) Action of Alkalies: It reacts with sodium hydroxide and potassium hydroxide to form series of salts.

disodium succinate ← succinic acid → dipotassium succinate

(ii) Action with Phosphorous Pentachloride: When treated with phosphorous pentachloride, it forms succinoyl chloride.

$$HOOC-CH_2-CH_2-COOH + 2PCl_5 \longrightarrow ClOC-CH_2-CH_2-COCl + 2POCl_3 + 2HCl$$

succinic acid — succinoyl dichloride

Succinic acid however on treatment with thionyl chloride gives succinic anhydride instead of succinoyl chloride

(iii) Action of heat: When heated above its melting points succinic acid sublimes and the rest being converted into the anhydride.

$$HOOC-CH_2-CH_2-COOH \xrightarrow{\Delta} \text{succinic anhydride} + H_2O$$

succinic acid — succinic anhydride

Excellent yield of succinic acid anhydride can however, be obtained by distilling succinic acid with acetic anhydride a acetyl chloride or phosphoryl chloride.

(iv) Action with Ethanol: Succinic acid when refluxed with absolute ethanol in presence of sulphuric acid gives diethyl succinate.

$$HOOC-CH_2-CH_2-COOH + H_3C-CH_2-OH \longrightarrow H_3C-CH_2-O-CO-CH_2-CH_2-CO-O-CH_2-CH_3$$

succinic acid — ethanol — diethyl succinate

(v) Electrolysis of sodium and Potassium salt: The electrolysis of a strong solution of its sodium or potassium salts yield ethylene.

$$KOOC-CH_2-CH_2-COOK \xrightarrow{\text{electrolysis}} \underbrace{CH_2=CH_2 + 2CO_2}_{\text{at anode}} + \underbrace{2KOH + H_2}_{\text{at cathode}}$$

dipotassium succinate — ethylene

(vi) Action with ammonia: When succinic acid is heated in a current of dry ammonia or with ammonium carbonate succinimide is produced.

succinic acid → succinaldehyde $\xrightarrow{+NH_3}$ succinimide + H_2O

Succinimide when treated with an alkaline solution of Br_2 at 0°C forms n – bromosuccinimide (NBS) which is a valuable reagent for allylic bromination

succinimide + Br_2 $\xrightarrow[0°C]{NaOH}$ n-bromosuccinimide + HBr

(vii) Action of Glycol: When heated with excess of ethylene glycol, succinic acid forms polyesters commonly referred to as alkyl esters. These esters are acidic in character due to succinic acid residues at the end of the chain.

succinic acid + ethylene glycol + succinic acid + ethylene glycol + ... →

α-(3-carboxypropanoyl)-ω-hydroxypoly[dioxy(1-oxopropane-1,3-diyl)oxy(1-oxopropane-1,3-diyl)]

Uses

Flavouring agent for food and beverages; Producing five heterocyclic compounds, Intermediate for dyes, perfumes, lacquers, photographic chemicals, alkyd resins, plasticizer, Metal treatment chemical, vehicle water cooling systems and coatings. Medicines of sedative, antispasmer, antiplegm, antiphogistic, anrhoter, contraception and cancer-curing.

GLUTARIC ACID

Formula: $C_5H_8O_4$
Also known as pentane-1, 5-dioic acid

Preparation

Glutaric acid is prepared from glutamic acid.

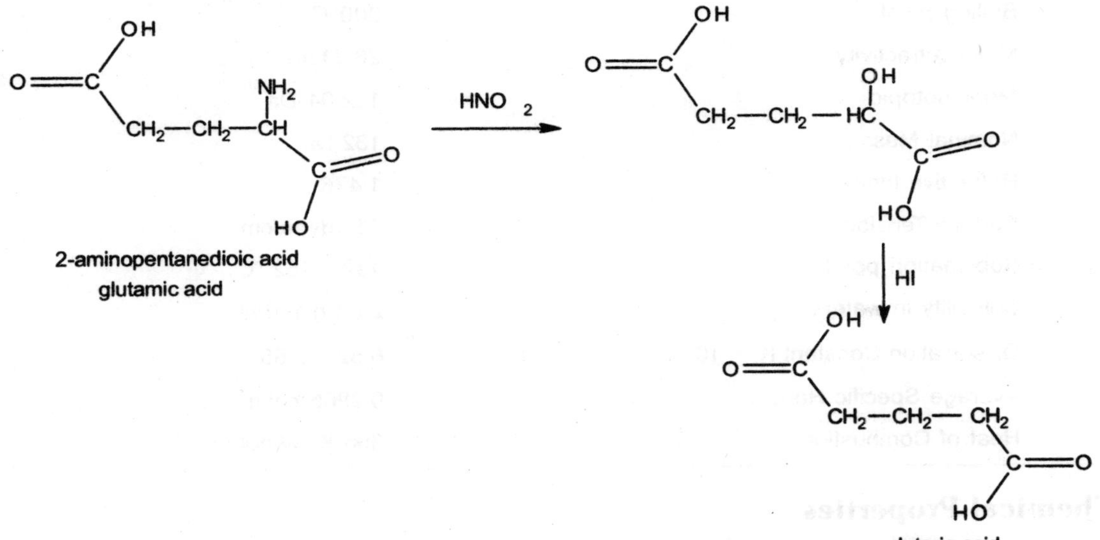

$$HOOC-CH_2-CH_2-CH(NH_2)-COOH \xrightarrow{HNO_2} HOOC-CH_2-CH_2-CH(OH)-COOH \xrightarrow{HI} HOOC-CH_2-CH_2-CH_2-COOH$$

2-aminopentanedioic acid
glutamic acid

glutaric acid

It is not a commercial product and hence it is made in the lab by following synthetic methods

(i) By hydrolysis of trimethylene cyanide:

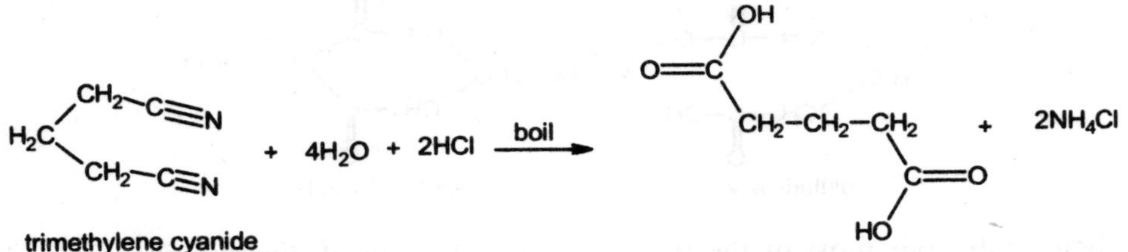

$$H_2C(CH_2-C\equiv N)_2 + 4H_2O + 2HCl \xrightarrow{boil} HOOC-CH_2-CH_2-CH_2-COOH + 2NH_4Cl$$

trimethylene cyanide

glutaric acid

(ii) By the oxidation of cyclopropane with nitric acid

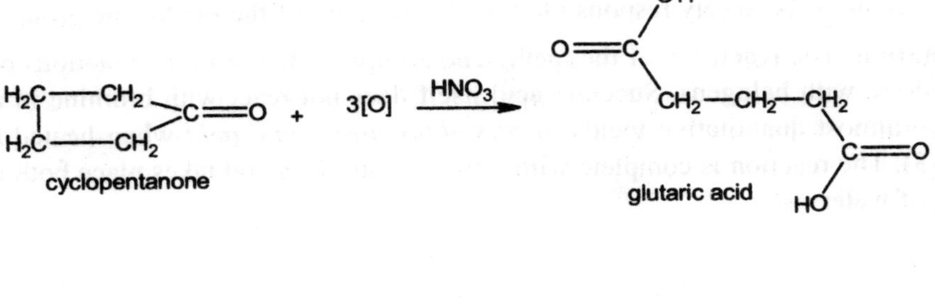

$$\text{cyclopentanone} + 3[O] \xrightarrow{HNO_3} HOOC-CH_2-CH_2-CH_2-COOH$$

cyclopentanone

glutaric acid

Physical Properties

Density	1.316 gm/cm^3
Polarizibility	11.23´10^{24}
Molecular mass	132.12 g/mol
Melting point	95°C to 98°C
Boiling point	200°C
Molar refractivity	28.34cm^3
Monoisotopic	132.04 Da
Nominal Mass	132 Da
Refractive Index	1.476
Surface Tension	56.1dyne/cm
Sublimation point	132 – 152 °C
Solubility in water	2.88 g/100ml
Dissociation Constant $K_1 \times 10^{-5}$	6.52 – 6.65
Average Specific Heat	0.2965 cal/g
Heat of Combustion	356 Kcal/mol

Chemical Properties

It forms anhydrates when heated

$$H_2C(CH_2-COOH)_2 \xrightarrow{\Delta} \text{glutaric anhydride} + H_2O$$

glutaric acid glutaric anhydride

Succinic acid undergoes most of the reactions characteristic of dicarboxylic acids. A common characteristic of succinic acid and its derivatives is the *succinoyl grouping*, -(O:)C-CH2-CH2-C(:O)-. This grouping, consisting of two adjacent methylene groups flanked by the two symmetrically attached unsaturated carbonyls, is largely responsible for the reactivity of the methylene groups.

Halogenation. The reactivity of the methylene groups is shovm in the reactions of succinic acid and its anhydride with halogens. Succinic acid itself does not react with bromine in an open vessel but does give almost quantitative yields of *mesodibromosuccinic acid* when heated at 100°C. in a sealed tube (3). The reaction is complete within two to four days and takes place both in the presence and absence of water:

$$2Br_2 + \text{succinic acid} \longrightarrow \text{2,3-dibromosuccinic acid} + 2HBr$$

There is no evidence that either the dl-acid or the monobromo acid is formed either were produced as an intermediate, it would probably be rapidly convert• the meso-dibromo acid under the conditions employed. When more than an eq1 lent amount of water is present, brominated hydrocarbons are produced which re the yield of the dibromo acid.

The anhydride, when heated with a molecular equivalent of bromine, principally the monobromo derivative, b.p. 130-133°C. With two moles of bro it yields the *dl-2,3-dibromosuccinic anhydride,* m.p. 118-119°C.

Diethyl succinate when treated with an excess of chlorine in sunlight is completely chlorinated, yielding small needles of *bis(pentachloroethyl) tdrachlorosuccinate,* $CCl_2OOCCCl_2CCl_2COOCCl_2CCl_3$.

Succinic acid reacts with an equal molecular amount of phosphorus pentachloride at 120-130°C. to yield succinoyl chloride.

Condensations with Ketones and Aldehydes. One of the most interesting reactions of the succinoyl, grouping is shown in the condensation of succinic di with aldehydes and ketones in the presence of sodium alkoxide. Unlike other pounds with reactive methylene groups, succinates do not form β-diketones by a Claisen type of reaction except in specific instances where polyketone produced as by-products in small amounts. Instead, half-esters of alkylidenesuccinic acids are usually the major products. These are formed by the *Stobbe condensation* which is peculiar to succinic esters and substituted succinic esters. From available evidence, the Stobbe reaction takes place in three steps: the reactants first an aldol or a ketol that loses alcohol to give a γ-Lactone ester known as a *paraconate* which in turn rearranges to the *alkylidenesuccinic acid half-ester.* A typical Stobbe condensation is that of acetone and diethyl succinate:

$$\text{acetone} + \text{succinic acid} \xrightarrow{NaOC_2H_5} \text{ethyl 3-hydroxy-3-methylbutanoate} \xrightarrow{-C_2H_5OH} \text{2,2-dimethyl-5-oxotetrahydrofuran-3-carboxylic acid}$$

Isomers of the alkylidenesuccinates are also obtained in varying amounts, depending upon the experimental environment and the structure of the ketone. *Alkenylsuccinic half-esters,* for example, are likely: to be the predominating products when the ketone contains a methylene group, as with methyl ethyl ketone:

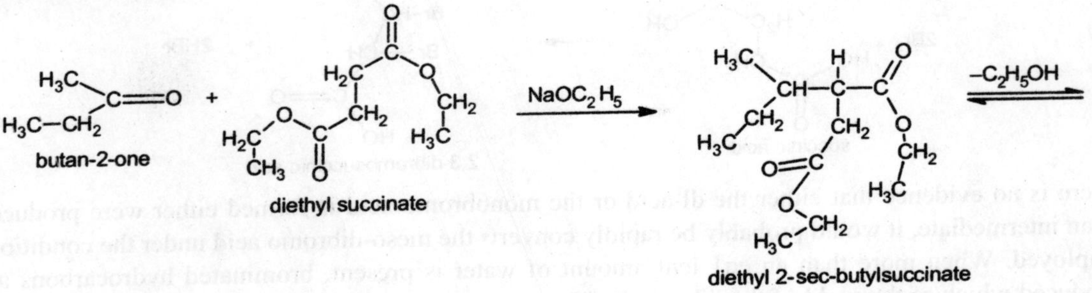

butan-2-one

diethyl succinate

$NaOC_2H_5$

$-C_2H_5OH$

diethyl 2-sec-butylsuccinate

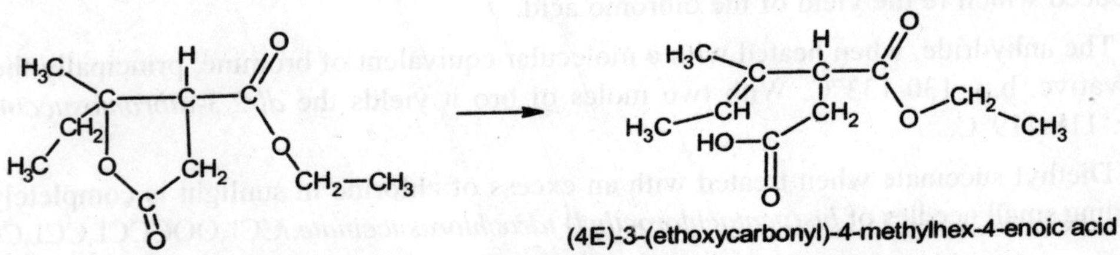

(4E)-3-(ethoxycarbonyl)-4-methylhex-4-enoic acid

A small amount of the (1-methylpropylidene)succinate is also formed in this reaction.

Decarboxylated products are obtained when the reaction is carried out at elevated temperatures. These can be formed from either the intermediate paraconates or from the alkylidenesuccinates (substituted itaconates), since both types of compounds lose carbon dioxide quite easily. Such a reaction occurs with 4-hydroxyvalerie acid -γ-lactone:

ethyl 2-ethyl-2-methyl-5-oxotetrahydrofuran-3-carboxylate

diethyl 2-[(1*E*)-1-methylprop-1-en-1-yl]succinate

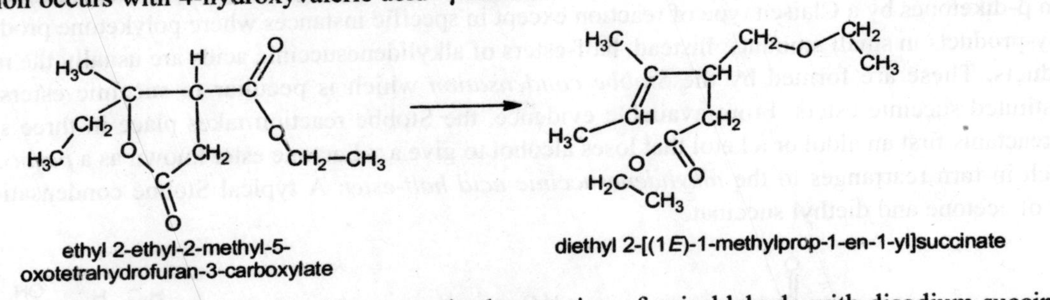

A similar loss of carbon dioxide occurs in the reaction of anisaldehyde with disodium succinate and acetic anhydride:

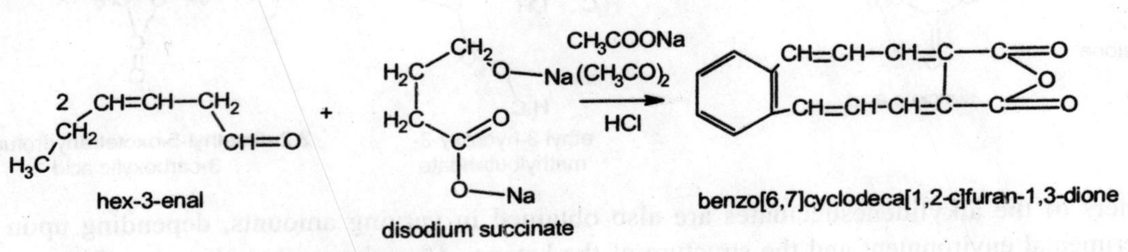

hex-3-enal

CH_3COONa

$(CH_3CO)_2$

HCl

disodium succinate

benzo[6,7]cyclodeca[1,2-c]furan-1,3-dione

These highly unsaturated anhydrides are known as *fulgides*. They are intensely colored. Small amounts of 6-phenyl-3,5-hexadienoic acid, $C_6H_5CH=CHCH=CH$

CH_2COOH, and 1 *,8-diphenyl-l ,3,5,7-octatetraene,* $C_6H_5CR=CHCH=CHCH=CH$

$CH=CHC_6H_5$, are formed also. The yield of the latter can be increased to 33% of the theoretical amount when lead oxide (lead acetate) is substituted for sodium acetate in the reaction. Dilactones are most likely formed during these preparations, giving the observed products. *H alfesters of dialkylidenesuccinic acids* may be prepared by allowing the diester of an alkylid~nesuccinic acid to react with a molecular equivalent of the aldehyde by the Stobbe reaction.

The Stobbe condensation is probably the most characteristic reaction of succinic acid esters. This is evident from the fact that only a very few failures are reported, and these can generally be attributed to steric hindrance. The keto esters, like ethyl acetoacetate and certain unsaturated ketones such as chalcone, react in a different manner. The former yields derivatives of furan acetic acid, possibly through the union of both reactants in their enol forms:

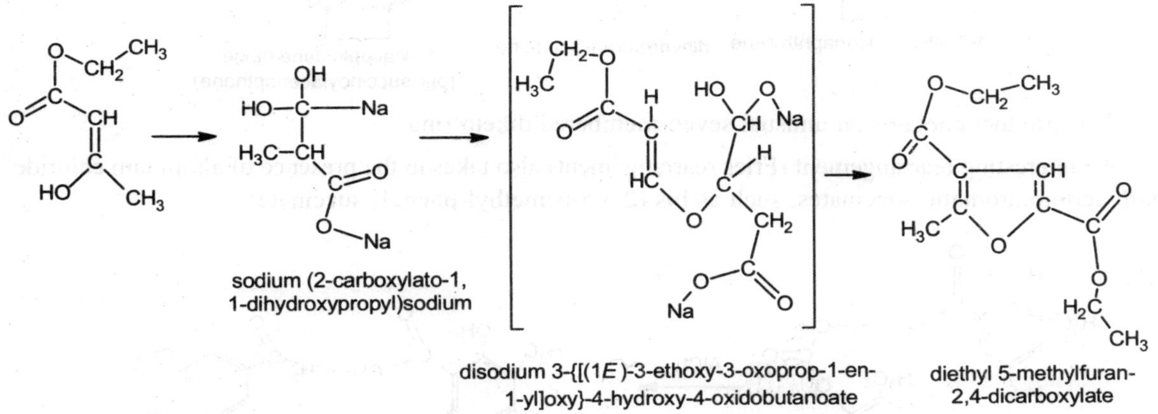

sodium (2-carboxylato-1, 1-dihydroxypropyl)sodium

disodium 3-{[(1*E*)-3-ethoxy-3-oxoprop-1-en-1-yl]oxy}-4-hydroxy-4-oxidobutanoate

diethyl 5-methylfuran-2,4-dicarboxylate

The reaction at best is rather complex since carbon dioxide, ethyl acetate, and acetone are also formed as by-products. Direct addition instead of condensation occurs in the case of chalcones:

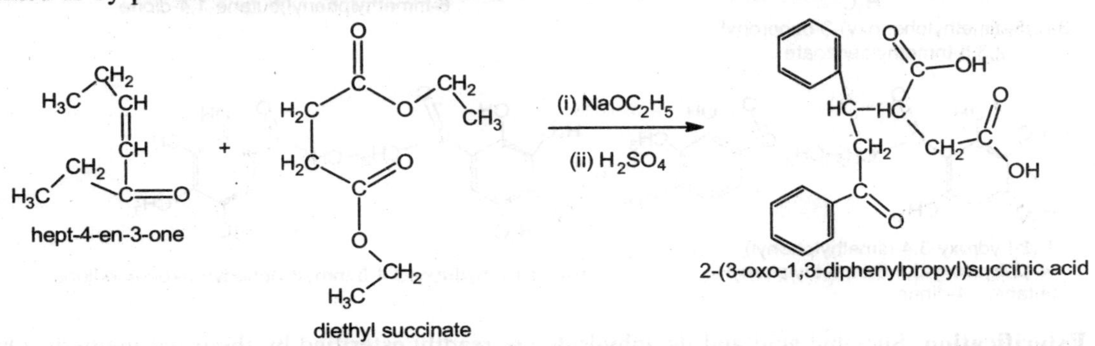

hept-4-en-3-one + diethyl succinate —(i) $NaOC_2H_5$, (ii) H_2SO_4→ 2-(3-oxo-1,3-diphenylpropyl)succinic acid

Friedel-Craft Reactions. Like other dibasic anhydrides, succinic anhydride readily undergoes addition to aromatic compounds in the presence. of anhydrous aluminum chloride (this type of reaction is known as *succinolation)* to form various derivatives of 4-oxobutyric acid (4), as for example in the reaction with inclan:

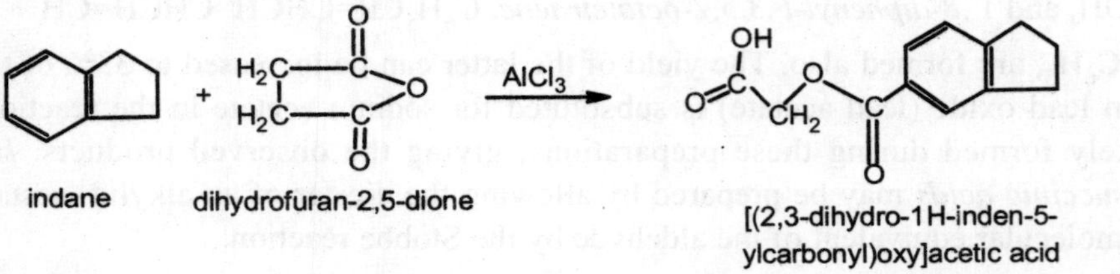

indane dihydrofuran-2,5-dione [(2,3-dihydro-1H-inden-5-ylcarbonyl)oxy]acetic acid

With acenaphthene, both carbonyl groups become attached to the hydrocarbon.

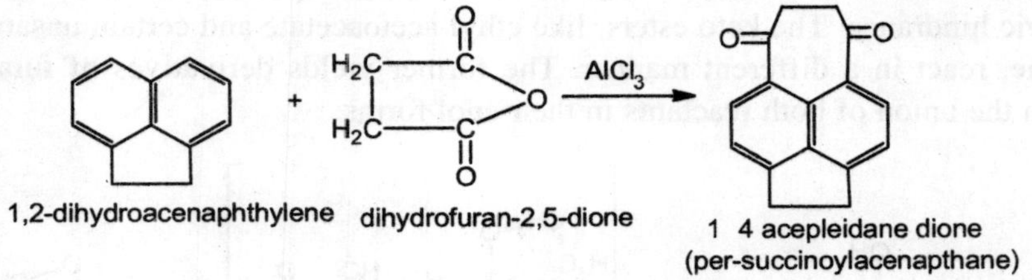

1,2-dihydroacenaphthylene dihydrofuran-2,5-dione 1 4 acepleidane dione (per-succinoylacenapthane)

This product contains an unusual seven-membered diketo ring.

An interesting rearrangement (Fries rearrangement) also takes in the presence of aluminum chloride with certain aromatic succinates, such as bis (2,3,5-trimethyl-phenyl) succinate:

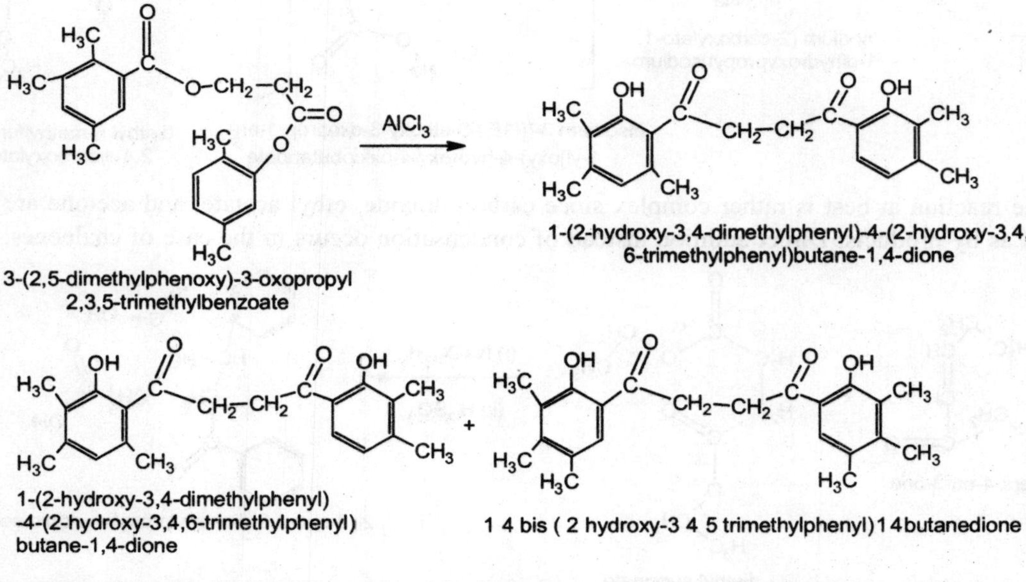

3-(2,5-dimethylphenoxy)-3-oxopropyl 2,3,5-trimethylbenzoate

1-(2-hydroxy-3,4-dimethylphenyl)-4-(2-hydroxy-3,4,6-trimethylphenyl)butane-1,4-dione

1-(2-hydroxy-3,4-dimethylphenyl) -4-(2-hydroxy-3,4,6-trimethylphenyl) butane-1,4-dione

1 4 bis (2 hydroxy-3 4 5 trimethylphenyl) 1 4 butanedione

Esterification. Succinic acid and its anhydride are readily esterified by the usual methods. One interesting procedure employs an unsaturated hydrocarbon for this purpose (26), to yield the secondary esters:

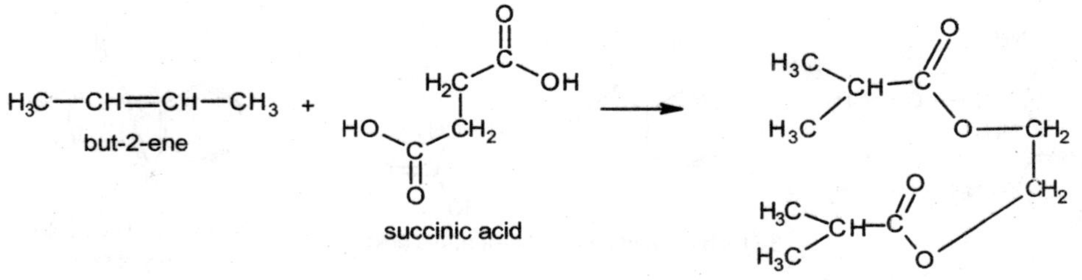

$H_3C-CH=CH-CH_3$ + succinic acid ⟶ ethane-1,2-diyl bis(2-methylpropanoate)

but-2-ene

succinic acid

ethane-1,2-diyl bis(2-methylpropanoate)

The polyesters of the various glycols can be depolymerized in the presence of stannous chloride to yield interesting poly-membered ring products. Poly (ethyl-ene succinate), for example, gives the 16-membered cyclic dimer of ethylene succinate, 1,4,9, 12-tetraoxacyclohexadecane-5,8, 13, 16-tetrane:

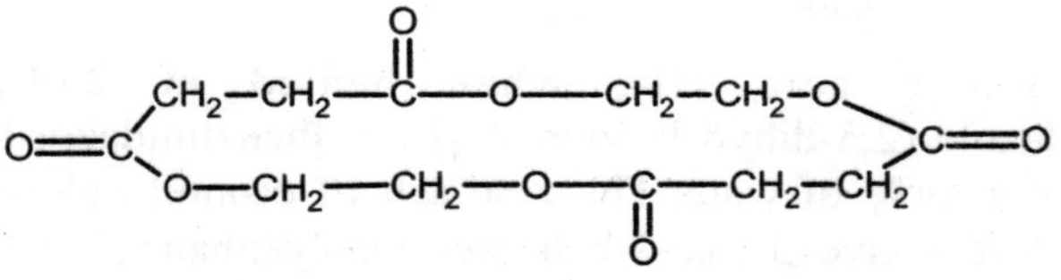

1,4,9,12-tetraoxacyclohexadecane-5,8,13,16-tetrone

Poly (propylene succinate) yields both a cyclic monomer and cyclic dimer.

Reactions with Amino Compounds. Several useful reactions also occur with compounds containing an amino group; in no case do they involve the methylene groups of succinic acid or its derivatives. Included is the reaction of succinic acid with ammonia and amines at elevated temperatures to give the various *succinimides* (R = hydrogen, alkyl, or aryl) :

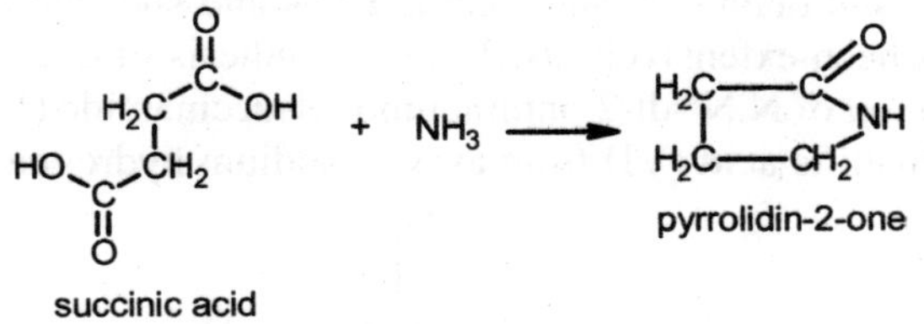

succinic acid + NH_3 ⟶ pyrrolidin-2-one

pyrrolidin-2-one

succinic acid

Diamines such as ethylenediamine or m-phenylenediamine react with two moles of succinic anhydride to yield *N,N'* -disuccinimides:

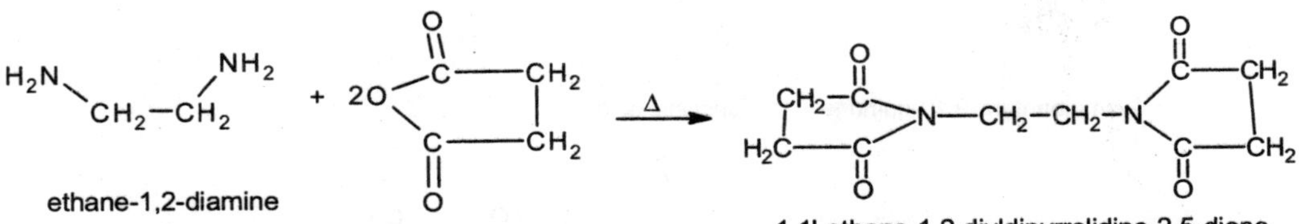

ethane-1,2-diamine + 2 dihydrofuran-2,5-dione $\xrightarrow{\Delta}$ 1,1'-ethane-1,2-diyldipyrrolidine-2,5-dione

ethane-1,2-diamine

dihydrofuran-2,5-dione

1,1'-ethane-1,2-diyldipyrrolidine-2,5-dione

By contrast, o-phenylenediarnine, in equimolecular proportion with succinic anhydride, gives a mixture of o,o'-diaminosuccinanilide (I) and 2-benzirnidazolepropionic acid (II), while two moles of the diarnine yield 1,2-di-2-benzimidazolylethane (2,2'-ethvlene bisbenzimidazole) (IV):

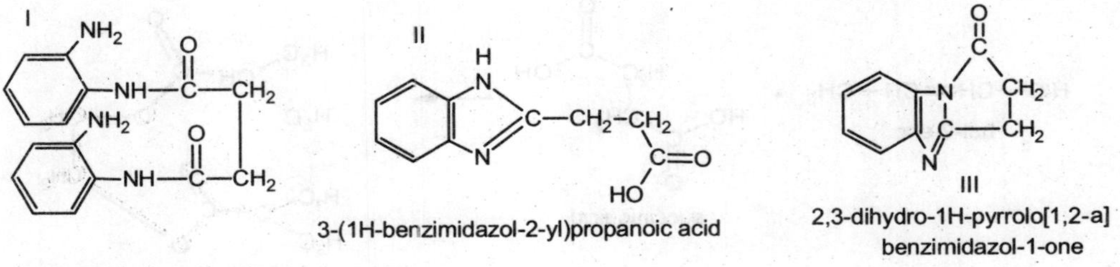

I

II

3-(1H-benzimidazol-2-yl)propanoic acid

III

2,3-dihydro-1H-pyrrolo[1,2-a] benzimidazol-1-one

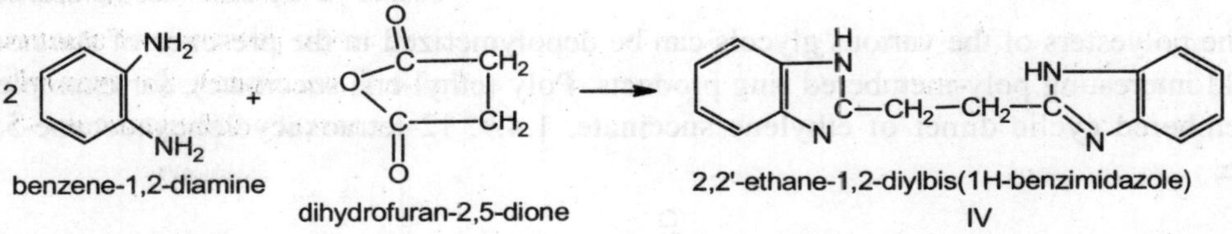

benzene-1,2-diamine

dihydrofuran-2,5-dione

2,2'-ethane-1,2-diylbis(1H-benzimidazole)

IV

2-Benzimidazolepropionic acid (II), when heated at 230-240°C., yields a-oxo-1,2-propylenebenzimidazole (2,3-dihydrl;.>pyrrolo[I.2-a]benzimidazol-I-one) (III) through further cyclization and loss of a mole of water.The zinc salt of o:-amlloselenophenol when treated with succinoyl chloride in pyridine gives 1,2-di-2-benzoselenazolylethane (2,2' ethylenebisbenzoselenazole):

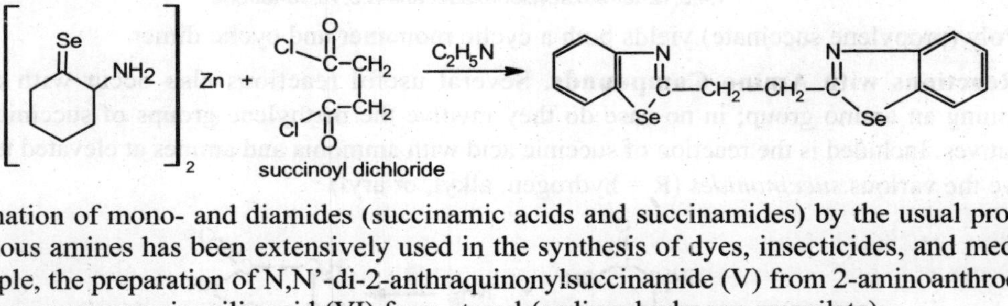

succinoyl dichloride

The formation of mono- and diamides (succinamic acids and succinamides) by the usual procedures with various amines has been extensively used in the synthesis of dyes, insecticides, and medicinals, for example, the preparation of N,N'-di-2-anthraquinonylsuccinamide (V) from 2-aminoanthroquinone and of *N* -p-arsonosuccinanilic acid (VI) from atoxyl (sodium hydrogen arsanilate):

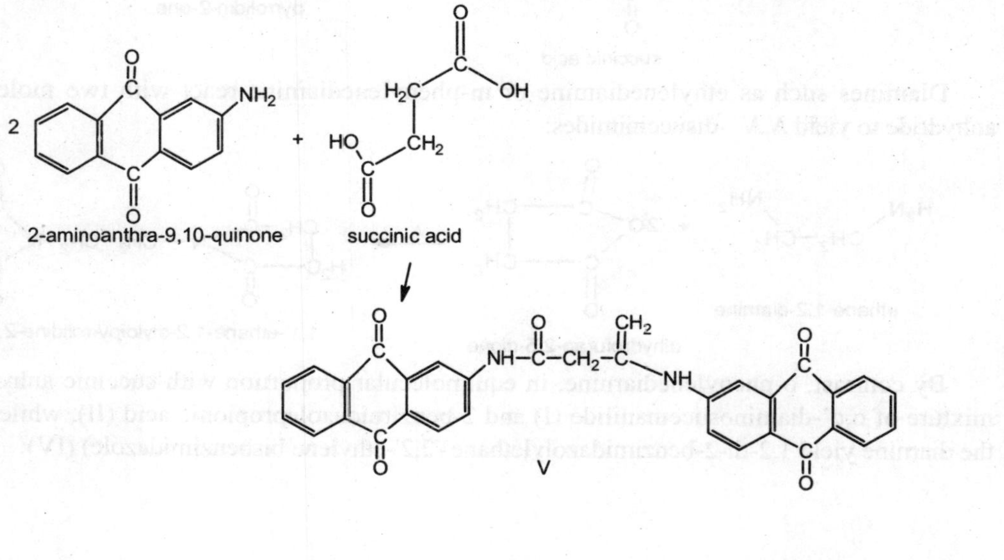

2-aminoanthra-9,10-quinone

succinic acid

V

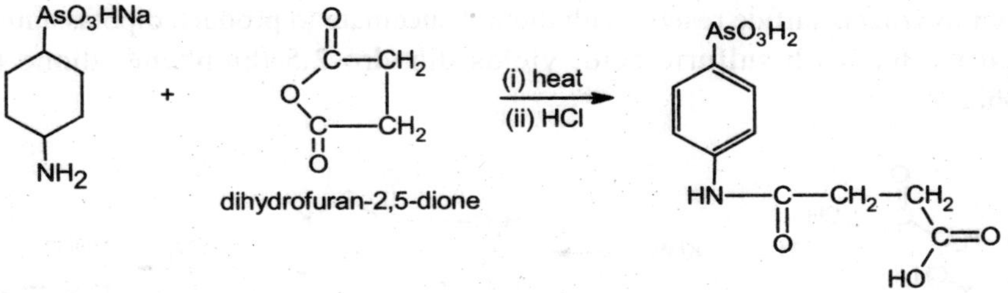

Benzamide, when heated with a molecular equivalent of succinic anhydride in benzene, yields a mixture of N,N'-dibenzoylsuccinamide and N-benzoylsuccinamic acid.

When various *N-substituted* succinamides are heated in dioxane at 250°C. under a hydrogen pressure of 200-300 atm. over a copper - chromium oxide cataJyst, fair yields of the respective pyrrolidines are obtained. N,N'-Bis(2-phenylethyl)-succinamide, for example, yields 65% of N-(2-phenylethyl)pyrrolidine and 13% of ethylbenzene:

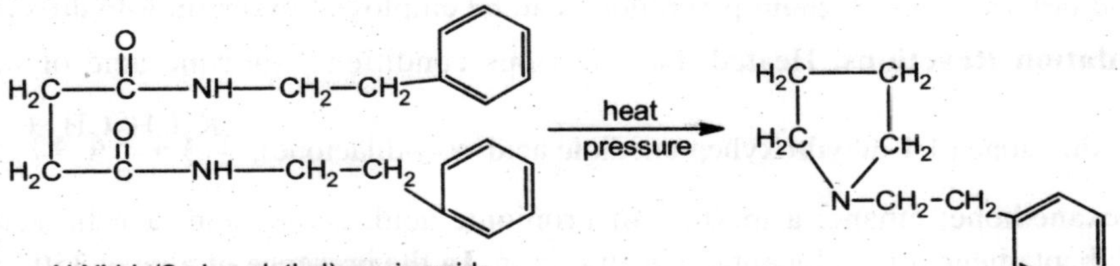

N,N'-bis(2-phenylethyl)succinamide

1-(2-phenylethyl)pyrrolidine

Reactions with Sulphur Compounds. Allyl mustard oil (allyl isothiocyanate) reacts with succinic anhydride to give *N-allylsuccinimide* by splitting off carbonyl sulphide:

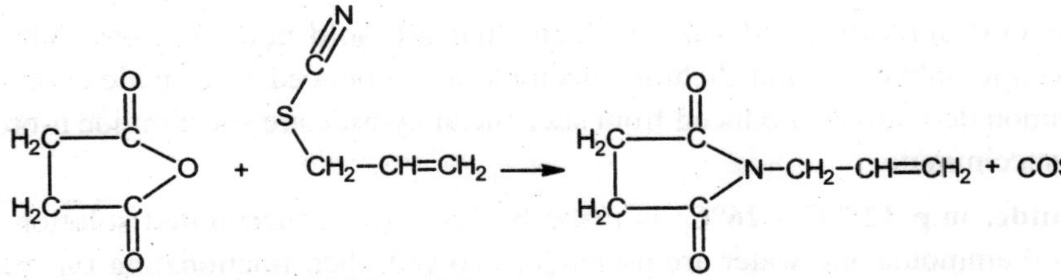

allyl thiocyanate

1-allylpyrrolidine-2,5-dione

Sulfur trioxide gives 2,3-disulphosuccinic acid as the result of its reaction with both methylene groups of succinic acid:

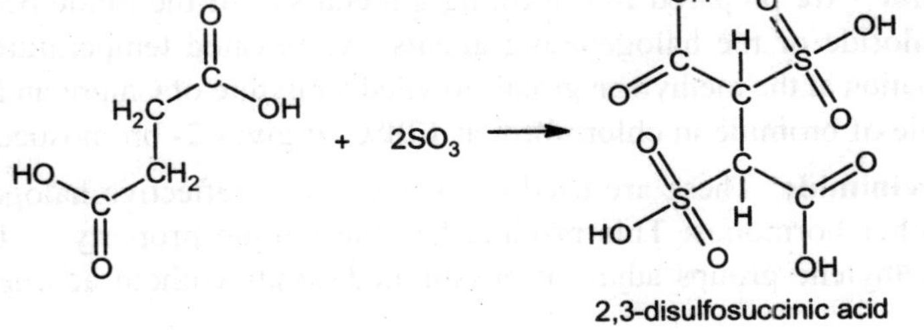

2,3-disulfosuccinic acid

Potassium hydrogen sulfide reacts 'with diethyl succinate to produce dipotassium dithiosuccinate, which, when treated with sulfuric acid, yields dihydro-2,5-thiophene- dione (dithiolsuccinic anhydrosulphide):

$$\text{diethyl succinate} + KHS \longrightarrow \text{dipotassium butanebis(thioate)} \xrightarrow{H_2SO_4} \text{dihydrothiophene-2,5-dione} + K_2S$$

diethyl succinate dipotassium butanebis(thioate) dihydrothiophene-2,5-dione

Oxidation. Hydrogen peroxide under various experimental conditions can be induced to yield either peroxysuccinic acid, $(CH_2COOOH)_2$, 2,2-dihydroxysuccinic acid, malonic acid, or a mixture of acetaldehyde, malonic and malic acids. Potassium permanganate gives either oxalic acid or a mixture of malic and tartaric acids. Sodium perchlorate can be employed to obtain 3-hydroxypropionic acid.

Degradation Reactions. Heated under various conditions, succinic acid or its salts yield -γ-ketopimelic dilactone (4,4-dihydroxyheptanedioic acid -γ,-γ-dilactone), $\underbrace{OC.CH_2.CH_2.C}_{O}, \underbrace{.CH_2.CH_2.CO}_{O}$

1,4-cyclohexanedione; ethane; a mixture of propionic acid, acetic acid, acrylic acid, and acrolein; oxalic acid; heptadiene; or cyclopentanone and furan. **In** the presence of zinc at 350—400°C. succinic acid gives high-melting hydrocarbons, and when subject to electrolysis, ethylene and acetylene.

Derivatives of Succinic Acid

There are several derivatives of succinic acid that are sold commercially. Many of these like dodecenylsuccinic anhydride and dichlorosuccinide are produced from maleic anhydride. These of the most common derivatives produced from succinic anhydride are succinimide n-bromosuccinimide, and n-chlorosuccinimide.

Succinimide, m.p 125°C - 26°C: is made by heating a concentrated solution of diammonium succinate, until ammonia and water are no longer evolved, then fractionating the molten product are collected , the material that boils at 285°C. It is also prepared, by heating succinic acid in a steam of ammonia or with area or formamide, or by treating aqueous acrylonitrile with alkali cyanide. It is used largely for the preparation of n-halogen substuited succinimide.

n- Succinimide: Are prepared first forming a metal salt of the imide before adding the halogen or using hydrochloride or the halogenating agents. At elevated temperatures (160°C) succinimide undergoes chlorination at the methylene groups to yiled a mixture of Chloro and 2, 3 dichloromaleimide, heated with a mole of bromide in chloroform at 128°C, it gives 2- bromosuccinimide.

N-bromosuccinimide: These are used extensively as a reflective halogenometry and oxidizing cortisones and other hormones. This product has the unique property of bringing about halogen substitution at methylene groups adjacent unsaturated bonds without adding to the ethylene group.

By varying reaction condition, it can be used as a reactant for sedative substitution, including the substitution of side chains attached to aromatics, for controlled dehydrogenation; for controlled dehydrogenation; for selective oxidation of alcohol groups to ketones or aldehydes. Those reactions have been applied to all types of compounds including olefines, carbonyl compounds aromatics hetrocyclines and even saturated substances.

N-Chlorosuccinimide: -Also called succinchloroimide) is a powerful germicide and development. It has been used widely as a disinfecting agent and to a limited extent as a halogenating agent similar to n-bromosuccinimide.

Another derivative is succinonitrile, which is obtained as a by product in the manufacturing of acrylonitrile. Succinonitrile has been for extracting aromatic compounds from petroleum distillate.

Succinaldehyde (butanedial), $OHCCH_2CH_2CHO$, is commercially available in limited quantitities as it is stable water soluble acetal, 2, 5 diethoxytetrahydrofurane formula weight 206.28 B.P 196°C at 5 mm Hg density 1.081.

ADIPIC ACID

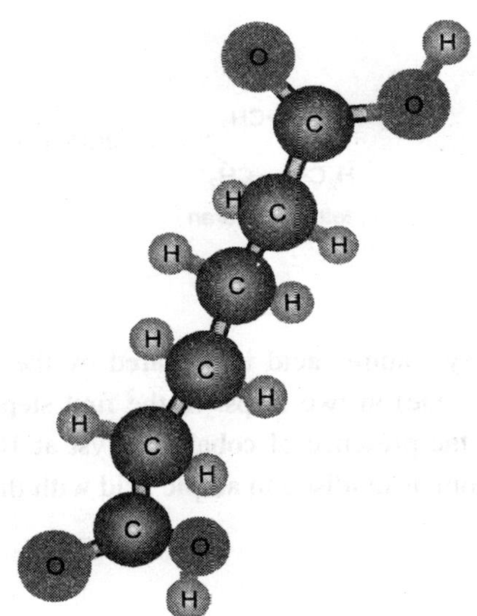

Formula: $C_5H_8O_4$

Also known as Hexane -1, 6-dioic acid

Introduction

This acid got its name form the word [Latin, adepes = fat] because it was obtained from the oxidation of fat.

Preparation

1. It can be prepared by the reaction of monosodiomaleic ester with ethylene bromide.

2. It is also obtained commercially by the action of cyclohexanal prepared from phenol or cyclohexanal with nitric acid in the presence of SeO_2 as a catalyst.

phenol + $3H_2$ → cyclohexanol —(HNO_3, SeO_2)→ cyclohexanone → adipic acid

3. It is also prepared in large scale by the reaction of tetrahydrofurane with carbon monoxide and water.

tetrahydrofuran + 2CO + H_2O → adipic acid

4. Now a day's adipic acid is prepared by the oxidation of cyclohexane (obtained by catalytic reduction of benzene) in two steps. In the first step cyclohexane in converted to cyclohexanone by air oxidation in the presence of cobalt catalyst at 100°C under a pressure of 150 psi. In the second step cyclohexanone is oxidised to adipic acid with the help of nitric acid.

benzene —(+ $3H_2O$, raney Ni)→ cyclohexane —($+O_2$, Co (catalyst))→ cyclohexanone —(+ $3O_2$, HNO_3)→ adipic acid

Physical Properties

Parachor	314.04 Cm^3
Melting Point	152 - 154°C
Boiling Point	337 C
Specific Gravity	1.36
Solubility In Water	Slightly Soluble
pH	3.45 (1%)
Vapor Density	5
Autoignition	420°C
Refractive Index	1.476
Dielectric Constant	52.4 Dyne/Cm
Flash Point	210°C
Stability	Stable Under Ordinary Conditions

Chemical Properties

Adipic acid is a crystalline solid, when heated in the presence of $Ba(OH)_2$ it is converted into cyclopentane.

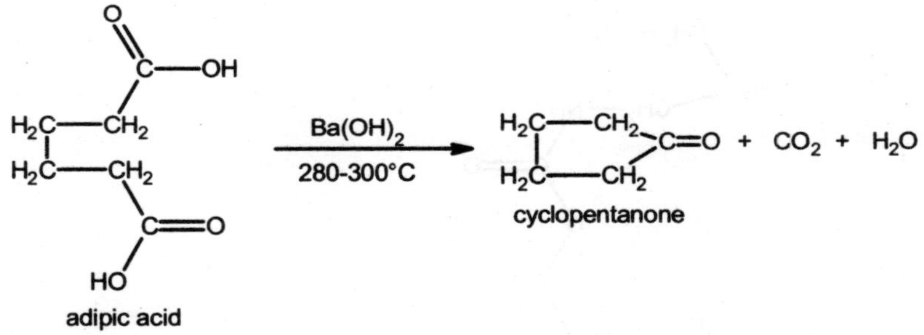

adipic acid $\xrightarrow[280\text{-}300°C]{Ba(OH)_2}$ cyclopentanone + CO_2 + H_2O

Upon heating with acetic anhydride, it gives linear polymeric anhydride

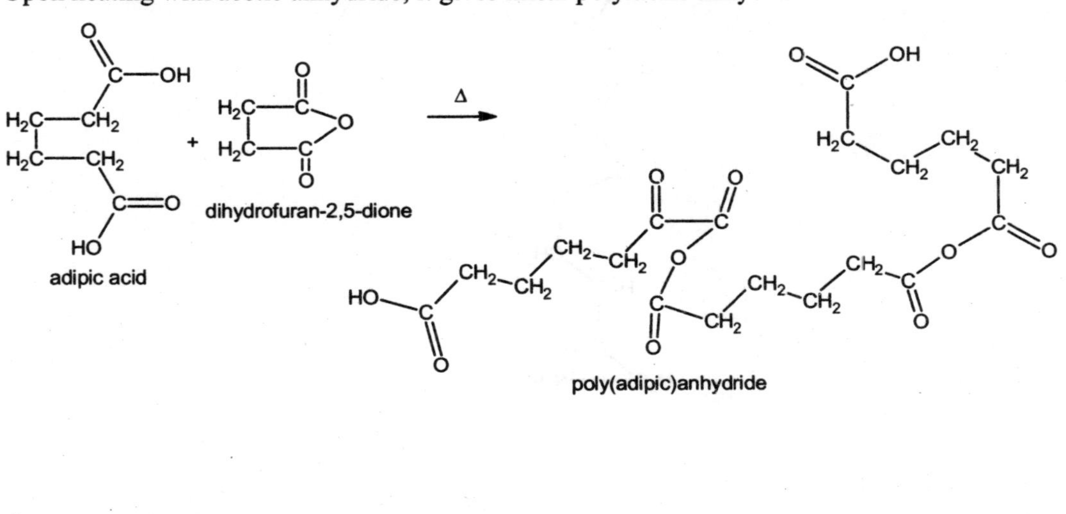

adipic acid + dihydrofuran-2,5-dione $\xrightarrow{\Delta}$ poly(adipic)anhydride

Adipic acid under goes condensation polymerization with 1,6 diaminohexane (hexamethylenediamine] to give a polyamide Nylon 6,6[see chapter polymerization]. It one of the most popular of the

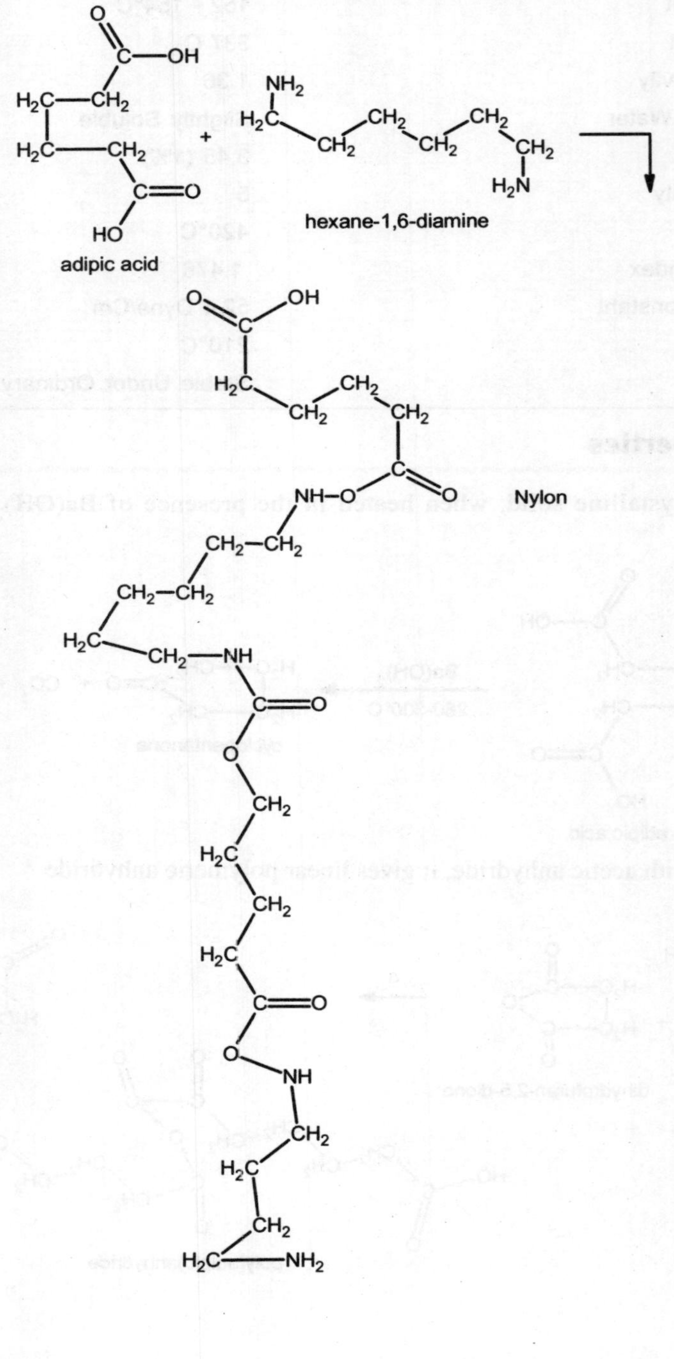

This why nylon is called nylon 6,6

adipic acid

hexane-1,6-diamine

Nylon

synthetic fibres and is so named in each repeating out its molecule contains six carbons in the amine fragment and six in the acid arrangement.

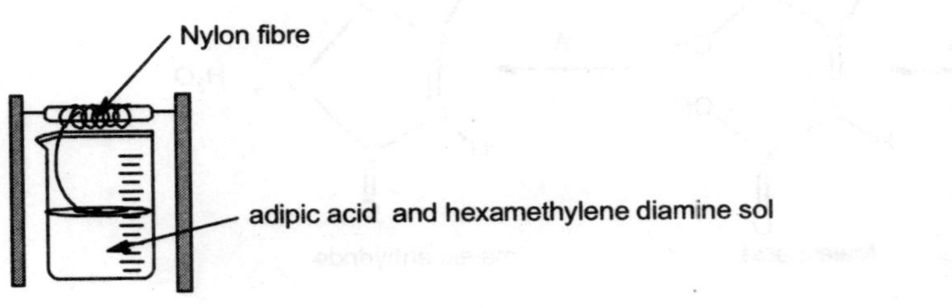

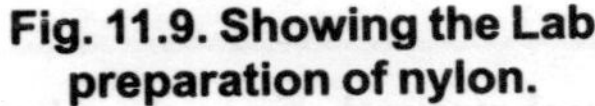
Fig. 11.9. Showing the Lab preparation of nylon.

Fig. 11.10. Showing the photograph of lab preparation of nylon.

UNSATURATED DICARBOXYLIC ACID

The two simplest and the important dicarboxylic acids are

Maleic acid fumaric acid

These acids from the classical example geometrical isomerism. Only one of them i.e. maleic acid when heated above its melting point readily yields a cyclic and hydride and hence this acid must be this *cis*-isomer. Therefore the other acid fumaric acid is the trans isomer. Fumeric anhydride is unknown. When fumaric acid sis heated for long periods of time, it is slowly transformed intro maleic anhydride by rotation about the double bond and elimination of a energy is absorbed shown by heat of combustion of the two acids, maleic acid 326.3 Kcal/mole; fumeric acid 320.1 Kcal/mole. This fumeric acid is more thermodynamically stable isomer as heat constant depicts, lower by 6.2 Kcal/mole

$$\text{Maleic acid} \xrightarrow{\Delta} \text{maleic anhydride} + H_2O$$

fumaric acid → (Δ) Maleic acid → (Δ) maleic anhydride + H_2O

Acidity of Maleic and Fumeric Acids

It might be thought at first sight that the two acids might be equal strength since they have identical structures. But actually maleic acid (K_a= 1.420′10^{-5}) is a much stronger acid than fumeric acid (K_a=93′10^{-5}). This is due to the fact that the maleate ion left after the removal of a position is stabilized by hydrogen bonding due to the close proximity of the two stabilized by resonance and also by hydrogen bonding fumarates is stabilized only by resonance. This greater stability of maleate anion relative to fumerate anion accounts for the higher acidity of maleic acid over fumeric acid.

Maleic acid ⇌ H^+ + Hydrogen bonded maleate ion

fumaric acid ⇌ H^+ + fumarate ion (no hydrogen bonding)

The second dissociation of fumaric acid occurs more readily than that of maleic acid (Ka_2 = 0.08 ′10^{-5}) and 3.6 ′ 10^{-5} respectively) since of is difficult to remove a proton the negatively charged cyclic system of maleate anion.

MALEIC ACID

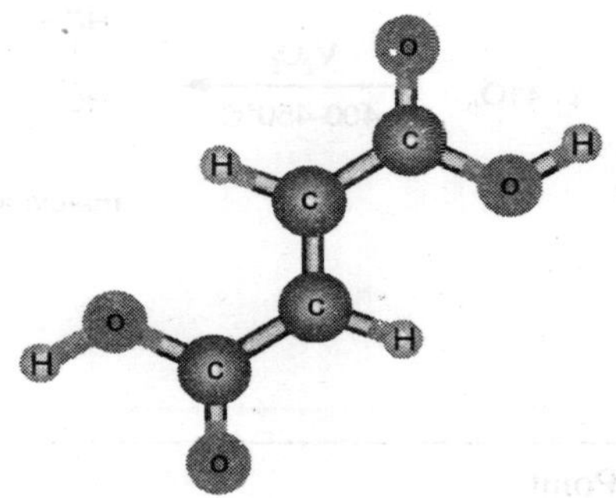

Formula: $C_4H_4O_4$

It does not occur in nature and hence it is a synthetic product

Preparation

(i) By heating malic acid rapidly at 258°C to form maleic anhydride which is then hydrolyzed with boiling water to give maleic acid.

$\xrightarrow[\Delta]{-H_2O}$

malic acid → maleic acid (intermediate) → maleic anhydride → maleic acid

(ii) By heating bromo succinic acid with ethanoic potash solution

$\xrightarrow{C_2H_5OH}$

2-bromosuccinic acid + potassium hydroxide → maleic acid

(iii) it is commercially prepared by the oxidation of benzene with air at the temperature of 400 - 450°C in presence of vanadium pentoxide (V_2O_5) as catalyst, followed by hydrolysis of maleic acid.

benzene + $4 1O_2$ $\xrightarrow[400\text{-}450°C]{V_2O_5}$ maleic anhydride $\xrightarrow{+H_2O}$ maleic acid

Properties

Melting Point	51 - 53 C
Boiling Point	202 C (Decomposes)
Specific Gravity	1.48
Solubility In Water	Hydrolysis
pH	
Vapor Density	3.4
Autoignition	477 C
Nfpa Ratings	Health: 3; Flammability: 1; Reactivity: 1
Refractive Index	
Flash Point	103 C
Stability	Stable Under Ordinary Conditions
Relative density (water = 1):	1.59
Solubility in water, g/100 ml at 25°C	78

Chemcal Properties

It forms two series of salts, ester amide and acid chloride.

sodium maleate diethyl maleate maleamide malyl chloride

(i) Action of Heat: When heated alone or with acetic anhydride it forms maleic anhydride

maleic acid $\xrightarrow[430°C]{\Delta}$ maleic anhydride

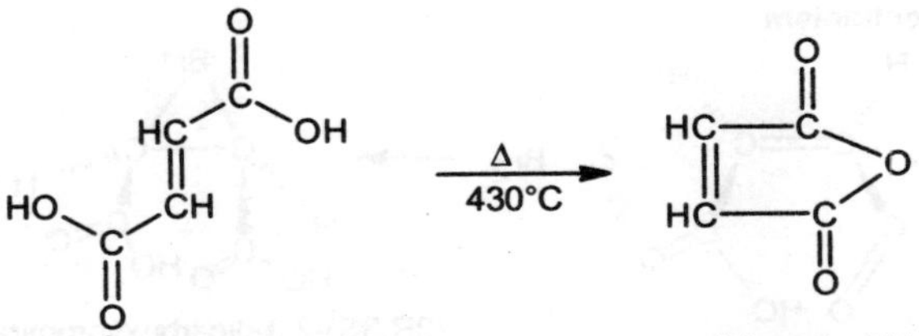

On prolonged heated at 150°C maleic acid isomerizes to fumeric acid

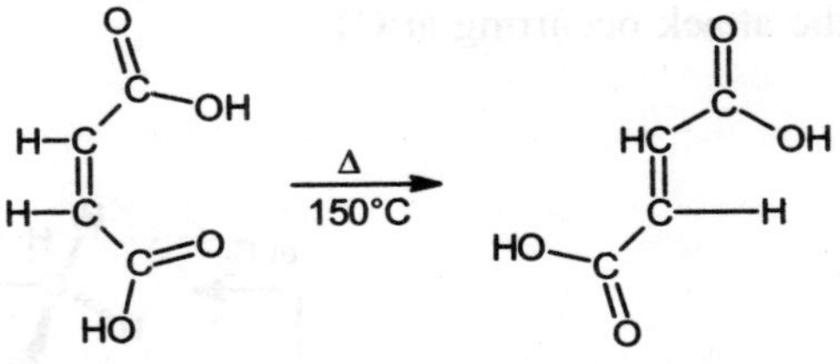

maleic acid $\xrightarrow[150°C]{\Delta}$ fumeric acid

(ii) Action with hydrogen bromide: Maleic acid reacts with hydrogen bromide to form bromosuccinic acid

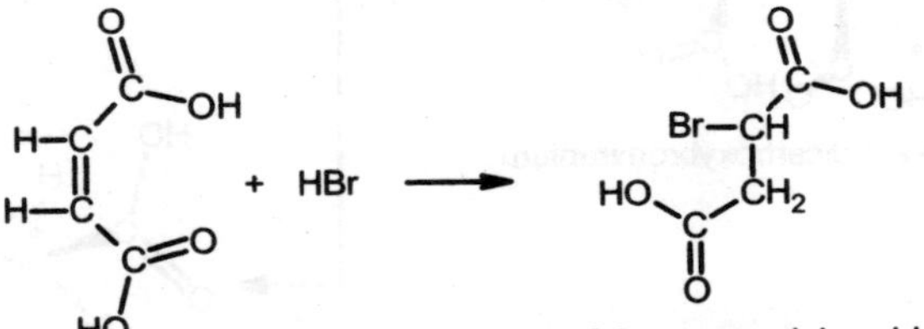

maleic acid + HBr ⟶ 2-bromosuccinic acid

(iii) Action with Bromine: Bromine dissolved in carbon tetrachloride or acetic acid adds to the double bond of maleic acid producing a mixture of and dibromosuccinic acid (racemic mixture)

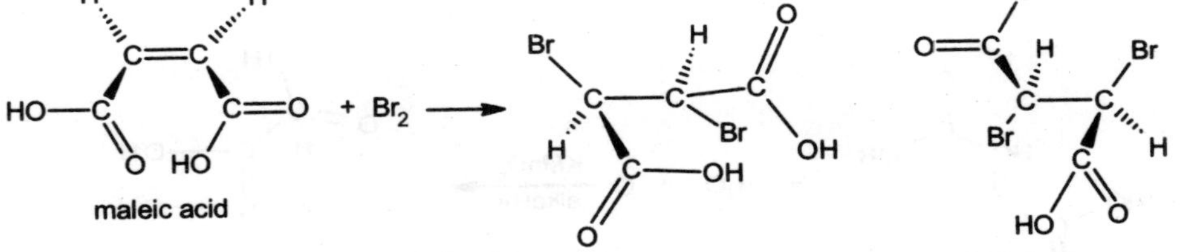

maleic acid + Br_2 ⟶ (2R,3R)-2,3-dibromosuccinic acid α-dibromosuccinic acid + (2S,3S)-2,3-dibromosuccinic acid l-dibromosuccinic acid

Mechanism

Step II: The bromonium ion undergoes nucliophilic attack by the bromide in two ways. When the attack occurs in C2 d-form of dibromosuccinic acid (I) is produced while l-dibromosuccinic acid (II) would result from the attack occurring at C1

(iv) Action with Alkaline $KMnO_4$: When heated with alkline potassium permanganate solution (or with osmium tetraoxide) it undergoes hydroxylation to produce mesotataric acid.

(b) Reduction: When reduced with hydrogen using Raney's Nickel as a catalyst, it gives succinic acid.

$$HOOC{-}CH{=}CH{-}COOH + H_2 \xrightarrow{\text{Raney's Nickel}} HOOC{-}CH_2{-}CH_2{-}COOH$$

maleic acid — succinic acid

Uses

(i) It inhibits rancidity in milk powders, oil and fats.

(ii) To prepare maleic anhydride for making varnishes and lacquers

It is also used in unsaturated polyester resins, lubricating oil additives, copolymers, fumaric acid, agricultural chemicals, malic acid, sulphosuccinic acid esters, alkenyl succinic anhydrides and alkyd resins.

FUMARIC ACID

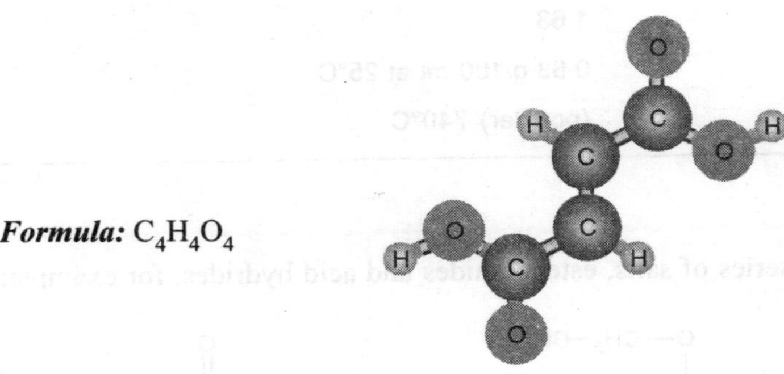

Formula: $C_4H_4O_4$

Also known as ethylene 1, 2 dicarboxylic acid. Fumeric acid occurs in many plants.

Perparation

1. By heating maleic acid for a long time at 150°C
2. By condensation of malonic acid with glycolinic acid in the presence of pyridine as a base.
3. Knovengal reaction, followed by decarboxylation at 130°C

oxoacetic acid + malonic acid —pyridine→ ethylene-1,1,2-tricarboxylic acid —$-CO_2$, Δ→ fumaric acid

4. Industrially fumaric acid is prepared by the action of maleic acid and hydrochloric acid or sodium hydroxide.

Physical Properties

Sublimation point:	200°C
Melting point	287°C (closed capillary, rapid heating);
Relative density (water = 1):	1.63
Solubility in water:	0.63 g/100 ml at 25°C
Auto-ignition temperature:	(powder) 740°C

Chemical Properties

(i) Formation of Salts: -It forms two series of salts, esters amides and acid hydrides, for example:

sodium fumarate diethyl fumarate fumaride

(ii) Reduction: - When reduced with dilute potassium permanganate, it forms di-nitrates of tartaric acids the addition takes place a *cis-* fusion

(iii) Action with HBr: Fumeric acid adds a molecule of hydrogen bromide to form bromosuccinic acid

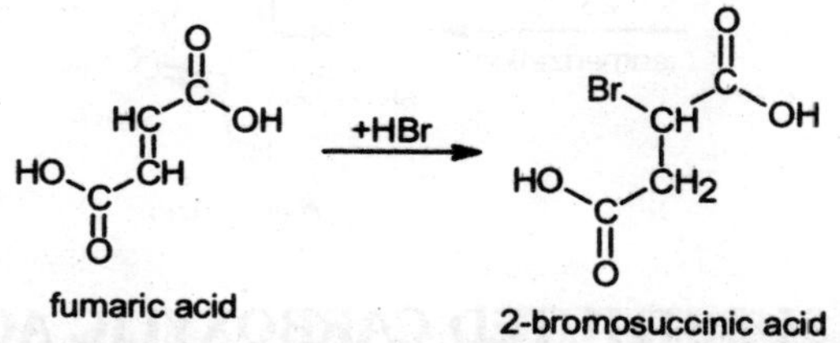

fumaric acid — +HBr → 2-bromosuccinic acid

(iv) Action with Bromine: When heated with Bromine in carbon tetrachloride or acetic acid or formic acid it gives monobromosuccinic acid.

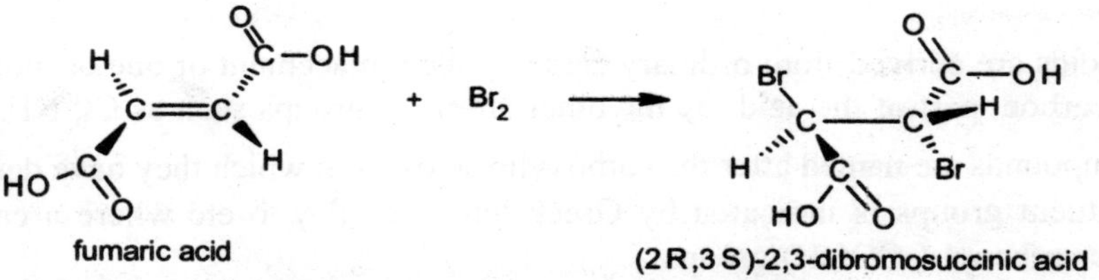

fumaric acid + Br_2 → (2R,3S)-2,3-dibromosuccinic acid

Mechanism: Like malic acid, fumaric acid also undergoes addition reaction with bromine. In terms of fusion steps of trans addition are: -

Step I

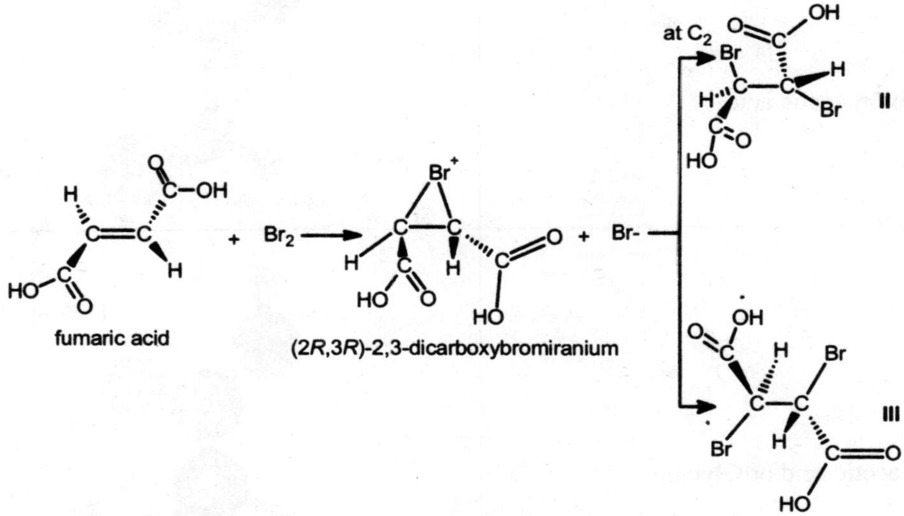

fumaric acid + Br_2 → (2*R*,3*R*)-2,3-dicarboxybromiranium + Br- → at C_2 II ; III

Now from II and III represent the same compound I as will be seen by rotating any one of the two through 180. The compound will not be optically active, since a plane of symmetry can be drawn through the molecule.

fumaric acid → (230°C, isomerization) → maleic acid → (Δ) → maleic anhydride

SUBSTITUTED CARBOXYLIC ACID

Introduction

Substituted acids are derived from ordinary acids by the replacement of one or more hydrogen atoms of the hydrocarbon, part of the acid, by the other atoms or groups such as Cl, NH_2OH etc

Such compounds are named after the carboxylic acids from which they have derived. The position of the substituent groups is indicated by Greek letters α, β γ, δ etc where a carbon atom is one nearest to the carbonyl (-COOH) groups.

The names and formula of the some of the important substituted carboxylic acid given below:

Monochloroacetic acid	H, O, O, C, C, H, H, Cl
Amino-acetic acid or Glycine	H, O, O, C, C, H, H, N, H

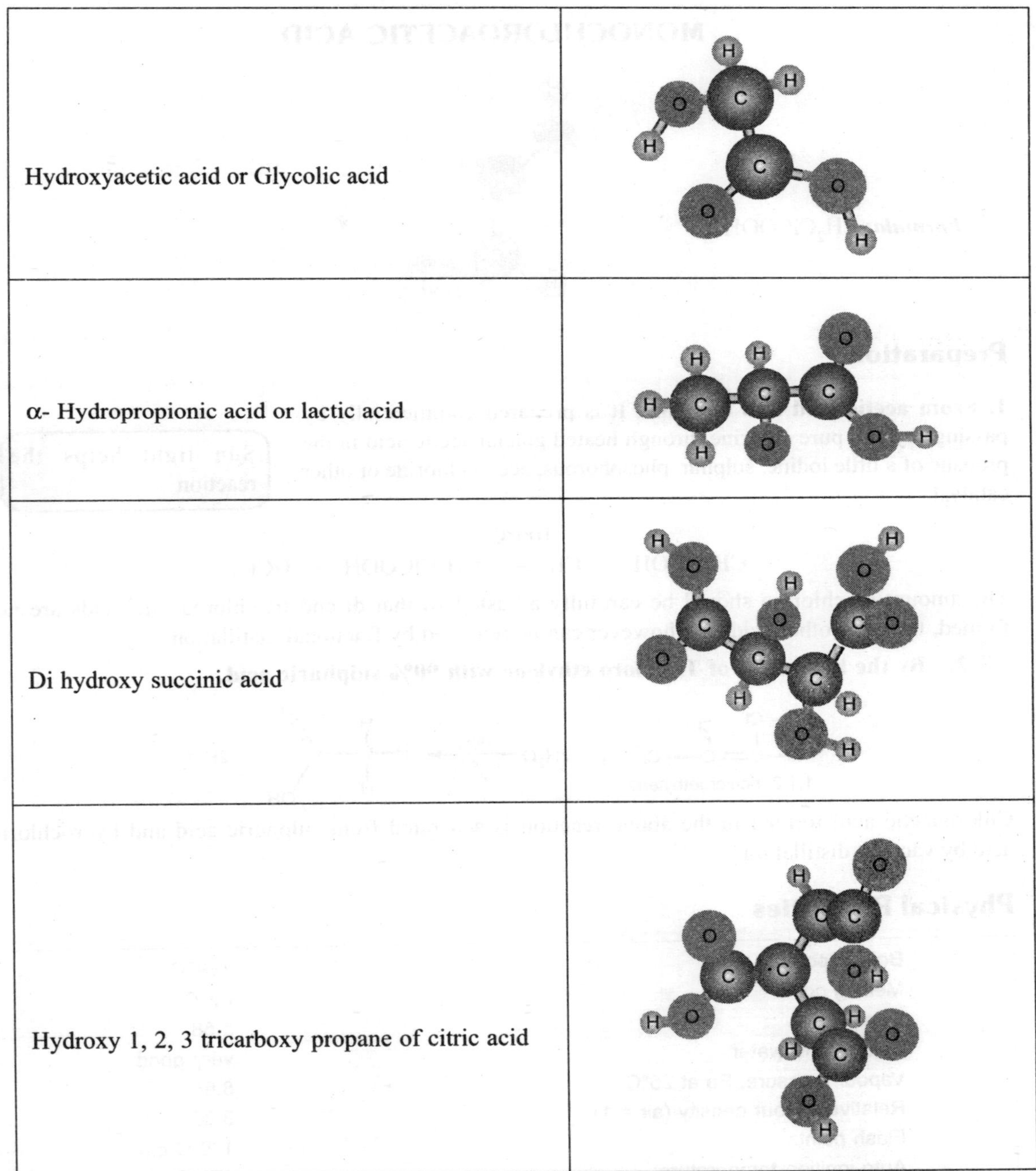

Hydroxyacetic acid or Glycolic acid	
α- Hydropropionic acid or lactic acid	
Di hydroxy succinic acid	
Hydroxy 1, 2, 3 tricarboxy propane of citric acid	

These compounds show the properties of the acid and also the properties of the substituent groups.

MONOCHLOROACETIC ACID

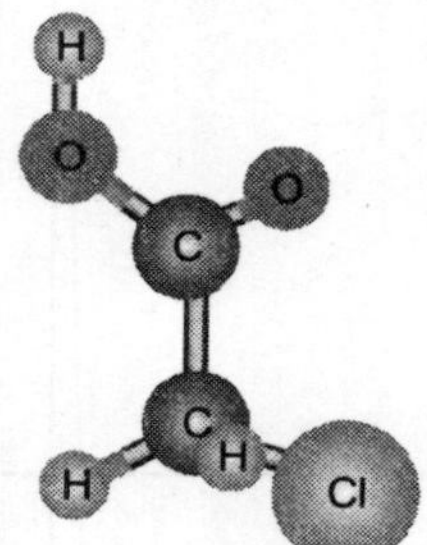

Formula: $CH_2ClCOOH$

Preparation

1. From acetic acid and chlorine. It is prepared commercially by passing dry and pure chlorine through heated glacial acetic acid in the propane of a little iodine, sulphur, phosphorous, acetyl chloride or other catalyst

Sun light helps the reaction

$$CH_3COOH + Cl_2 \xrightarrow{100°C} CH_2ClCOOH + HCl$$

The amount of chlorine should be carefully adjusted so that di and tri chloroacetic acids are not formed, traces of other acids are however can be removed by fractional distillation

2. By the hydrolysis of Trichloro ethylene with 90% sulphuric acid.

$$H-C(Cl)=C(Cl)-Cl + 2H_2O \xrightarrow[H_2SO_4]{90\%} H-CH_2-C(=O)OH + 2HCl$$

1,1,2-trichloroethylene

Chloroacetic acid formed in the above reaction is separated from sulphuric acid and hydrochloric acid by vacuum distillation.

Physical Properties

Boiling point:	189°C
Melting point:	63°C
Density:	1.58
Solubility in water:	very good
Vapour pressure, Pa at 25°C:	8.68
Relative vapour density (air = 1):	3.26
Flash point:	126°C c.c.
Auto-ignition temperature:	470°C
Explosive limits, vol% in air:	8
Octanol/water partition coefficient as log Pow:	0.34

Halogen substituted acids are stronger than the parent carboxylic the influence of the different halogens being in the order Cl>Br>I and carrying also with the number and position of the halogen atoms present.

Chemical Properties

It gives all the reaction of the acetic acid and alkyl halides

(A) Reactions of the carboxyl group

1. Neutralization: Salts are formed

$CH_2ClCOOH + NaOH \rightarrow CH_2ClCOONa + H_2O$

Choloroacetic acid — sodium chloroacetae

2. Esterification: Esters are produced

$CH_2ClCOOH + C_2H_5OH \rightarrow CH_2ClCOOC_2H_5 + H_2O$

Choloroacetic acid — Ethyl chloroacetate

3. With Phosphorous Pentachloride: Acid chloride is produced

$CH_2ClCOOH + PCl_5 \rightarrow CH_2ClCOCl + HCl + POCl_3$

Choloroacetic acid — Chloroacetyl Chloride

(B) Reactions of Chlorine

1. With Potassium Cyanide: Cyanoacetic acid is produced

$CH_2ClCOOH + KCN \rightarrow CH_2CNCOOH + HCl$

Cyanoacetic acid

This reaction is used for the preparation for the preparation of malonic acid

2. With Ammonia: Amine compounds are produced

$CH_2ClCOOK + 2NH_3 \rightarrow CH_2NH_2COOK + NH_4Cl$

Potassium Chloroacetate — potassium amino acetate

3. With Silver Nitrate: Nitro compounds are produced

$CH_2ClCOOH + AgNO_3 \rightarrow CH_2NO_2COOH + AgCl$

Nitroacetic acid

4. Hydrolysis with moist silver oxide: Hydroxyacetic acid is formed

$CH_2ClCOOH + H_2O \rightarrow CH_2OHCOOH + HCl$

5. With Potassium Iodide: Iodic acetic acid is formed

$CH_2ClCOOH + KI \rightarrow CH_2ICOOH + KCl$

USES

Chloroacetic acid is used in the inorganic synthesis and in the preparation of indigo

AMINO ACID

Amino Acids are those in which the hydrogen atoms of the hydrocarbon radical is replaced by the monovalent amino group NH_2. They are of great physiological importance and occur in animal body and plant tissue.

They are the decomposed product of albuminoidal substances and proteins. They are essential for the healthy growth of animals. Amino acetic acid is the most typical member of this class.

AMINO ACETIC ACID

Formula: NH_2CH_3OH

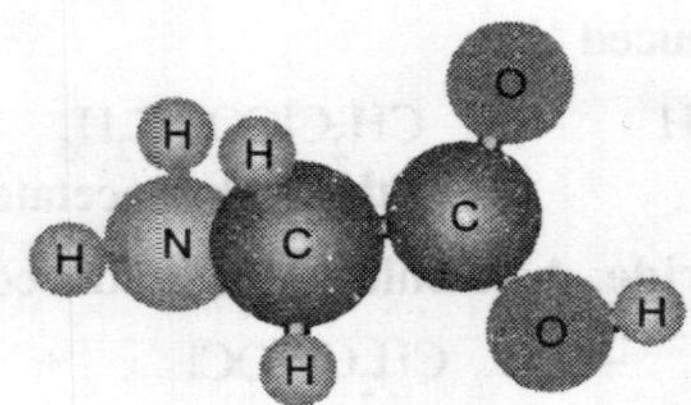

Also known as Glycocoll

Occurrence

It occurs in the urine of ox or hence as benzyl Glycine or Hippuric acid

$$H_3C-CH_2-\overset{\overset{\displaystyle O}{\|}}{C}-\overset{\overset{\displaystyle H}{|}}{N}-CH_2-\underset{\underset{\displaystyle HO}{|}}{C}{=}O$$

Preparation

1. Laboratory Method

Theory: By the action of liquor ammonia on chloroacetic acid

Procedure: A solution of chloroacetic acid in water in slowly added to ammonia solution. The mixture is allowed to stand for a few hours and then evaporated to a small amount. A current of steam may be passed through the liquid to recover excess of ammonia. The resulting solution is heated and treated with a slight excess of copper carbonate. This solution is followed, concentrated and cooled. Blue coloured crystals of cupric amino acetate $(NH_3CH_2COO)_2CuH_2O$ are obtained. They are requested by filtration, washed with dilute alcohol, dissolved in water and hydrogen sulphide is passed through then till the blue colour of the solution has disappeared. Copper sulphide (CuS) is filtered off and the filtrate of amino acetic acid are obtained.

$$(NH_2CH_2COOH)_2Cu + H_2S \rightarrow CuS + 2NH_2CH_2COOH$$

cupric amino acetic acid — amino acetic acid

2. From Protein: Hydrolysis of sulphuric acid (dilute)

Glue is boiled with sulphuric acid till completely hydrolyzed. From the resulting solution glycocoll is recovered through the formation of cupric amino acetate as derived below: -

3. From Formaldehyde: By the treatment with HCN and NH_3 followed by hydrogen (Streaker's reaction).

$$H-CHO \xrightarrow{HCN} H-C(OH)(H)-C\equiv N \xrightarrow[-H_2O]{NH_3} H-C(NH_2)(H)-C\equiv N \xrightarrow{2H_2O} H-C(NH_2)(H)-COOH + NH_3$$

formaldehyde; hydroxyacetonitrile; aminoacetic acid

4. From Hippuric Acid: By hydrolysis with dilute acids.

$$CH_3CH_2-CO-NH-CH_2-COOH + H_2O \longrightarrow H_3CCH_2-COOH + H_2N-CH_2-COOH$$

(propionylamino)acetic acid
hippuric acid; propionic acid; aminoacetic acid

Benzoic acid is filtered off glycine is crystallized out from the filtrate

Physical Properties

Physical State	White Crystals, Odorless
Melting Point	245°C (Decompose)
Specific Gravity	1.6
Solubility in Water	25g/100 Ml
pH	5.97 (Isoelectric Point)
Polarizability	6.50×10^{24} Cm^3
Refractive Index	1.460
NFPA Ratings	Health: 0 Flammability: 1 Reactivity: 0
Flash Point	145°C
Stability	Stable Under Ordinary Conditions
Prachor	162.5 Cm^3

Chemical Properties

Basic Properties

All amino acids found in proteins have this basic structure, differing only in the structure of the R-group or the side chain.

The simplest, and smallest, amino acid found in proteins is glycine for which the R-group is a hydrogen (H).

R-group or side chain
α-carbon
α-hydrogen
$^{+}H_3N$ — CH — C=O
α amino group
Carboxyl group

L-isomer

In proteins, only the L-isomer is found normally.

As you travel onward (from the carbonyl carbon to the amino group), the R group of L-amino acids will be on the left as shown in the molecular graphic on the right

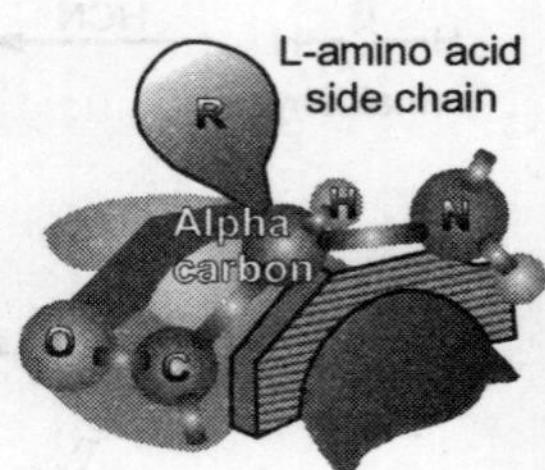

Since the amino acetic acid contains both the groups (-COOH) and the amino ($-NH_2$) group, it gives all the reactions of acetic acid and primary amine.

(A) Reactions of Carboxyl group:

1. Neutralization: Salt are formed,

$$NH_2CH_2COOH + KOH \rightarrow NH_2CH_2COOK + H_2O$$

Amino acetic acid → Potassium amino acetate

2. Esterification: Esters formed,

Hydrogen chloride is passed through a mixture of amino acetic acid and ethanol. The product containing ethyl amino acetate hydrochloride is then decomposed by alkaline to get free ester.

$$2HN_2CH_2COOH + 2C_2H_5COOH \xrightarrow{2HCl} 2HClNH_2CH_2COOC_2COOC_2H_5 + 2H_2O$$

ethyl amino acetate hydrochloride

$$2HClNH_2CH_2COOC_2COOC_2H_5 \rightarrow NH_2CH_2COOC_2H_5 + CO_2 + H_2O + 2NaCl$$

ethyl amino acetate

3. With phosphorous pentachloride, PCl_5: Acid chloride is formed

$$NH_2CH_2COOH + PCl_5 \rightarrow NH_2CH_2COOCl + POCl_3 + HCl$$

Amino acetyl chloride

4. With Copper carbonate: Blue precipitate of cupric amino acetate is produced

$$2NH_2CH_2COOH + CuCO_3 \rightarrow (NH_2CH_2COOH)_2Cu + CO_2 + H_2O$$

cupric aminoaceate (blue ppt)

The reaction is used for the isolation and purification of glycine.

(B) Reaction of the Amino Group

5. with hydrogen chloride: - Salt is formed

$NH_2COOHCH_2$ + HCl → $HClNH_2COOH$
Glycine hydrochloride

6. With Nitrous Acid: - Hydroxy acetic acid is produced

$NH_2COOHCH_2$ + HNO_2 → $HOCH_2COOH$ + H_2O
Hydroxyacetic acid

7. With acetyl chloride: (Acetylation): Acetyl glycine is produced.

$H_2N-CH_2-C(=O)OH + H_3C-C(=O)Cl \longrightarrow H_3C-C(=O)-NH-CH_2-C(=O)OH + HCl$

aminoacetic acid, acetyl chloride, (acetylamino)acetic acid

8. With methyl iodide (Alkulation): Methyl glycine is obtained (decomposition)

$H_2N-CH_2-C(=O)OH + H_3C-I \longrightarrow H_3C-NH-CH_2-C(=O)OH + HI$

aminoacetic acid, iodomethane, (methylamino)acetic acid

9. Action of Heat: Glycine anhydride is formed

$2\ H_2N-CH_2-C(=O)OH \xrightarrow{-2H_2O}$ glycine anhydride ($H_3C-NH-C(=O)$ / $O=CH-NH-CH_2$)

aminoacetic acid, glycine anhydride

10. Action of heat with barium oxide or soda lime: Methylamine is produced

$H_2N-CH_2-C(=O)OH + BaO \longrightarrow H_3C-NH_2$

aminoacetic acid, methanamine

Uses

Flavour enhancers and maskers, pH buffers and stabilizers, an ingredient in pharmaceutical products, food and personal care products and as a chemical intermediate.

It helps to improve glycogen storage utilized in the synthesis of hemoglobin, collagen, and glutathione, and facilitates the amelioration of high blood fat and uric acid levels.

HYDROXY ACETIC ACID
OR
GLYCOLIC ACID

Formula: $C_2H_3O_2$

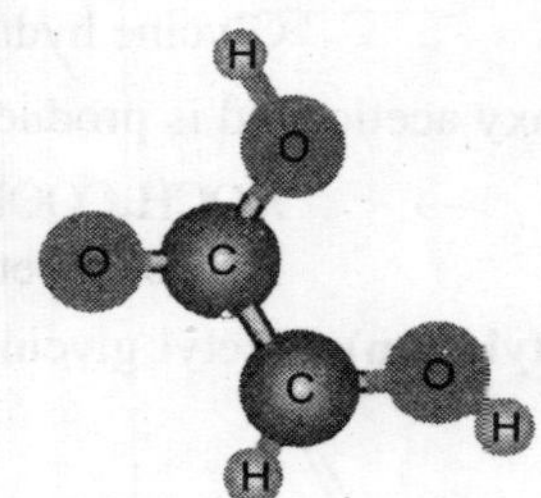

Occurrence

It is present in the cane juce, unripe grapes and tomato puree.

Preparation

1. From Chloroacetic acid: By hydrolysis

$$\underset{\text{chloroacetic acid}}{Cl{-}CH_2{-}C(=O){-}OH} + H_2O \xrightarrow{\text{boil}} \underset{\text{hydroxyacetic acid}}{HO{-}CH_2{-}C(=O){-}OH} + HCl$$

2. From glycine: By the action of Nitrous acid

$$\underset{\text{aminoacetic acid}}{H_2N{-}CH_2{-}C(=O){-}OH} + HNO_2 \longrightarrow \underset{\text{hydroxyacetic acid}}{HO{-}CH_2{-}C(=O){-}OH}$$

3. From the Glyoxalic acid: By reduction with sodium and water

$$\underset{\text{oxoacetic acid}}{HO{-}C(=O){-}HC{=}O} + 2H \longrightarrow \underset{\text{hydroxyacetic acid}}{HO{-}CH_2{-}C(=O){-}OH}$$

4. From Oxalic acid: By reduction with Zn and sulphuric acid

$$\underset{\text{oxalic acid}}{HOOC-COOH} + 4H \longrightarrow \underset{\text{hydroxyacetic acid}}{HO-CH_2-COOH}$$

Electrolytic reduction of oxalic is used for preparation glycolic acid on the commercial scale

Manufacture: The reaction is conducted in cell containing lead cathode surrounded by a solution of oxalic acid in dilute sulphuric acid (see fig 54). The lead anode is placed

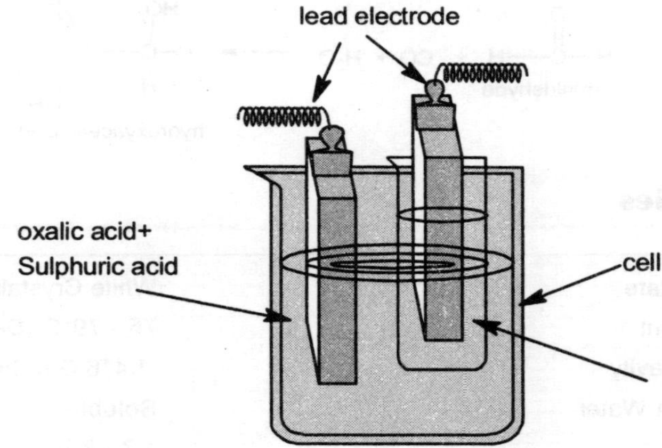

Fig. 11.11. Showing the Reduction of Oxalic Acid.

The product containing glycolic acid with excess oxalic acid and sulphuric acid, which is neutralized with calcium carbonate. Calcium glycolate (soluble), calcium oxalate (insoluble) and calcium sulphate (insoluble) are formed. Glycolic acid is finally recovered from calcium glycolate by treating it with the calculated amount of sulphuric acid.

$$\underset{\text{calcium glycolate}}{(CH_2OHCOO)_2Ca} + H_2SO_4 \rightarrow \underset{\text{glycolic acid}}{2CH_2OHCOOH} + \underset{\text{calcium sulphate (ppt)}}{CaSO_4\downarrow}$$

The solution of glycolic acid is concentrated and allowed to cool with colourless, crystalline hydroscopic solid is obtained.

5. From Ethylene Glycol: By partial oxidation with dilute nitric acid

$$\underset{\text{ethylene glycol}}{H_2C(OH)-H_2C(OH)} \longrightarrow \underset{\text{hydroxyacetic acid}}{HO-CH_2-COOH} + H_2O$$

6. From Formaldehyde: By the treatment with Hydrogen cyanide followed by hydrolysis

$$H-\overset{O}{\overset{\|}{C}}-H \xrightarrow{HCN} H-\underset{\underset{\underset{N}{|||}}{C}}{\overset{H}{\overset{|}{C}}}-H \xrightarrow{2H_2O} HO-CH_2-\overset{O}{\overset{\|}{C}}-OH$$

formaldehyde → acetonitrile → hydroxyacetic acid

or by heating foemaldehyde, carbon monoxide and water under pressure in the presence of a catalyst (acetic acid and sulphuric acid).

$$H-\overset{O}{\overset{\|}{C}}-H + CO + H_2O \longrightarrow H-\underset{H}{\overset{HO}{C}}-\overset{O}{\overset{\|}{C}}-OH$$

formaldehyde → hydroxyacetic acid

Physical Properties

Physical State	White Crystals
Melting Point	75 - 79°C (Decomposes)
Specific Gravity	1.416 Gm/Cm^3
Solubility in Water	Soluble
pH	1.7 - 2.5
Polarizibity	5.71×10^{-24} Cm^3
Surface Tension	61.3 Dyne/Cm^3
Refractive Index	1.449
Stability	Stable under Ordinary Conditions

Chemical Properties

It possess a dual character of a carboxylic acid and a primary alcohol

A. Reaction of the Carbonyl Group

1. Neutralization: It forms salts

$$HOCH_2COOH + NaOH \rightarrow HOCH_2COONa + H_2O$$

Sodium glycolate

2. Esterification: It forms ester

$$HOCH_2COOH + C_2H_5OH \xrightarrow[Dry]{HCl} HOCH_2COOC_2H_5$$

3. With Ammonia: It forms amide

$HOCH_2COOH + NH_3$ → $HO—CH_2—COONH_2$

Hydroxy acetic acid → Ammonium glycolate

$HO - CH_2 - COONH_4$ → $HO - CH_2COONH_2 + H_2O$

Ammonium Glycolate → Glycol amide

B. Reaction of the Primary Alcoholic (-CH_2OH) group

4. With hydrogen chloride: it forms chloroacetic acid

$$HO - CH_2 - COOH + HCl \rightarrow ClCH_2COOH + H_2O$$

5. With acetyl chloride: It forms acetyl glycolic acid

HO–CH₂–C(=O)–OH + H₃C–C(=O)–Cl → H₃C–C(=O)–CH₂–C(=O)–HO + HCl

hydroxyacetic acid   acetyl chloride   3-oxobutanoic acid

6. With Hydrogen (reduction): It gives acetic acid

$$HOCH_2COOH + 2HI \rightarrow CH_3COOH + H_2O + I_2$$

7. Oxidation: It gives oxalic acid

$H_2C(OH)–C(=O)OH \xrightarrow[-H_2O]{O} HC(=O)–C(=O)OH \xrightarrow[-H_2O]{O} HO–C(=O)–C(=O)–OH$

hydroxyacetic acid   oxoacetic acid   oxalic acid

C. Reactions involving both carbonyl and Alcohol groups

8. With Sodium: Disodium glycolate is produced

$HO–CH_2–C(=O)–OH \xrightarrow{2Na} O=C(OH)–CH_2–O–Na + H_2$

hydroxyacetic acid   sodium carboxymethanolate

9. With Phosphorous Pentachloride: Chloroaldehyde chloride is produced

$HO–CH_2–C(=O)–OH + 2PCl_5 \rightarrow Cl–C(=O)–CH_2–H_2C–Cl + 2POCl_3 + 2HCl$

hydroxyacetic acid   3-chloropropanoyl chloride

10. Action of heat: Inner anhydride and glycolide is formed

$$2\ HO{-}H_2C{-}COOH \xrightarrow[{[-H_2O]}]{100^{\circ}C} \text{glycolide}$$

hydroxyacetic acid

hydroxyacetic acid

glycolide

Uses

Due to its excellent capability to penetrate skin, glycolic acid finds applications in skin care products, most often as a chemical peel performed by a dermatologist in concentrationns of 20%-80% or at-home kits in lower concentrations of 10%. It is used to improve the skin's appearance and texture. It may reduce wrinkles, acne scarring, hyperpigmentation and improve many other skin conditions. Once applied, glycolic acid reacts with the upper layer of the epidermis, weakening the binding properties of the lipids that hold the dead skin cells together. This allows the outer skin to "dissolve" revealing the underlying, healthier, smoother, brighter-looking skin.

Glycolic acid is also a useful intermediate for organic synthesis, in a range of reactions including: oxidation-reduction, esterification and long chain polymerization. Among other uses this compound finds employment in the textile industry as a dyeing and tanning agent, in food processing as a flavouring agent and as a preservative. Glycolic acid is often included into emulsion polymers, solvents and additives for ink and paint in order to improve flow properties and impart gloss.

LACTIC ACID

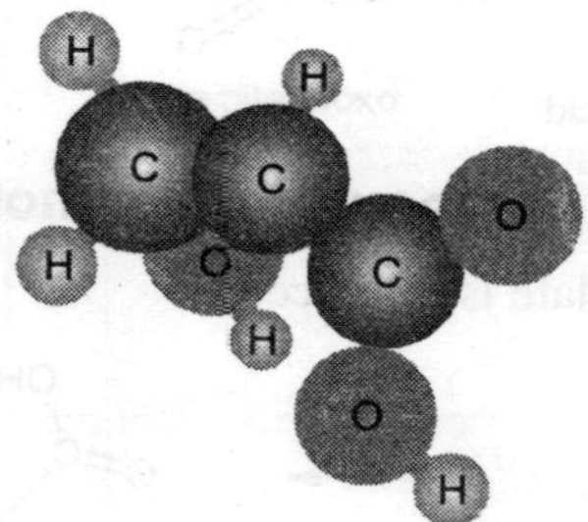

Formula: $C_3H_6O_2$

Occurrence

Lactic acid (Latin Lactis = milk) occurs in sour milk, curd and cheese where it is formed by the lactic fermentation of milk sugar –lactose. It also occurs in the muscles of the body and blood.

Preparation

It resembles glycolic acid in preparation

1. From α-chloro-propinic acid: By hydrolysis

$$CH_3-CH(Cl)-COOH + H_2O \xrightarrow{boil} CH_3-CH(Cl)-COOH + HCl$$

2-chloropropanoic acid   2-chloropropanoic acid

2. From α-Amino-propionic acid (alanine): by the action of nitrous acid

$$CH_3-CH(NH_2)-COOH + HNO_2 \longrightarrow CH_3-CH(OH)-COOH + N_2 + H_2O$$

2-aminopropanoic acid   2-hydroxypropanoic acid

3. From Pyruvic acid: By reduction of pyruvic acid

$$CH_3-CO-COOH + 2H \longrightarrow CH_3-CH(OH)-COOH$$

pyruvic acid   lactic acid

4. From Methyl glycol: By partial oxidation

$$CH_3-CH=CH_2 \xrightarrow{Cl_2} CH_3-CHCl-CH_2Cl \xrightarrow[aqueous]{2NaOH} CH_3-CH(OH)-CH_2OH \xrightarrow{O} CH_3-CH(OH)-COOH$$

prop-1-ene   1,2-dichloropropane   propylene glycol   lactic acid

5. From Acetaldehyde: By treatment with hydrogen cyanide followed by hydrolysis.

$$CH_3-CHO \xrightarrow{HCN} CH_3-CH(OH)-C\equiv N \xrightarrow[-NH_3]{2H_2O} CH_3-CH(OH)-COOH$$

acetaldehyde   2-hydroxypropanenitrile   lactic acid

Laboratory Preparation and Manufacture

From sugar by Fermentation: To some dilute solution (15% - 20%) of glucose or cane sugar is added, old chese or sour milk which supplies lactic bacilli. The temperature is maintained between 45°C - 55°C to avoid the formation of butyric acid. As the reaction proceeds calcium carbonate is gradually added to neutralize the lactic acid as soon as it is formed. This is necessary because the presence of free acid hinders the growth and working of lactic ractions

$$\underset{\text{cane sugar}}{C_{12}H_{22}O_{11}} + H_2O \rightarrow \underset{\text{glucose}}{C_6H_{12}O_6} + \underset{\text{fructose}}{C_6H_{12}O_6}$$

$$\underset{\text{glu cos e}}{C_6H_{12}O_6} \xrightarrow{\text{Lactic Fermentation}} \underset{\text{lactic acid}}{2CH_3CHOH.COOH}$$

$$2CH_3CHOH.COOH + CaCO_3 \rightarrow (CH_3CHOH.COO)Ca + H_2O + CO_2$$

Calcium lactate so formed is treated with the required amount of H_2SO_4

$$\underset{\text{calcium lactate}}{(CH_3CHOH.COO)_2Ca} + H_2SO_4 \rightarrow CH_3COOH.COOH + \underset{\text{lactic acid}}{CaSO_4}$$

The precipitate of calcium sulphate is filtered off and the filtrate of the lactic acid is made concentrated by distillation under pressure and by cooling below 18°C white crystalline solid is obtained.

Physical Properties

Molecular weight	90.08
Melting point	16.8°C
Boiling point	82°C at 0.5 mm Hg122°C at 14 mm Hg
Dissociation constant, K_a at 25°C	1.37×10^{-4}
Heat of combustion, DH_c	1361 KJ/mole
Specific Heat, C_p at 20°C	190 J/mole/°C

Chemical Properties

It is similar to glycolic acid in reactions:

(A) As an acid

1. **Neutralized**: It produces salt

$$\underset{\text{lactic acid}}{2CH_3CHOHCOOH} + Ca(OH)_2 \rightarrow \underset{\text{calcium lactate}}{(CH_3CHOH.COO)_2Ca} + H_2O$$

2. **Esterification**: It forms esters

$$2CH_3CHOHCOOH + C_2H_5OH \xrightarrow{HCl/Dry} CH_3COOHCOOC_2H_5 + H_2O$$

3. With Ammonia: It forms amide

$$\underset{\text{lactic acid}}{CH_3CHOHCOOH} \xrightarrow{NH_3} CH_3CHOH.COONH_4 \xrightarrow{Heat} H_2O + \underset{\text{lactic amide}}{CH_3CHOHCONH_2}$$

(B) As a secondary Alcohol

4. With hydrogen chloride: It forms α-Chloro propionic acid

$$\underset{\text{Lactic acid}}{CH_3CHOHCOOH} + HCl \rightarrow \underset{\text{a- Chloro propionic acid}}{CH_3CHClCOOH} + H_2O$$

5. With acetyl chloride (acetylation): It forms acetyl derivatives

$$\underset{\text{lactic acid}}{CH_3CH(OH)COOH} + \underset{\text{acetyl chloride}}{ClCOCH_3} \longrightarrow \underset{\text{acetyl lactic acid}}{HOOC{-}CH(CH_3){-}CO{-}CH_3} + HCl$$

6. With hydrogen Iodide(reduction): It gives propionic acid

$$\underset{\text{lactic acid}}{CH_3CH(OH)COOH} \xrightarrow{2HI} \underset{\text{propionic acid}}{HOOC{-}CH_2{-}CH_3} + H_2O + I_2$$

7. Oxidation: With hydrogen peroxide and potassium permanganate

It gives pyruvic acid

$$\underset{\text{lactic acid}}{CH_3CH(OH)COOH} \xrightarrow{O} \underset{\text{2-oxopropanoic acid}}{HOOC{-}C(=O){-}CH_3} + H_2O$$

8. With hot and dilute sulphuric acid: It decomposes to acetaldehyde and formic acid

$$CH_3CH(OH)COOH \xrightarrow[\text{Sulphuric acid at 180°C}]{\text{heated with}} H_3C{-}CHO + HCOOH$$

lactic acid → acetaldehyde + formic acid

9. With phosphorus pentachloride: α-chloropropionyl chloride is formed

$$CH_3CH(OH)COOH + 2PCl_5 \longrightarrow CH_3CH(Cl)COCl + POCl_3 + HCl$$

lactic acid → 2-chloropropanoyl chloride

10. Action of heat:

$$2\,CH_3CH(OH)COOH \xrightarrow{-2H_2O} \text{lactamide (cyclic dimer)}$$

2-hydroxypropanoic acid → lactamide

Uses

Lactic acid (a-hydroxypropionic acid) is a chemical compound that plays a role in several biochemical processes.

Lactic acid is a carboxylic acid with a chemical formula of $C_3H_6O_3$. Its structure is reflected in its systematic name, 2-hydroxypropanoic acid. In solution, it can lose a proton from the COOH carboxy group, turning into the **lactate** ion $CH_3CHOHCOO^-$.

There are two optical isomers of lactic acid (and of lactate) since the central carbon atom is bound to four different groups. The first isomer is known as L(+)-lactic acid or (S)-lactic acid and the second is D(-)-lactic acid or (R)-lactic acid.

Lactic acid accumulates in skeletal muscles during extensive anaerobic exercise, causing temporary muscle pain. Lactic acid is quickly removed from muscles when they resume aerobic metabolism. Delayed onset muscle soreness usually becomes apparent more than 24 hours after exercising and is not caused by lactic acid buildup.During one form of anerobic glycolysis or fermentation, L-lactate is produced from pyruvate via the enzyme lactate dehydrogenase. This conversion also oxidizes one molecule of NADH to NAD^+, and this is the reason for the conversion: NAD^+ has to be regenerated so that glycolysis can continue.This lactic acid fermentation occurs in red blood cells since they don't have mitochondria, and in skeletal muscle during intense exertion when sufficient amounts of oxygen cannot be supplied fast enough. This lactate is released into the bloodstream. The typical lactate concentration in the blood is 1-2 mmol/L. The liver takes up about 60% of the lactate and reoxidizes it to pyruvate, which is then reconverted to glucose in a process known as gluconeogenesis. The glucose enters the bloodstream and can be used by the tissues. This glucose → lactate → glucose cycle, originally described by Carl and Gerty Cori, is known as the Cori cycle. About 40% of the lactate is taken up by well oxygenated muscle cells and oxidized to pyruvate, which is then directly used to fuel the citric acid cycle.Lactic acid fermentation is also performed by lactic acid bacteria. These bacteria can operate in the mouth; the acid they produce is responsible for the tooth decay known as caries.

Lactic acid fermentation performed by lactic acid bacteria is responsible for the sour taste of old milk and is used in the production of dairy products such as cheese, yogurt, and kefir. Lactic acid fermentation also gives the sour taste to fermented vegetables such as traditionally cultured sauerkraut and pickles and many fermented starches such as poi.

Lactic acid can be used as a food additive where it acts as an acidity regulator. In the food industry it is produced by heating and fermenting carbohydrates in milk whey, potatoes, cornstarch, or molasses. It is used in sweets, dressings, soft drinks (sometimes beer), infant formulas, and confectionary. It is denoted by E number E270.

Lactic acid is also the result of malolactic fermentation, a process used in winemaking to convert sharp-tasting malic acid into the gentler lactic acid.

Optical Activity of Lactic Acid

Lactic acid exists a special type of isomerism known as optical isomerism

Optical Activity: Ordinary white lactic acid consists of vibration in all phases perpendicular to the path of light. There are certain minerals such as tourmaline, Nicol's prism (A) see fig 11.12. The rays of light that emerge from the prism begin to vibrate only in one phase. The light is known a polarized light. If another similar prism (B) is now placed parallel to the first one, the polarized light will able to pass through and will be visible to the eye placed beyond the second prism.

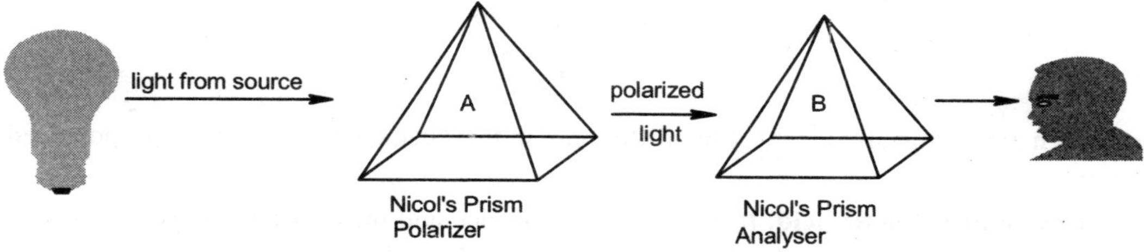

Fig. 11.12. Showing the Polarization Process by Nicol's Prism

Prism (A) is called a *'Polariser'* and prism B *Analyser* and a combination of two such prisms with an arrangement for passing light and a telescope for observing the rotation of the second prism

(B) is known as Polarimeter or **Polariscope.** If however, the second prism is crossed by rotating it through 90°, the light will be cut off and the eye will see no light. If now a tube full of lactic acid solution is placed between these two prisms, some light will pass through the second prism. This means that *lactic acid has the power to rotate the plane of polarization. thus enabling the polarized light to pass through the prism B.* Such a substance which can rotate the plane of polarized light v is known as **Optically Active** substance. Can sugar glucose tartaric acid, etc., behave similarly towards polarized light. The property is know as **Optical Activity**

On studying the constitution of optically active substances it has been conclusively proved that the molecule of all such compounds contain at least one carbon atom which is connected to four different atoms or group of atoms. Such a carbon atom is called **Asymmetric** Carbon atom.

```
     d
     |
a —  C — b
     |
     e
```

Optical activity is due to the presence of at least one **Asymmetric carbon atom in the molecule.**

It has further been studied that certain substances rotate the plane of polarization to the right while others rotate It to the left. Substances which can rotate the plane of polarization towards right are called **Dextro or** d-rotatory where as those which rotate the plane to the left are called **Laevo** or *l*-rotatory. The dextro and laevo rotation of a substance may also be indicated by + and – sign.

To explain optical activity Le Bell and Van Hoff, (1874) suggested a model for carbon atom. According to it carbon atom .lies at the centre of a regular tetrahedron and its 4 valences are directed towards 4 corners of tetrahedron so that angle between any two is same i *e.* 109°28'. If in such a model we link 4 different groups to carbon atom, the molecule (model) so prepared is asymmetric. *i.e. this model has no plane along which we can divide it into 2 identical halves.* We may also express it by saying that *this model and its mirror image can never be placed in identical positions.* The model represent one type of molecules say dextro and image represents the other type *i.e.* laevo.

Isomerism of Lactic Acid

Lactic acid contains an *Asymmetric carbon atom* shown in bold type and therefore occurs in three isomeric

```
        CH3
        |
  HO —  C — H
        |
        C — OH
        ||
        O
```

forms which are chemically identical but differ only in their behaviour to the plane of polarised light. They are:

(a) **Dextro *or* d-Lactic acid.** It is found in meat juice and muscles and is also known as *Sarco lactic acid.* It rotates the plane of polarised light towards the right.

(b) **Laevo or I-Lactic acid.** It is obtained from cane sugar by fermentation with *Bacillus Acidi Laevolactici.* This specimen rotates the plane towards the left. *It is regarded as the mirror image of dextra-lactic acid.*

```
   CH3               CH3
    |                 |
 H—C—OH         HO—C—H
    |                 |
HO—C                 C—OH
    ||                ||
    O                 O
```

(c) Racemic or dl-Lactic acid. It is formed by synthesis or from sugar solution by fermentation with Lactic Bacilli It does not effect the plane of pplarised light since it is a mixture of equal quantities of d- and l- forms. This form is optically inactive but capable of separation (resolution) into its active components by suitable methods.

TARTARIC ACID

Formula: $C_4H_6O_6$

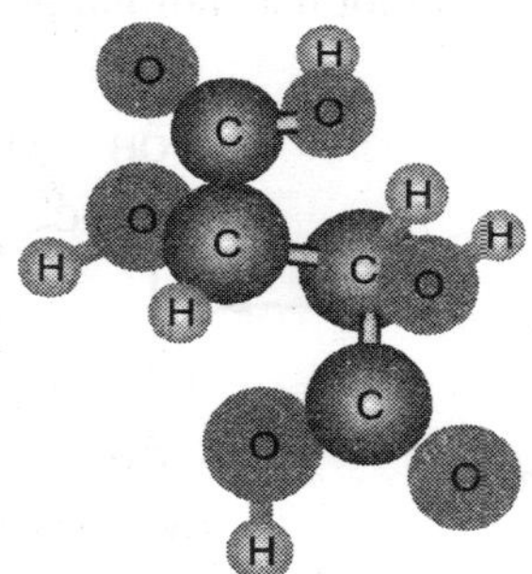

Tartaric acid is a white crystalline organic acid. It occurs naturally in many plants, particularly grapes and tamarinds, and is one of the main acids found in wine. It is added to other foods to give a sour taste, and is used as an antioxidant.

Tartaric acid was first isolated from potassium tartrate, known to the ancients as tartar, circa 800 AD by the alchemist **Jabir ibn Hayyan**, who was also responsible for numerous other basic chemical processes still in use today. The modern process was developed in 1769 by the Swedish chemist Carl Wilhelm Scheele. The chirality of tartaric acid was discovered in 1832 by Jean Baptiste Biot, who observed its ability to rotate polarized light. Louis Pasteur continued this research in 1847 by investigating the shapes of tartaric acid crystals, which he found to be asymmetric. Pasteur was the first to produce a pure sample of levotartaric acid.

Exactly, it is a dihydroxy derivative of dicarboxylic acid - 2,3-dihydroxysuccinic, or 2,3-dihydroxybutanedioic acid, with formula HO_2C-CH(OH)-CH(OH)-CO_2H ($C_4H_6O_6$). Naturally-occurring tartaric acid is chiral, meaning that tartaric acid molecules are non-superimposable on their mirror-images. It a useful raw material in organic chemistry for the synthesis of other chiral molecules. The naturally occurring form of the acid is **L-(+)-tartaric acid** or **dextrotartaric acid**. The mirror-image (enantiomeric) form, **levotartaric acid** or **D-(-)-tartaric acid**, and the achiral form, **mesotartaric acid**, can be made artificially. Note, that the *dextro* and *levo* prefixes are not related to the D/L configuration (which is derived from the reference D- or L-glyceraldehyde), but to the orientation of the optical rotation, (+) = dextrorotatory, (-) = levorotatory. Sometimes, instead of capital letters, small italic *d*, *l* are used. They are abbreviations of *dextro-* and *levo-*, and nowadays should not be used. Levotartaric and dextrotartaric acid are enantiomers, mesotartaric acid is a diastereomer of both of them. Rarely occurring optically inactive form of tartaric acid, 1:1 mixture of the *levo* and *dextro* forms, **DL-tartaric acid** (which must not be confused with mesotartaric acid) was called **racemic acid** (from Latin *racemus* - "a bunch of grapes"). This word later changed its meaning, becoming a general term for 1:1 enantiomeric mixtures - racemates.

Manufacture

From Argol: The argol is the raw material for the manufacture of tartaric acid. Argol is purified by recrystallization from boiling water. Pure potassium hydrogen tartrate – known as cream of tartar is obtained. This is dissolved in hot water and the original tartrate is converted in calcium tartrate (ppt) and the other half remains in solution as normal potassium tartrate

2 potassium hydrogen tartrate + $CaCO_3$ → calcium tartrate + dipotassium Hydrogen tartrate + CO_2 + H_2O

potassium hydrogen tartrate | calcium tartrate | dipotassium Hydrogen tartrate

To the product so obtained excess calcium chloride solution is added, when potassium tartrate is also changed into calcium tartrate.

dipotassium Hydrogen tartrate + $CaCl_2$ → calcium tartrate + 2KCl

dipotassium Hydrogen tartrate | calcium tartrate

The precipitate of calcium tartrate is now separated by filtration and concentrated under reduced pressure when crude tartaric acid crystallizes out. It is purified by decolourization with animal charcoal and recrystallization

Syntheis

Tartaric acid is synthesized from calcium and hydrogen as described below:

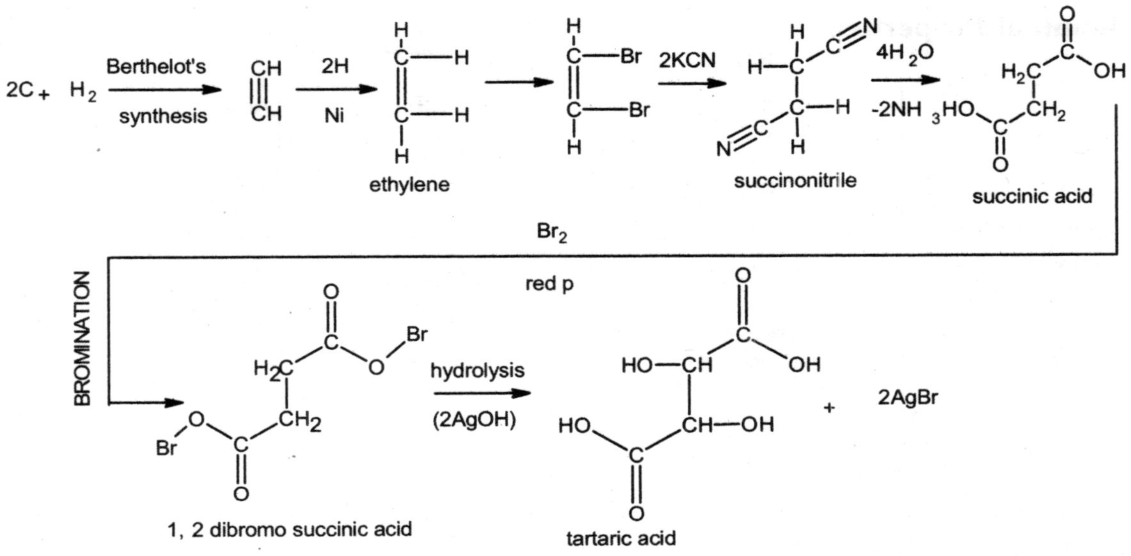

2C + H_2 —Berthelot's synthesis→ CH≡CH —2H, Ni→ $H_2C=CH_2$ (ethylene) → BrHC=CHBr —2KCN→ NC–CH₂–CH₂–CN (succinonitrile) —$4H_2O$, $-2NH_3$→ HOOC–CH₂–CH₂–COOH (succinic acid)

BROMINATION —Br_2, red p→ 1, 2 dibromo succinic acid —hydrolysis (2AgOH)→ tartaric acid + 2AgBr

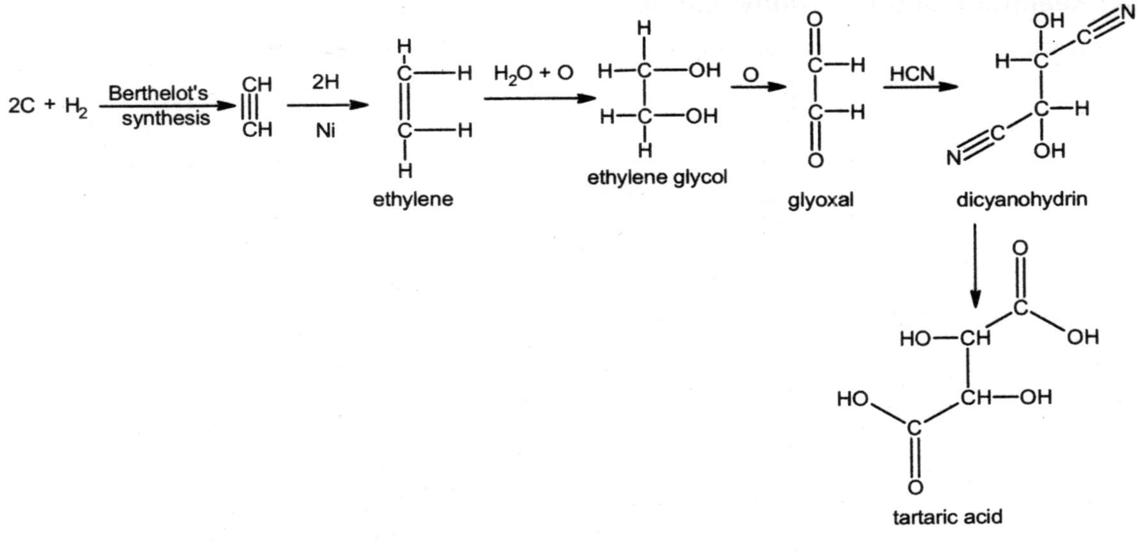

2C + H_2 —Berthelot's synthesis→ CH≡CH —2H, Ni→ ethylene —H_2O + O→ ethylene glycol —O→ glyoxal —HCN→ dicyanohydrin → tartaric acid

Physical Properties

Melting point: 206°C	Flash point: 210°C o.c.
Relative density (water = 1): 1.79	Auto-ignition temperature: 425°C
Solubility in water, g/100 ml at 20°C: 20.6	Octanol/water partition coefficient as log Pow: -0.76 (calculated)

Chemical Properties

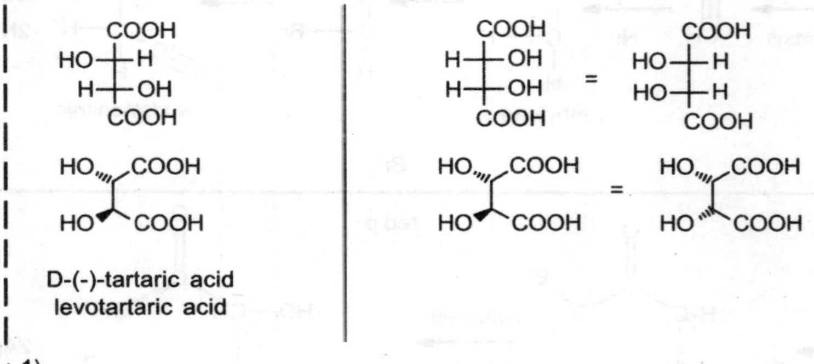

(natural) tartaric acid
L-(+)-tartaric acid
dextrotartaric acid

D-(-)-tartaric acid
levotartaric acid

(1 : 1)
DL-tartaric acid
"racemic acid"

It is a dibasic and gives all the reactions of acids and secondary alcohol

(A) Reactions of the carbonyl Group

1. It reacts with alkalies and alcohols forming two series of salts and easters

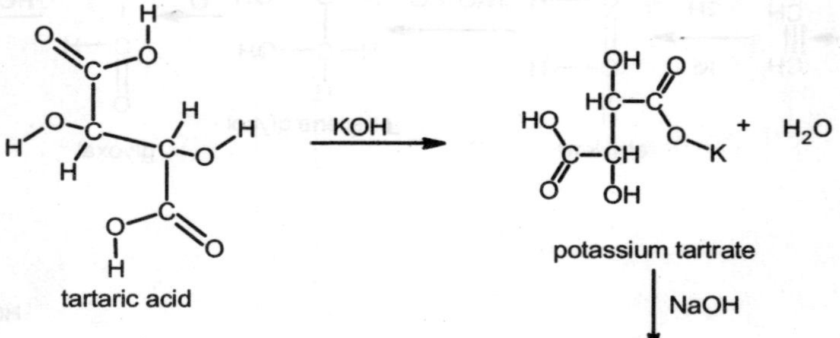

tartaric acid $\xrightarrow{KOH}$ potassium tartrate + H_2O

potassium tartrate $\xrightarrow{NaOH}$ sodium potassium tartrate
Rochelle salt

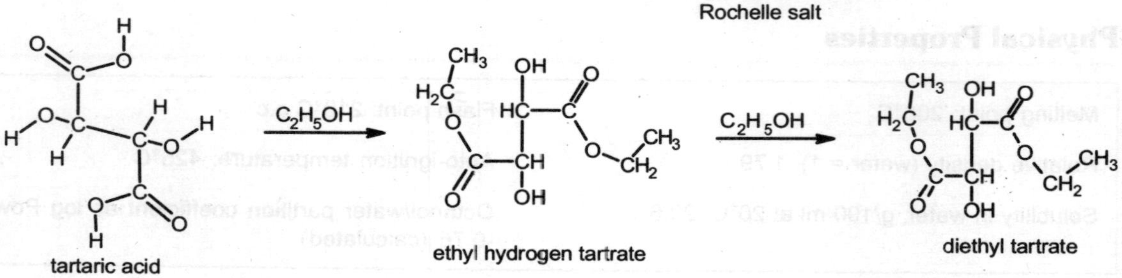

tartaric acid $\xrightarrow{C_2H_5OH}$ ethyl hydrogen tartrate $\xrightarrow{C_2H_5OH}$ diethyl tartrate

2. With ammonia: It forms amide

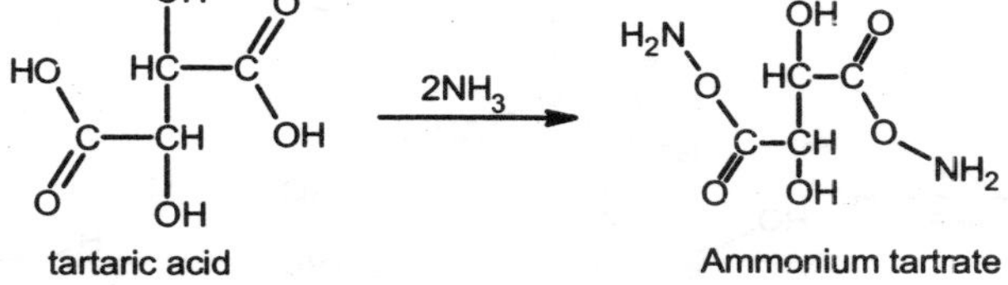

tartaric acid —$2NH_3$→ Ammonium tartrate

(B) Reactions of Secondary Alcoholic Groups:

3. With Hydrogen Bromide: It forms dibromosuccinic acid

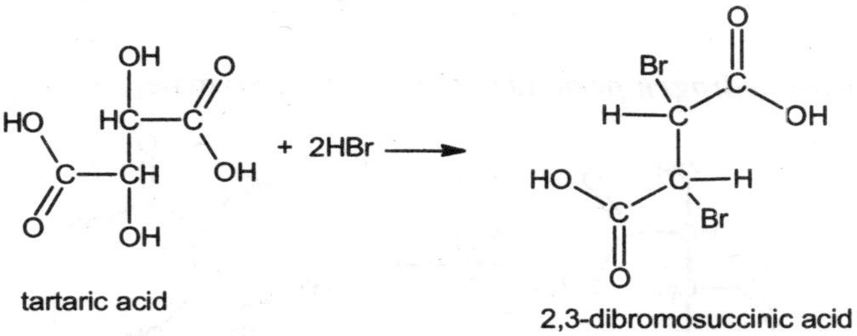

tartaric acid + 2HBr → 2,3-dibromosuccinic acid

4. With Acetyl Chloride: It forms acetyl derivative

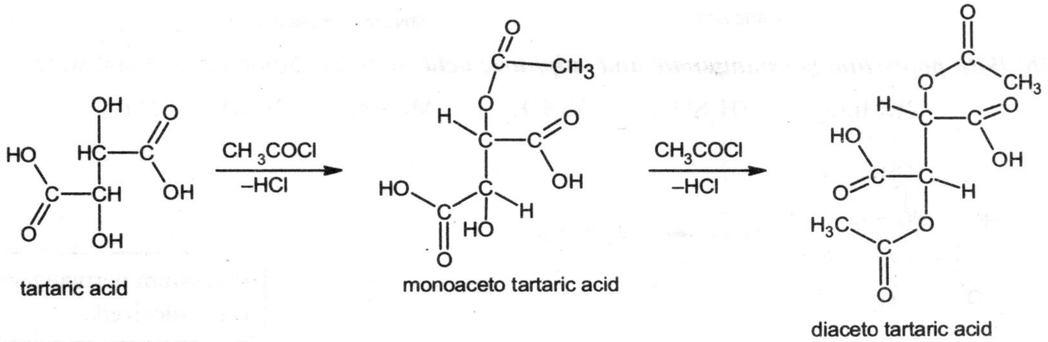

tartaric acid —CH_3COCl, –HCl→ monoaceto tartaric acid —CH_3COCl, –HCl→ diaceto tartaric acid

5. With acetic Anhydride: In the presence of a trace of conc. Sulphuric acid, tartaric acid gives diacetyl anhydride

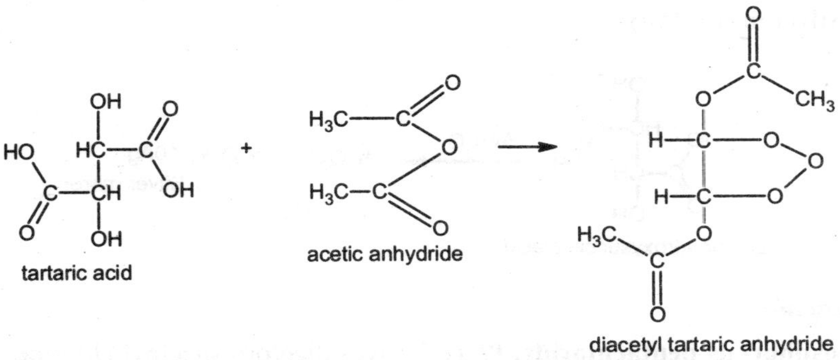

tartaric acid + acetic anhydride → diacetyl tartaric anhydride

8. With Hydrogen Iodide: It gives succinic acid

tartaric acid $\xrightarrow[\text{heat}]{2HI}$ malic acid $\xrightarrow{2HI}$ succinic acid + H_2O + I_2

+ H_2O + I_2

7. Oxidation: ***(a) with hydrogen peroxide*** gives di hydroxy maleic acid

tartaric acid $\xrightarrow{H_2O_2}$ dihydroxy maleic acid

(b) With potassium permanganate and sulphuric acid: It gives carbon dioxide and water

$$2KMnO_4 + 3H_2SO_4 \rightarrow K_2SO_4 + 2MnSO_4 + 3H_2O + 5O$$

tartaric acid + 5O ⟶ CO_2 + $3H_2O$

Potassium permanganate is decolorized

Tartaric acid and titrates are therefore reducing agents. They reduce ammonia silver nitrate to metallic silver (silver mirror test)

2,3-dihydroxysuccinic acid $\xrightarrow{5Ag_2O}$ $4CO_2$ + $3H_2O$ + 10Ag
Silver mirror

(c) Other reaction:

8. With phosphorous pentachloride, PCl_5: It gives dichloro succinyl chloride

$$\text{2,3-dihydroxysuccinic acid} + 4PCl_5 \longrightarrow \text{2,3-dichlorosuccinoyl dichloride} + 4POCl_3 + 4HCl$$

2,3-dihydroxysuccinic acid

2,3-dichlorosuccinoyl dichloride

9. Action of Heat: ***(a) On heating alone of with con sulphuric acid***: charring takes place and emit and odour of burnt sugar.

$$\text{tartaric acid} \longrightarrow 6H_2O + 3CO_2 + 5C$$

tartaric acid

$$[H_2SO_4 + C \longrightarrow H_2O + SO_2 + CO]$$

(b) On destructive distillation it gives pyruvic acid:

$$\text{tartaric acid} \longrightarrow H_3C{-}CO{-}COOH + CO_2 + H_2O$$

tartaric acid

pyruvic acid

(c) On heating for a long time at about 150°C it forms hydride:

$$\text{tartaric acid} \longrightarrow \text{tartric anhydride}$$

tartaric acid

tartric anhydride

10. With potassium chloride: It gives white precipitate of potassium hydrogen tartrate (test for tartaric acid or potassium)

$$\text{tartaric acid} + KCl \longrightarrow \text{potassium hydrogen tartrate} + HCl$$

tartaric acid

potassium hydrogen tartrate

11. With calcium chloride: A solution of neutral tartrate solution is treated with calcium chloride a white ppt is produced, which is dissolved is acetic acid (test and distinction from oxalate)

$$\text{tartaric acid} + CaCl_2 \xrightarrow[-2H_2O]{2NaOH} \text{calcium tartrate} + 2NaCl$$

tartaric acid

calcium tartrate

Uses

Tartaric acid in wine

Tartaric acid may be most immediately recognizable to wine drinkers as the source of "wine diamonds," the small potassium bitartrate crystals that sometimes form spontaneously on the cork. These "tartrates" are harmless, despite sometimes being mistaken for broken glass, and are prevented in many wines through cold stabilization. The tartrates that remain on the inside of aging barrels were at one time a major industrial source of potassium bitartrate.However, tartaric acid plays an important role chemically, lowering the pH of fermenting "must" to a level where many undesireable spoilage bacteria cannot live, and acting as a preservative after fermentation. In the mouth, tartaric acid provides some of the tartness that is currently out of fashion in the wine world, although citric and malic acids also play a role. The modern practice of extended hang time, where grapes are allowed to sit on the vine nearly until they become raisins, can dramatically reduce the taste of tartaric acid in a wine, leaving it smoother but also potentially less compatible with food.

Tartaric acid is used chiefly in the form of its salts, e.g., cream of tartar and Rochelle salt Tartaric acid is used as a flavoring agent in foods to make them taste sour. The potassium salt of tartaric acid (potassium bitartrate or potassium hydrogen tartrate) is weakly acidic, and is known as "cream of tartar". Since it is a dry acid, cream of tartar is used in baking powders (along with sodium bicarbonate) to produce carbon dioxide gas when water is added. Other acids used in baking powder are fumaric acid and phosphoric acid.

Applications of Tartaric Acid

Tartaric acid is a muscle toxin. When administered to experimental animal, tartaric acid causes muscle damage. In fact, as little as 12 grams have been fatal to a human (3). (One gram is about the weight of a cigarette.)

The main natural source of tartaric acid is yeast (2). The process in the body is similar to the wine brewing process, where tartaric acid forms sludge and has to be removed from the final product (2). Wine is essentially sugar fermented by yeast into alcohol and byproducts.

Of course, humans do not produce this material. However, when yeast in the intestinal tract are fed sugar from the diet, they produce tartaric acid—just like the yeast in the wine-making process does.

Tartaric acid is also extremely elevated in many patients with fibromyalgia who also experience muscle and joint pain. Tartaric acid was initially discovered by Dr. Shaw, Director of the Great Plains Laboratory, in the urine of two autistic brothers with muscle weakness so severe that they could not stand up (4).

Where the Yeast is & Why it Causes Problems

Tartaric acid is an analog (or close chemical relative) of malic acid. Malic acid is a key intermediate in the Krebs cycle, a biochemical process used for the extraction of most of the energy from our food.

Presumably tartaric acid is toxic because it inhibits the biochemical production of the normal compound, malic acid. Tartaric acid is a known inhibitor of the Krebs cycle enzyme fumarase (Figure 11.1) which produces malic acid from fumaric acid (5).

A large percentage of patients with fibromyalgia respond favourably to treatment with malic acid (6). I presume that supplements of malic acid are able to overcome the toxic effects of tartaric acid by supplying deficient malic acid. Treatment with the antifungal drug Nystatin kills the yeast and values for tartaric acid steadily diminish with antifungal treatment (Figure 11.3).

Fifty percent of patients with fibromyalgia often suffer from hypoglycemia (7)—low blood sugar—even though their diet may have adequate or even excessive sugar. The reason may be due to the inhibition of the Krebs cycle by tartaric acid.

The Krebs cycle is the main provider of raw materials such as malic acid that can be converted to blood sugar (Figure 11.3) when the body uses up its supply. If sufficient malic acid cannot be produced, the body cannot produce the sugar glucose which is the main fuel for the brain. The person with hypoglycemia feels weak and their thinking is foggy because there is insufficient fuel for their brain.

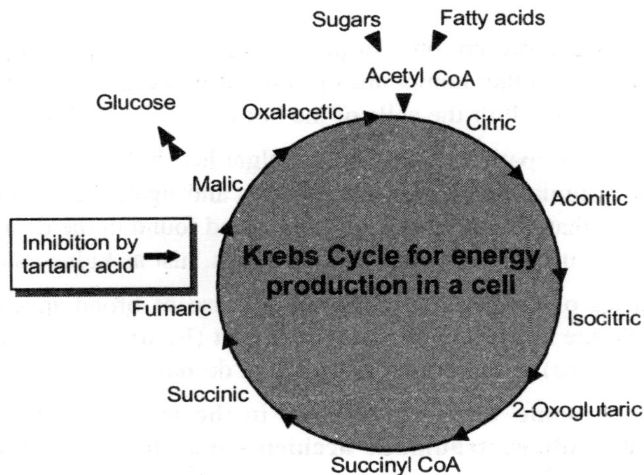

Fig. 11.13. Site of tartaric acid inhibition of the Krebs Cycle, the major energy producing mechanism of the cell.

In addition to the inhibition of energy production, tartaric acid prevents the production of malic acid, which is a key intermediate in the production of glucose in the process of gluconeogenisis, the principal fuel for the brain.

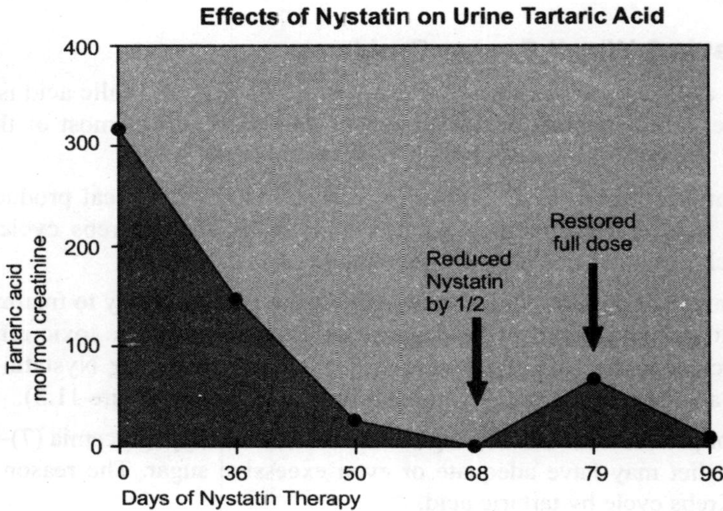

Fig. 11.14. Patient with high tartaric acid was started on the antifungal drug, Nystatin, and tested while on this drug. Even after 68 days, tartaric increased when the dose was reduced in half and then decreased again when the full dose was restored.

Contributing Factors

Most of the time the yeast are present only in the intestinal tract, not in the blood and other organs. However, the tartaric acid and other compounds produced by yeast in the intestine are absorbed into the bloodstream and may enter all of the cells of the body.

Dr. St Anand noticed that patients with fibromyalgia had high amounts of dental tartar on their teeth and speculated that similar deposits in the muscles and ligaments might be causing the pain of fibromyalgia. It's possible that tartaric acid is the compound found in the dental tartar and that crystals of this substance may be causing the muscle and joint pain, just as kidney stones cause kidney pain.

The two main causes of the yeast overgrowth are use of broad-spectrum antibiotics and the high sugar and carbohydrate content of the American diet (Figure 11.3). These antibiotics kill most of the normal bacteria (germs) in the intestinal tract, but do not kill organisms such as yeast (8-15).

As a matter of fact, some yeast grow faster in the presence of antibiotics. **The reason fibromyalgia commonly follows traumatic accidents may be related to the use of antibiotics to treat trauma.**

Sugar consumption is the second major factor in causing yeast-related illnesses. The average American consumes 10 times more sugar than Americans in the time of George Washington(about

150 lb per year). In a study done in mice, those mice receiving sugar in their water had 200 times more yeast in their intestine than mice receiving plain water (16).

Other factors that cause yeast overgrowth may include stress, birth control pills, viral infections, and a weak immune system.

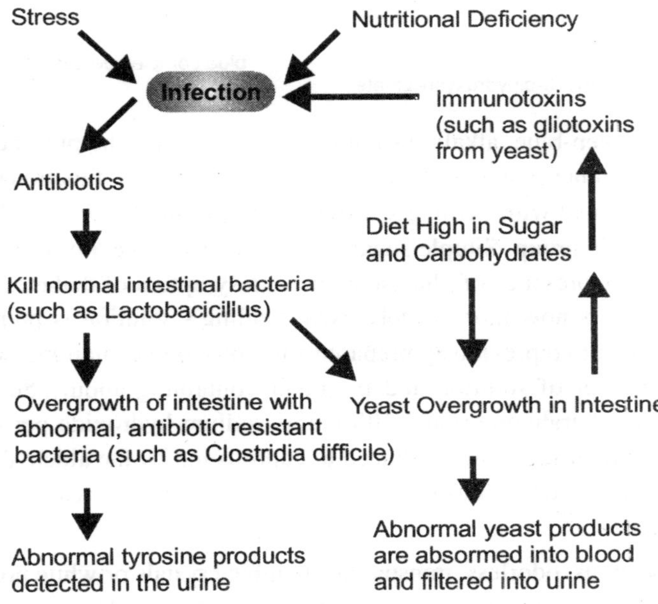

Fig. 11.15. Factors that Contribute to Yeast Overgrowth

Derivatives of Tartaric Acid

Cream of tartar: White crystalline powder. Chemically it is potassium hydrogen tartrate, $KC_4 H_5 O_6$, the acidic potassium salt of tartaric acid. It is used as the leavening agent in baking powders. An impure form, called tartar or argol, forms naturally during the fermentation of grape juice into wine and crystallizes in the wine casks.

Rochelle salt: Colorless to blue-white orthorhombic crystalline salt with a saline, cooling taste. It is also called Seignette salt after Pierre Seignette, an apothecary of La Rochelle, France, who was the first to make it (c.1675). Chemically, it is potassium sodium tartrate, $KNa (C_4 H_4 O_6) \cdot 4H_2O$. It is soluble in water and slightly soluble in alcohol, melts at about 75°C, has specific gravity 1.79, and exhibits double refraction. It is used in medicine as a mild purgative, often in the form of Seidlitz powders. It is an ingredient of Fehling's solution . It is used in silvering mirrors. Crystals of Rochelle salt are easily grown and are used in piezoelectric devices, e.g., crystal microphones and phonograph pickup cartridges.

dipotassium 2,3-dihydroxysuccinate → blue complex derivative + $2H_2O$

Fehling's Solution: Deep-blue, alkaline solution used to test for the presence of aldehydes (e.g., formaldehyde, HCHO) or other compounds that contain the aldehyde functional group, –CHO. The substance to be tested is heated with Fehling's solution; formation of a brick-red precipitate indicates the presence of the aldehyde group. Simple sugars (e.g., glucose) give a positive test, so the solution has been used to test for the presence of glucose in urine, a symptom of diabetes; Benedict's solution, which gives the same test, is now more widely used. Fehling's solution is prepared just before use by mixing equal volumes of two previously prepared solutions, one containing about 70 grams cupric sulfate pentahydrate per liter of solution and the other containing about 350 grams Rochelle salt (potassium sodium tartrate tetrahydrate) and 100 grams sodium hydroxide per liter of solution. The cupric ion (complexed with tartrate ion) is reduced to cuprous ion by the aldehyde (which is oxidized) and precipitates as cuprous oxide (Cu_2O); for this reason, sugars that react with Fehling's solution are called reducing sugars.

Tartar emetic: Poisonous, odorless, transparent rhombic crystals or white powder with a metallic, sweetish taste. Chemically, it is potassium antimony tartrate, $KSbC_4H_4O_7 \cdot 1/2\ H_2O$. It is used as a mordant in dyeing. Medically, it was formerly used as an emetic and expectorant, to produce sweating, and in the treatment of several diseases, but had frequent toxic side effects.

CITRIC ACID

Formula: $C_6H_8O_7$

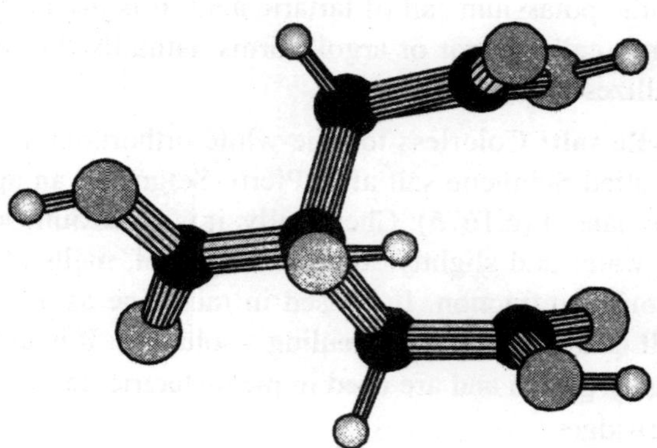

History and Introduction

Citric Acid was discovered by a Swedish Chemist, Carl Wilhelm Scheele, in 1784. Scheele was born in Stralsund, Pomerania [Germany] on December 9, 1742 and died on May 21, 1786 in Köping, Sweden. The isolation of citric acid was just one of the few substances that he discovered. Some of his other significant conquests were the identification of oxygen (rarely given credit over Priestley), arsenic acid (1775), molybic acid (1783), malic acid, oxalic acid, gallic acid, glycerine, chlorine oxide, barium oxide, hydrogen sulfide, hydrogen cyanide, and hydrogen fluoride. Scheele also published a book called *Abhandlung von der Luft und dem Feuer* (1777; *Chemical Observations and Experiments on Air and Fire*). He happened to have encountered and isolated this compound from an experiment involving lemon juice.(Citrus Limon = Latin lemon).

Citric Acid is found in almost all plants and in many animal tissues and fluids. It is a key element in the physiological oxidation of fats, proteins, and carbohydrates to carbon dioxide and water. It is obtained or manufactured commercially by fermentation of cane sugar or molasses in the presence of a fungus, Aspergillus Niger. As well, it can be collected by extraction from lemon juice, lime juice, and pineapple canning residues. Although most people believe that its natural abundance is found in citric fruits, most of it is produced by refineries.

Manufacture

1. From lemon Juice: Lemon juice contains 7 – 8% of citric acid. The juice is boiled. Protein and aluminous matters is precipitated and filtered off. The filtrate is neutralized with chalk or lime and boiled; calcium citrate is precipitated at the bottom. It is filtered, washed with hot water and decomposed with sulphuric acid. Calcium sulphate is filtered off and the filtrate is concentrated to obtain large transparent crystals of citric acid.

$$\underset{\text{(lemon juice)}}{2C_6H_8O_7} + 3CaCO_3 \rightarrow \underset{\text{calcium citrate}}{(C_6H_5O_7)_2Ca_3} + 3H_2O + 3CO_2$$

$$\underset{\text{calcium citrate}}{(C_6H_5O_7)_2Ca_3} + 3H_2SO_4 \rightarrow 3CaSO_4 + \underset{\text{citric acid}}{2C_6H_8O_7}$$

2. From glucose by citric acid fermentation: Citric acid can be prepared on an industrial scale by the fermentation of glucose with Aspergillus Niger or citronyces citrium. At fungus at 20 - 30°C. the reaction completes in 9 – 10 days.

$$\underset{\text{glucose}}{3C_6H_{12}O_6} + 9O_2 \rightarrow \underset{\text{citric acid}}{2C_6H_8O_7} + 6CO_2 + 10H_2O$$

From fermented liquid citric acid is recovered and purified as lemon juice, a 60% yield of citric acid is obtained by the method.

Synthesis: From its elements C and H

$2C + H_2 \xrightarrow[\text{synthesis}]{\text{berthelot}} HC{\equiv}CH$ (acetylene) $\xrightarrow[HgSO_4]{H_2SO_4} H_3C{-}CH{=}O$ (acetaldehyde) $\xrightarrow{O} H_3C{-}COOH$ (acetic acid) $\xrightarrow{CaO} Ca(COOCH_3)_2$ (Calcium Acetate) $\xrightarrow{\text{distil}}$ $CH_2{=}CH{-}CH_3$ (prop-1-ene) $\xrightarrow{Cl_2}$ $CH_2Cl{-}CH_2{-}CH_2Cl$ (1,3-dichloropropane) $\xrightarrow{ICl}$ $CH_2Cl{-}CHCl{-}CH_2Cl$ (1,2,3-trichloropropane) $\xrightarrow{3H_2O}$ $CH_2OH{-}CHOH{-}CH_2OH$ (glycerol) $\xrightarrow[\text{heat}]{2HCl}$ $CH_2Cl{-}CHOH{-}CH_2Cl$ (glycerol 1, 3 dichlorohydrin) $\xrightarrow[K_2Cr_2O_7]{O}$ $CH_2Cl{-}CO{-}CH_2Cl$ (1,3-dichloroacetone) $\xrightarrow{HCN}$ $H_3C{-}C(OH)(C{\equiv}N){-}CHCl{-}H$ (dichloroacetone cyanohydrin) $\xrightarrow{2KCN}$ 1, 2,3 tricyano glycerol $\xrightarrow{6H_2O}$ citric acid

Physical Properties

Decomposes below boiling point at 175°C	Flash point: 100°C
Melting point: 153°C	Explosive limits, vol% in air: 0.28-2.29
Solubility in water, g/100 ml at 20°C: 59	Octanol/water partition coefficient as log Pow: -1.7

Chemical Properties

It contains 3 carbonyl groups and one alcoholic group and therefore shows the following important reactions.

(A) Reactions of Carbonyl Groups

(a) Neutralization: It is a tribasic acid and therefore forms three series of salts and esters citric acid —KOH→ potassium citrate —KOH→ dipotassium citrate —KOH, $-H_2O$→ tri potassium citrate

(b) Esterification: Similarly diethyl and trimethyl citrates are produced.

citric acid —C_2H_5OH→ monoethyl citrate

(3) Amides: Citric acid also forms amides when its ammonium salt is heated.

(B) Reactions of the Alcoholic Group

(4) With acetyl chloride: It gives acetyl derivatives.

citric acid —CH_3COCl, $-HCl$→ mono aceto citric acid

This reaction (acetylation) proves the presence of one alcoholic (-OH) group in a molecule of citric acid.

4. With Hydrogen Iodide (Reduction): It gives tricarballyic acid

citric acid —2HI→ tricaballyic acid + H_2O + I_2

5. Oxidation: Oxidizing agents decompose citric acid yield a variety of products. Citric acid gives silver Mirror test slowly and only on boiling. Potassium permanganate is decolorized.

6. With Phosphorous Pentachloride: It forms Chloro citryl chloride

$$\text{citric acid} + 4PCl_5 \longrightarrow \text{chlorocitryl chloride} + 4POCl_3 + 4HCl$$

citric acid — chlorocitryl chloride

7. Action of Heat: On long heating above at 170°C is charred and gets hydrated

$$\text{citric acid} \xrightarrow[(-H_2O)]{175°C} \text{aconitric acid}$$

citric acid — aconitric acid

8. On heating with conc. sulphuric acid: Gives acetone dicarboxylic acid. Slow charring takes place on continued heating.

$$\text{citric acid} \longrightarrow \text{acetone dicaboxylic acid} + CO_2 + H_2O$$

$$\text{acetone dicaboxylic acid} \xrightarrow{\text{heat strongly}} CH_3-CO-CH_3 + 2CO_2$$

citric acid — acetone dicaboxylic acid — acetone

9. With Calcium chloride: Citric acid solution made alkaline with ammonia gives white precipitate of calcium citrate only on heating.

$$2C_8H_8O_7 + 2CaCl_2 \xrightarrow{6NH_4OH} (C_6H_5O_7)_2Ca_3 + 6NH_4Cl + 6H_2O$$

10. Citric acid can also prevent the ppt of the $Cu(OH)_2$ or $Fe(OH)_2$ from solution of their salts (similar to tartaric acid)

Uses

What is the first image that enters your mind when you hear the words "acidic acid"? Well, one word screams in your face, ACID. AS well, citrate is a citric acid salt that is commonly associated with this compound. It is formed when metals combine with citric acid. Here is a list of common uses of acidic acid or where it may be found.

Foods

Yes, the use of citric acid is primarily used in the food industry. The industry amounts to approximately 70% of its general use. It provides the tart taste in various fruits such as lemons, limes, oranges, pineapples, grapefruit, and gooseberries. Another great use of acidic acid is found in the line of sodas created by Pepsi. You know it, Britney Spears and Bob Dole are fans of this compound. After checking up on Coke and their ingredients... it wasn't labeled on the can... could it possibly be... PEPSI is more acidic. Of course! *AhEm... Coke is the best!* One of the unknown daily uses of citric acid is for meat tenderizing. It is used in many food processors as an acidulant in their products in order to control acidity. This allows the meat to loosen from the weak acid present in citric acid.

Detergents and Cleaners

This group represents 18% of the general use of acidic acid. This acid is capable of preserving colour and flavour in canned and frozen fruits and fish. Take a look next time you pop a microwave dinner or open those canned peaches. You'll notice that acidic acid is there. As well, it's found in inks, mordant (dye-fixative). buffering agents, refrigerant, dispering agents, acidifiers, and textile painting. One of the more interesting uses is in bath "bombs". It provides that initial sizzle that appears when it reacts with water.

Pharmaceuticals and Cosmetics

You ever have problems going to the bathroom? Or do you just want to play a prank on a friend or foe? Laxitives is the answer to both of those questions. Yes, even lactic acid is found in that wonderful creation. Another interesting use is with antioxidants. This helps alcoholics recover by prompting an irritation when they consume alcohol. It allows the normal effects of alcohol to conteract their effects. Last of all, its used in anticoagulants. This allows for slower blood clotting when necessary. Overall, the pharmaceutical and cosmetic division of the uses of citric acid amount to 6%.

Industrial and Chemical Processing Uses / Other

This group represents another 6% of the uses of citric acid. Most of them are foreign to the common individual, but here is a nice anyways: alkyd resins, flasticizers, chelating agent, hemostyptic, lithotriptic, sequestering agent, and cathartics.

Citrates

One of the most common uses of citrates is ferric ammonium citrate, which is used to make blueprint paper. The ferric ammonium citrate works when light is exposed to a drawing on top of it. Where there are not lines, the paper will be able to react and turn into ferrous salt that leads into another reaction that creates a blue background.

Side Effects

Frequent and large doses may lead to corrosion of the teeth and produce an irritant action.

Formation of Complex Ions

sodium cupracitrate

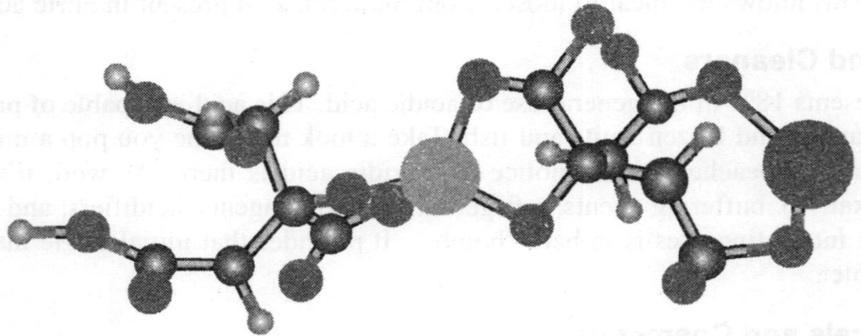

Benedict's reagent: **Like tartaric acid, citric acid also forms soluble complex salts and thus prevents the precipitation of hydroxides of heavy metals. Benedict's solution which contains copper sulphate solution and sodium carbonate and sodium citrate provides an excellent complex in similar to that of the complex formed by tartaric acid.**

The formation of this complex lowers the concentration of Cu (II) ions to such content that it does not form a chelate (of the type shown) with citric acid the brick red copper II oxide is precipitated when Cu(II) when Cu(II) are reduce to Cu(I) ions by an aldehyde or any other oxidizing substance.

Benedict's reagent is more soluble than Fehlling's solution as it is does not react with the creatinine or uric acid and therefore used for the detection of the sugar in urine.

Distinctive Tests for Oxalic Tartaric and Citric Acid

Action of	Oxalic Acid	Tartaric Acid	Citric Acid
Heat on solid	Decompose without charring carbon monoxide carbon dioxide and formic acid	Chars at once smell of burnt sugar is observed	Melts without charring on long heating it gradually chras
Hot and conc. Sulphuric acid on solid	No charring carbon dioxide and carbon dioxide are evolved. no evolution of sulphur dioxide	Chars immediately odour of burnt sugar is produced carbon monoxide carbon dioxide and sulphuric acid are evolved	Slow charring on continuous heating carbon monoxide, carbon dioxide and sulphur dioxide is evolved
Silver nitrate on neutral solution	White precipitate soluble in excess ammonium hydroxide, silver mirror is formed on warming.	White precipitate soluble in dilute ammonia on warming silver mirror is produced	Same as tartaric acid but the process is very slow
Potassium permanganate and sulphuric acid on hot solution	Decomposition and carbon dioxide is produced	Decomposition and carbon dioxide is produced	Decomposition and carbon dioxide is produced

This is for Revision for the Students

1. Reaction of Glycine

Synthesis Reaction

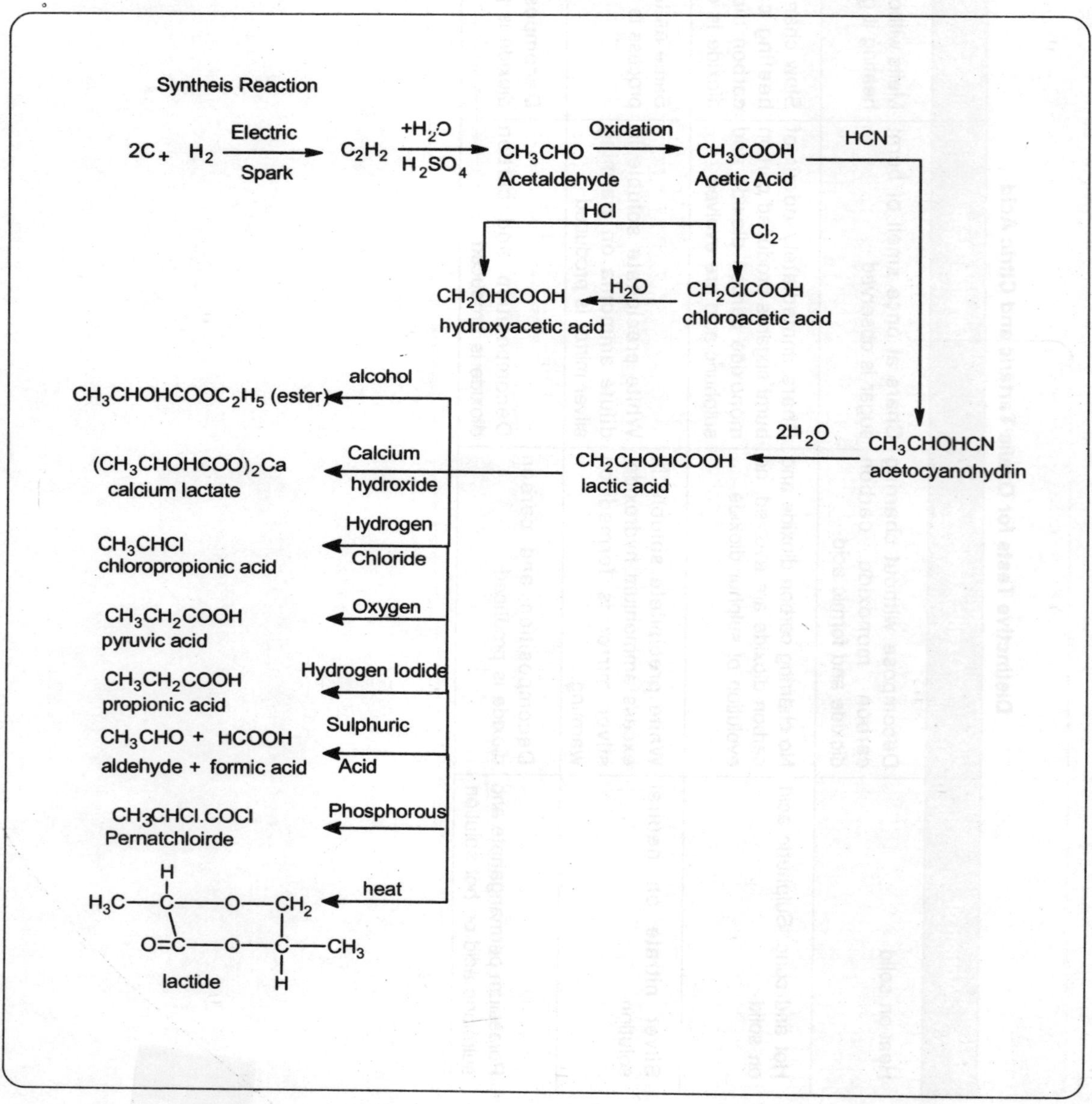

2. Tartaric Acid (Synthesis and Reaction)

$$C_2H_2 \xrightarrow[Ni]{2H} \underset{\text{ethylene}}{CH_2{=}CH_2} \xrightarrow[\text{bayer's reagent}]{H_2O + O} \underset{\text{ethylene glycol}}{H_2C(OH){-}H_2C(OH)} \xrightarrow{O} \underset{\text{glyoxal}}{OHC{-}CHO} \xrightarrow{2HCN} \underset{\text{dicyanohydrin}}{N{\equiv}C{-}CH(OH){-}CH(OH){-}C{\equiv}N} \longrightarrow \underset{\text{tartaric acid}}{HOOC{-}CH(OH){-}CH(OH){-}COOH}$$

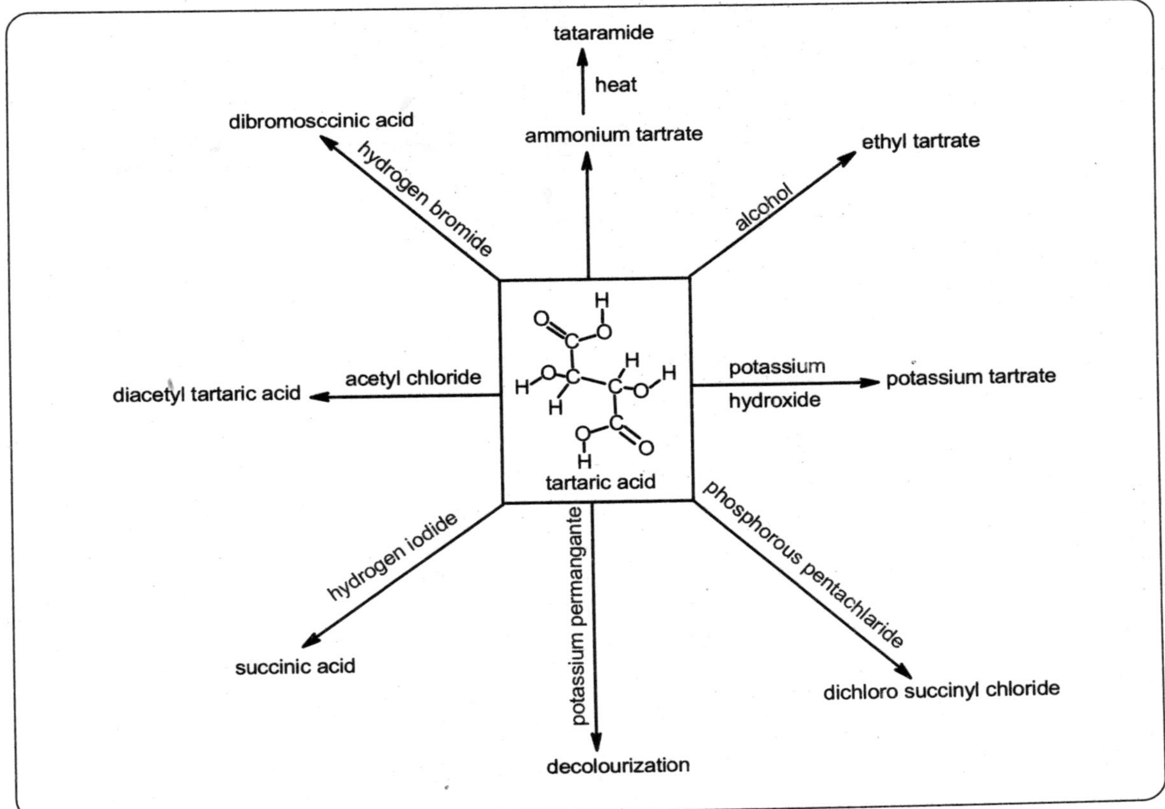

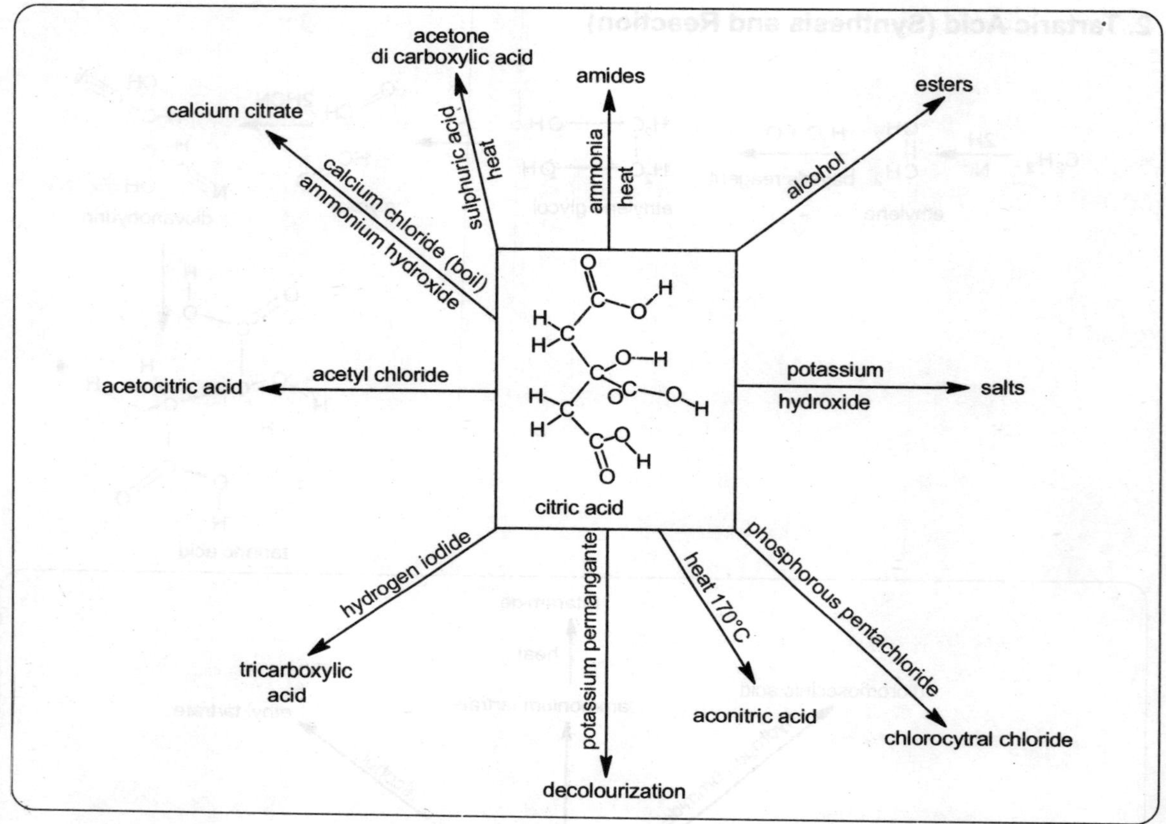
acetone
di carboxylic acid
amides
esters
calcium citrate
sulphuric acid
heat
ammonia
heat
alcohol
calcium chloride (boil)
ammonium hydroxide
acetocitric acid
acetyl chloride
potassium
hydroxide
salts
citric acid
hydrogen iodide
potassium permangante
heat 170°C
phosphorous pentachloride
tricarboxylic
acid
aconitric acid
chlorocytral chloride
decolourization